Diercke Praxis

Activity Book

advanced level (Einführungsphase)

Editor:
Wolfgang Latz

Authors:
Frank Böning
Andreas Bremm
Ursula Brinkmann-Brock
Rolf Brück
Erik Elvenich
Guido Hoffmeister
Dr. Norma Kreuzberger
Christine Kreuzberger
Wolfgang Latz
Dr. Lars Schmoll
Björn Schray
Wolfgang Stark
Silke Weiß

Translators:
John Attfield
Joanne Moar

westermann

ⓞ **Optional assignment**
In this case, the student may choose to work on a section of the assignment in line with his or her interests.

Ⓢ **Supplementary assignment**
These assignments consolidate existing knowledge and can be worked on additionally to the foregoing assignments.

key terms
→ **special key terms** with explanation

Druck A[1] / Jahr 2015
Alle Drucke der Serie A sind im Unterricht parallel verwendbar.

Redaktion: Dietmar Falk (Berlin)
Verlagslektorat: Malin Schneider-Pluppins (Witten), Lektorat Eck (Berlin), Steffen Stierhof
Umschlaggestaltung und Layoutkonzept: JANSSEN KAHLERT Design & Kommunikation GmbH (Hannover)
Satz und Seitengestaltung: Yvonne Behnke (Berlin)
Druck und Bindung: westermann druck GmbH, Braunschweig

ISBN 978-3-14-**114958**-6

Note about the materials

To a large extent, *DIERCKE PRAXIS* comprises original materials. These come from a multiplicity of analogue and digital publications by a wide range of very different authors, institutes and research groups. Therefore, statements made in different materials about specific factual elements may diverge from one another, and may even be contradictory. For example, the boundaries of the prairies in the USA are delineated differently by different scientists. Even such apparently definite data as population figures or the economic strength of a geographical region may vary from one source to another. The authors have consciously refrained from eliminating such differences, as this would represent a falsification of the sources..

Geography = Earth Studies

Geography in the Secondary Level II

Geography – a new subject?

No! You already had "Erdkunde" (Earth Studies) up to the 9th grade. And the lessons in Secondary Level II will build on those of the earlier years. In S II you will encounter many topics that you are already familiar with – but now you will be working much more on your own. For this you will use the skills you acquired in S I, and in particular you will dive deeper into the specialist subject areas of geography.

What are these subject areas?

What will you cover in the introductory phase (M2) and later in the qualification phase (M1)?

M2 Subjects in the introductory phase

Subjects in the qualification phase

- Agricultural production between conflicting demands of nutrition and feeding a growing world population
- Market- and export-oriented agro-business as a fit solution for the future?
- Changing economic regions – factors of influence and their effects
- Promotion of economic zones – necessary for global competition between industrialised regions?
- Global disparities – the challenge of unequal development in geographical areas
- Population development and migration as a cause of spatial problems
- Similar problems, similar solutions? Strategies and instruments to reduce disparities in areas at different stages of development
- Services and their importance for peripheral and underdeveloped areas
- Cities as a complex living space between tradition and progress
- Metropolisation and marginalisation – inevitable processes in the context of worldwide urbanisation?
- The city as a space with good quality of life for all? – Problems and strategies for future-oriented urban planning
- Modern cities – centres for the service sector alone?
- Goods and services – always available?
  The importance of logistics and freight transport

Source: Ministry for Schools and Further Education, NRW

M1 Subjects of geography teaching in SII

Geography

"Geography, as the 'science of the Earth', analyses spatial structures and the processes relevant to them in the fields of natural and social sciences. [...]
Physical geography is concerned with the dynamic forces and their interactions that determine the shape of the Earth's surface, and thus includes geology, geomorphology, vegetation and climate in particular.
In human geography, the focus is on the study of social and economic systems such as population, the economy and cities."

Source: www.ph-heidelberg.de/geographie, 29.1.2014

M3 Geography – the science of the Earth

Geography [...] is a classic science that deals with the landscape sphere of the Earth.
The landscape is studied in an integrative way; this means that physical, biotic and anthropogenic facts are seen as a causal network which has developed into its present state in the course of time.
Its future development tendencies are also the subject of investigation in geography. [...]

Source: H. Leser, Wörterbuch allgemeine Geographie, 2011

M4 Geography – a definition

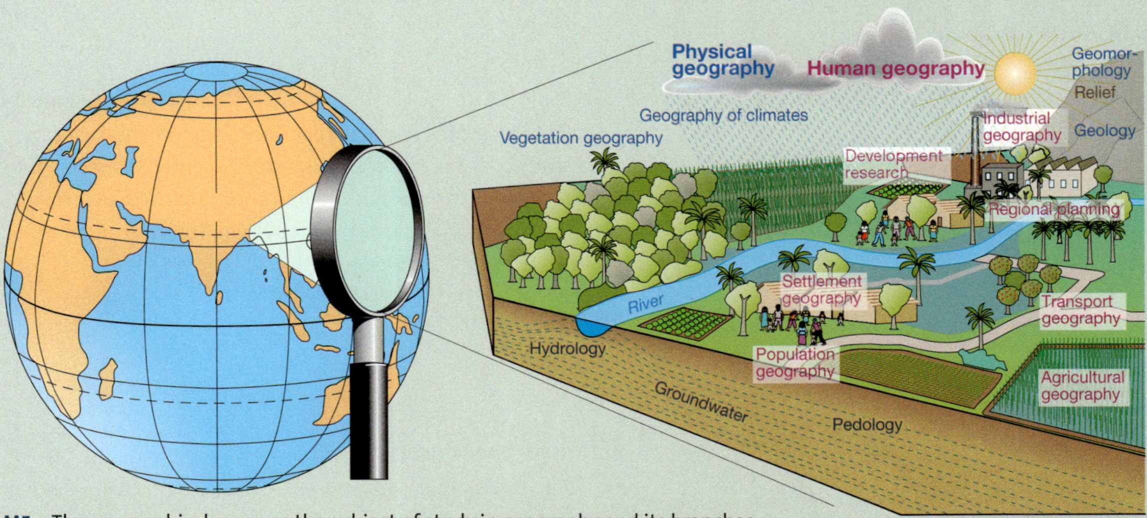

M5 The geographical space as the subject of study in geography and its branches

The frame of reference and the field of research in geography is a geographical space. The size of the space can differ very widely: the Earth, a continent, a country, a landscape, or perhaps just a stream. Geography studies not only the factors operating in the geographical space (e.g. climate, vegetation, population, the economy), but also and above all their interactions.

Ⓞ **1. A** Geography is a science of geographical space. Explain (M3 – M5).
B Allocate the photographs to the different branches of geography (M2/ M5).
C Name, in each case, three of your own concrete examples of topics that are covered in physical geography and in human geography (M5).
D Mention five current news reports that fall within the scope of geography (M5).
2. Find more detailed information about one of the subjects in the qualification phase and explain it using examples (M1).
Ⓢ **3.** "System Earth" – what does it mean? Explain (M4).

Our Earth – worthy of protection!

Astronauts who have had the opportunity to admire the Earth from afar have always been deeply impressed by the beauty, but also the apparent vulnerability, of our home planet.
We are far too seldom conscious of the fact that we are only a tiny part of the great ecosystem that is the Earth.
What does this mean for our day-to-day actions?
What does this mean for our conduct in a geographical area, for our conduct in our home place, on the Earth?

1. Often, people who have travelled into space campaign in a very special way for sustainable policies and sustainable action. Explain this.
2. Sustainability is not limited to the environment. Explain each of its three components with a concrete example (M4).
(O) **3.** An intervention in the ecosystem (the causal network) of a landscape has a multiplicity of consequences (M1/ M4).
A Present an example of this in a causal network.
B Draft a newspaper report about a possible intervention.
(S) **4.** The most important goal of geography teaching is to convey spatially-related action skills. The most important principle here is that of sustainability.
Explain (→ definitions).
(S) **5.** One planet says to the other:
"You look so sick, what do you have?"
Says the other: "I have humans!"
The first one replies: "Don't worry, it will soon go away!"
Explain this joke.

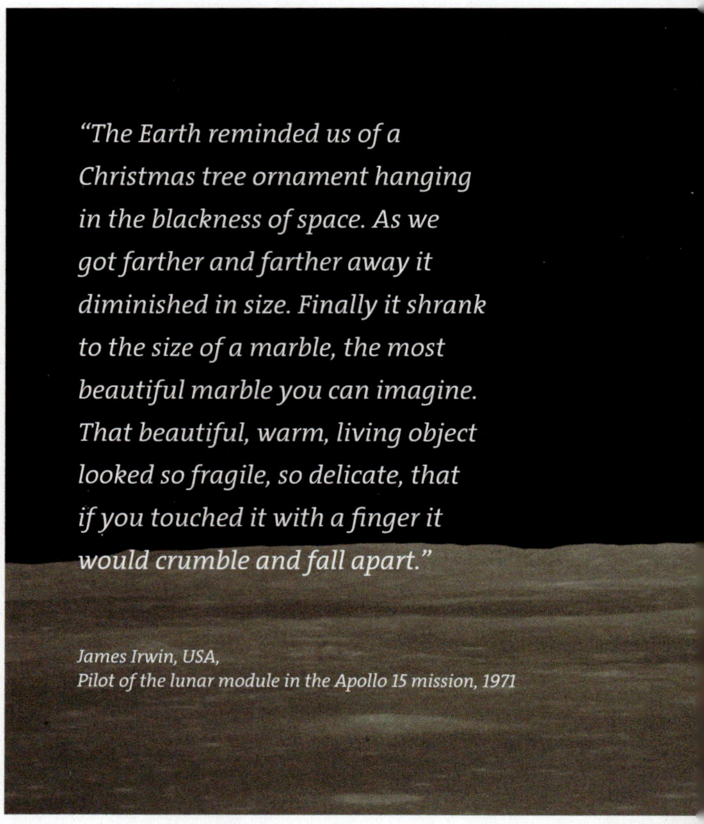

M2 "The Earth is rising"

→ Causal network
The landscape is studied in an integrative way; this means that physical, biotic and anthropogenic facts are seen as a causal network which has developed into its present state in the course of time. Its future development tendencies are also the subject of investigation in geography (...).
Source: H. Leser, Wörterbuch der Allgemeinen Geographie, 2005

Ecosystem
→ Causal network, Sustainability

→ Sustainability
Sustainability means to equally consider environmental aspects, social and economic aspects. Thus, future-oriented management means: We have to leave our children and grandchildren an intact ecological, social and economic system. The one cannot be achieved without the other.
Source: German Council for Sustainable Development (RNE), www.nachhaltigkeitsrat.de, 29.1.2014

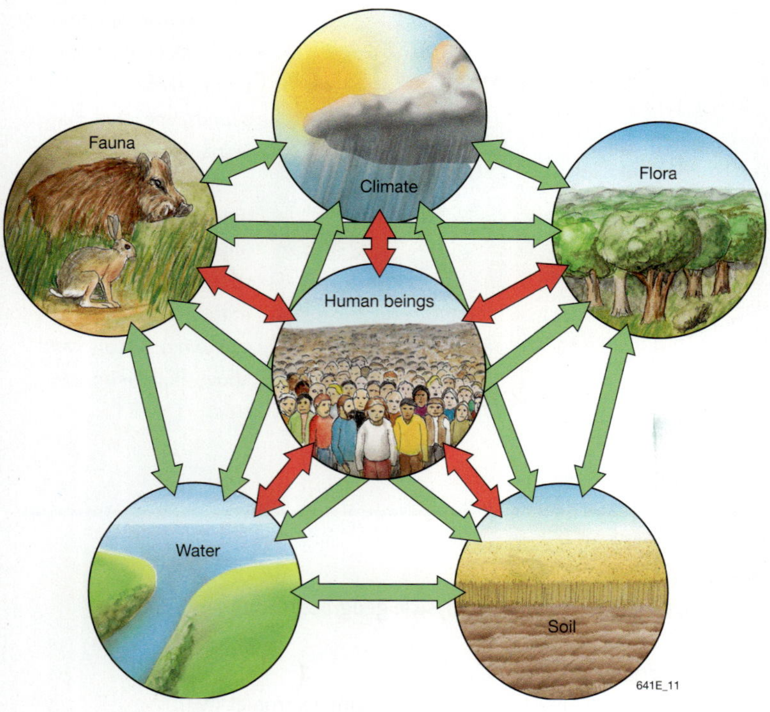

M1 Model of an ecosystem – the "landscape as a causal network"

Acting for the future ...

The United Nations declared the years 2005 to 2014 as the Decade of Education for Sustainable Development. (...)
This aims to open up educational opportunities for all people that will enable them to acquire knowledge and values as well as to learn forms of behaviour and lifestyles that are necessary for a future worth living and for shaping a society fit for the future.
Source: www.kmk.org

M3 Report to the conference of education ministers, 13.12.2012

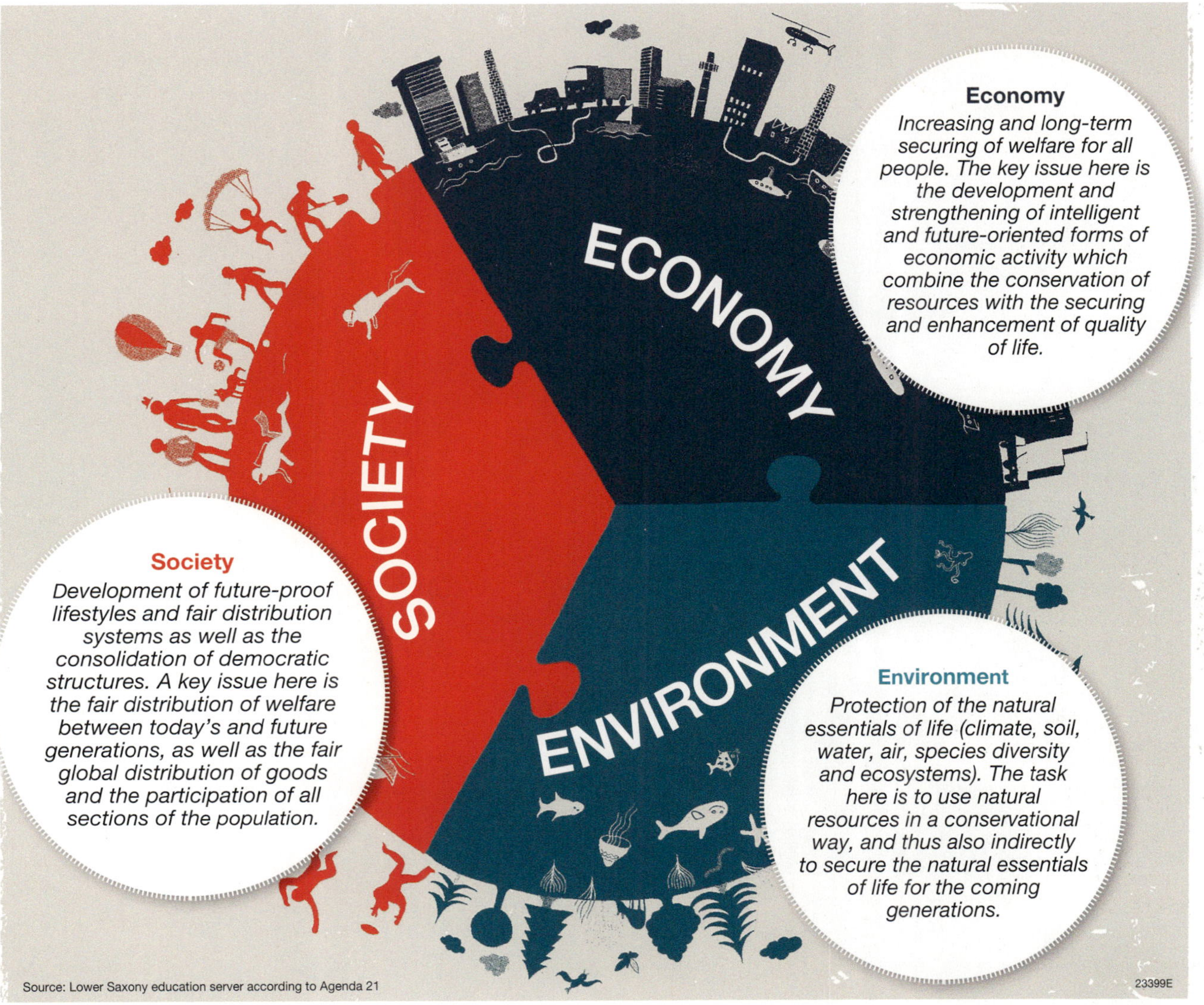

M4 Sustainability at all levels

Between Inhabited and Uninhabitable

The human habitat in different landscape zones

The Dades Valley in Morocco. In this valley, landscapes characterised by aridity alternate with green, fertile oases.

The zonal classifications of Earth

Landscape zones and their utilisation as human habitats

There are many different natural subregions (or landscapes) on Earth. The interaction of geofactors such as climate, soil, water and vegetation determine the characteristics of each landscape. Together, landscapes with similar conditions form the landscape zones – the natural regions of the Earth.

But what is it exactly that determines the individual landscape zones? How do geofactors work together? What potential do the different landscape zones offer human beings to live and cultivate? Which favourable and unfavourable factors determine life in the different landscape zones? Can human beings overcome natural boundaries and live and cultivate in regions that are essentially inhospitable?

M3 Deciduous and mixed forest in the temperate latitudes

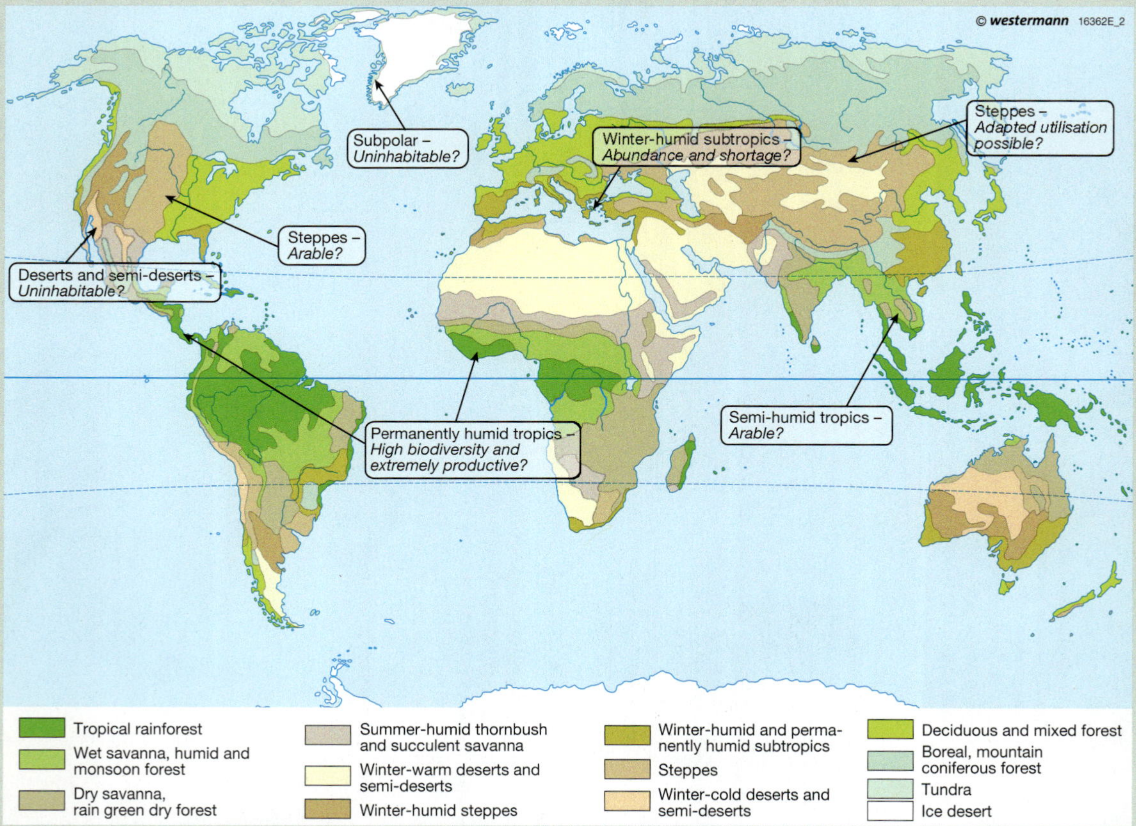

M1 Vegetation zones of the Earth (generalised) – largely consistent with the landscape zones

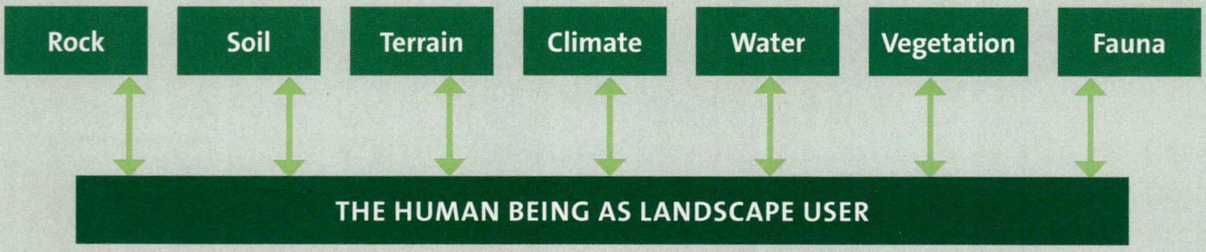

M2 Geofactors and human activity

M4 Tropical rainforest

M6 Ice desert in the polar zone

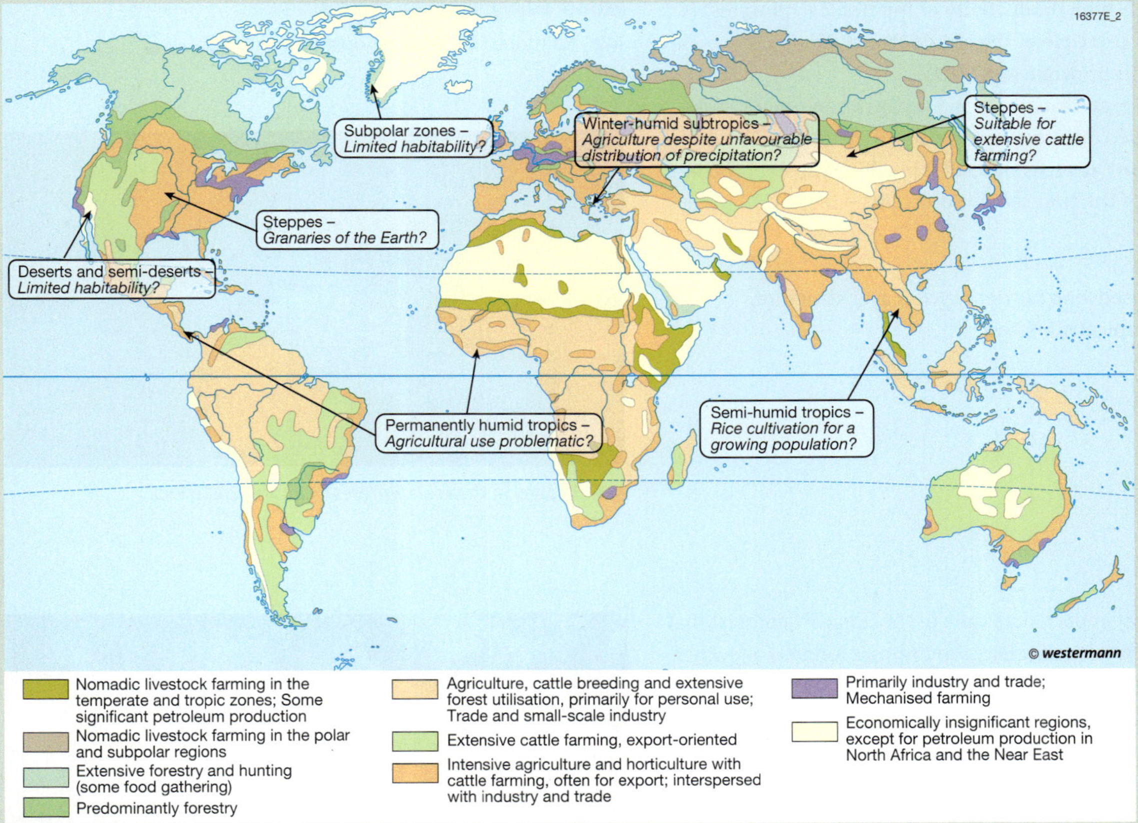

M5 Land use zones on Earth

→ **Inhabited/ Uninhabitable**
Those regions of the Earth that are permanently colonised and cultivated by human beings are referred to as being inhabited (ecumene). Regions that are not permanently colonised by human beings belong to the uninhabitable regions. They are too cold, too dry or too high in altitude.

1. Using examples show how climate and vegetation determine the characteristics of a landscape (M1).
2. Consider which regions on Earth could be particularly
   a) inhospitable (M1, M5),
   b) suitable for human habitation (M1, M5).
3. Name possible solutions for human beings to overcome aridity or severe cold.

Heat and precipitation in excess – the permanently humid tropics

The tropical rainforest in Costa Rica – high biodiversity and extremely productive

The tropical rainforest – A popular holiday destination

Ecotourism is a growth industry. Travel companies have reported double-figure growth rates in this branch. Diverse destinations are available – among others the tropical rainforest. Tour operators advertise with spectacular images and attractive excursions. Online feedback pages are full of enthusiastic comments. What makes the tropical rainforest so unique?

1. Locate Golfito, Costa Rica (atlas).
2. List possible motivations for people to travel to the tropical rainforest (M1 – M5).
3. Describe the situation on the forest floor and above 25 m (M4 – M6).
4. ⓞ The extremely high productivity of the tropical rainforest is based on the principle of the closed loop nutrient cycle.
   - **A** Explain this principle (M6/ M7).
   - **B** Explain the particular role the mycorrhizae fungus plays in the closed loop nutrient cycle (M6 – M9).
5. Ⓢ Produce a short advertising leaflet in which you explain the uniqueness of the tropical rainforest (M1 – M9).

Biodiversity, Closed loop nutrient cycle, Layer structure, Mycorrhizae, Tropical rainforest

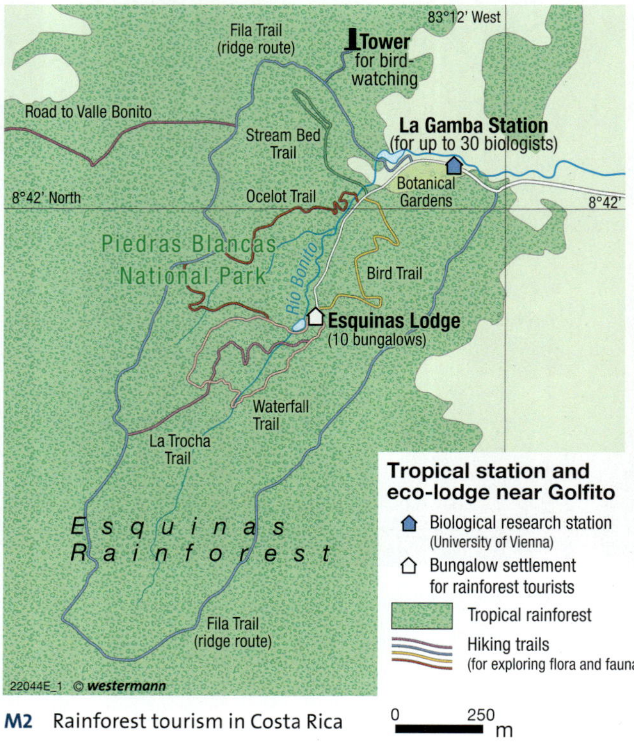

M2 Rainforest tourism in Costa Rica

M3 Lodges in tropical rainforest, Golfito, Costa Rica

The first and most important feature of the tropical rainforest that must be named is its biodiversity. For example, on only a few hectares in the Amazon Basin more species of plant and insects have been found than are known to exist in the entire European flora and fauna. 70 percent of all plant species in the rainforest are trees. It is estimated that there are between 10,000 and 30,000 different species of tree. [...]

Up to 100 different species of tree were counted on one hectare alone. One plant family particularly characteristic to the tropics, the palm tree, is now known to have about 3,500 species. Many scientists estimate that about half of all the world's plant and animal species live in the permanently humid tropical rainforest zone. Many of these have never been recorded or described.

Source: K. Müller-Hohenstein, F.-D. Miotke: Die geoökologischen Zonen der Erde (1997), In: Girndt, T. et al., Raummodul Lateinamerika, 2012

M1 Biodiversity of the tropical rainforest

M4 Treetop walkway at a height of about 25 m – This is the zone of the highest biodiversity in the tropical rainforest

M5 View from the treetop walk into the upper tree layer

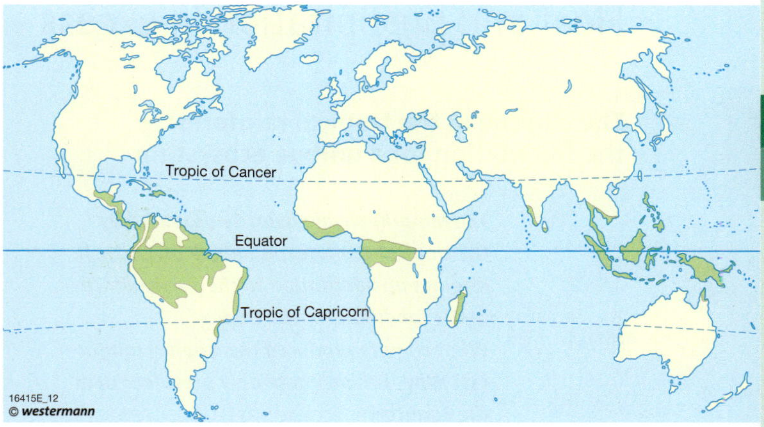

M8 Distribution of tropical rainforest

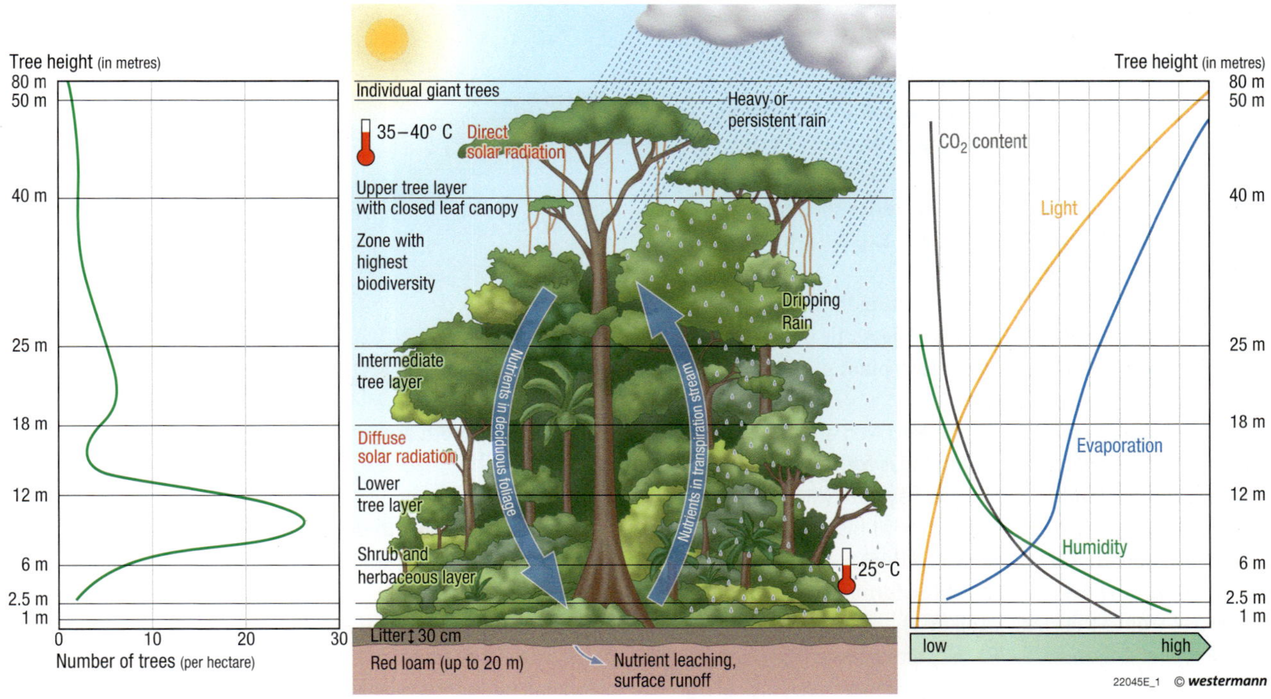

M6 The layer structure and nutrient cycle of the tropical rainforest

How can the thin humus layer of the tropical rainforest produce such lush growth?

The roots of the tropical rainforest's tree giants grow very strong but not very deep (max. 40 cm). They draw all their nutrients (e.g. phosphate, nitrogen, calcium, potassium etc.) from the forest's thin humus layer. This is made possible by a very dense, widely branching root system.

Additionally the thin root hairs form a symbiotic relationship with mycorrhizae fungi. These fungi live in the soil and have countless, microscopically fine branching hairs, with which they absorb the nutrients in the soil and release these to the tree (nutrient traps). As opposed to plants, fungi cannot carry out photosynthesis. They are therefore dependent on the nutrients (e.g. sugar, starch) that the tree provides them.

Almost 99 percent of the nutrients are thus filtered from the decomposing leaf litter, the humus layer and seepage water and fed straight back into the nutrient cycle.

M7 The closed loop nutrient cycle

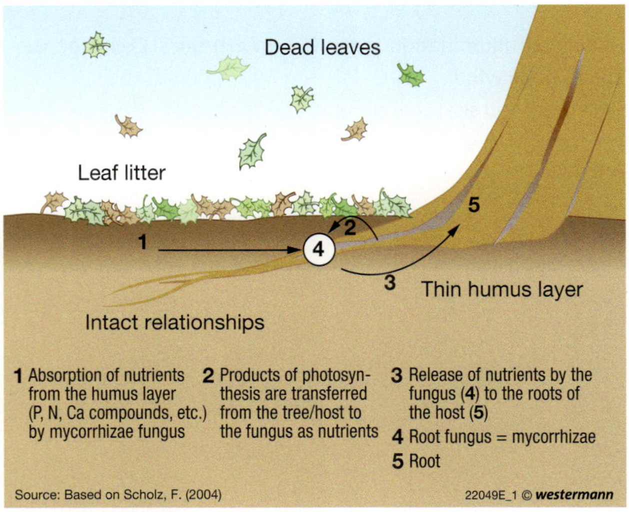

M9 The function of the mycorrhizae in the closed loop nutrient cycle

Heat and precipitation in excess – the permanently humid tropics

The climate in the tropical rainforest – the foundation for a unique ecosystem

The remarkable biodiversity produced by the tropical rainforest is unique on Earth. It displays an adaptation to the particular climatic conditions of the tropics. What climate produced the tropical rainforest? Why is this climate only prevalent near the Equator?

Ⓞ **1. A** Describe the global distribution of the tropical rainforest (atlas).
B Produce a table to list countries (ordered according to continent) that have tropical rainforest. Include the latitudinal coordinates of each country.

2. Describe the climate of the permanently humid tropics (M2 – M5).

3. M4 shows a diurnal climate. Characterise the properties of this climate (M4/ M5, → definition). Explain why we speak of a seasonal climate.

4. Compare the radiation and temperature in the tropics with our latitudes, approx. 50°N and in the Arctic Circle. Explain the role the climate plays in the lush plant growth of the tropical rainforest (M1 – M5).

Ⓞ **5.** The land and sea breeze system (M6/ M9) and the experiment (M10) demonstrate how wind is formed.
A Explain the land and sea breeze system (M6).
B Outline and explain the situation at night when the land cools more quickly than the sea (M6).

6. Transfer your knowledge about the formation of wind from exercise 5 onto the Hadley cell (M7).

Ⓢ **7.** Christopher Columbus took a very direct route from Europe to the Caribbean. Consider why Columbus made particularly speedy progress from La Gomera to the Bahamas (atlas, M8/ M9). Draw his route on a map sketch and fill in the relevant wind currents (internet, atlas).

M3 Saturated air above evergreen tropical rainforest

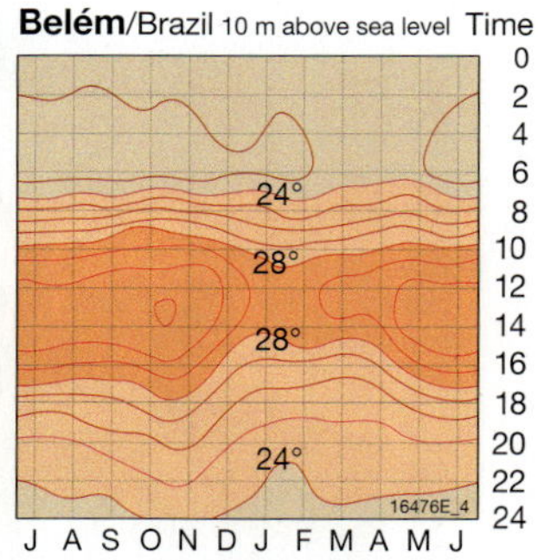

M4 Thermoisopleth diagram for Belém

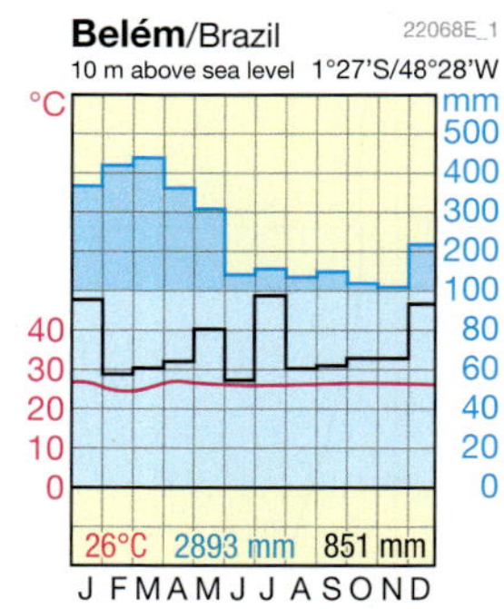

M5 Climate diagram for Belém, Brazil

→ Thermoisopleth diagram, Diurnal and seasonal climate

Thermoisopleth diagrams show annual and daily variations in temperature. It is thus possible to differentiate climates with large temperature fluctuations between night and day from climates with large seasonal fluctuations. Equal temperature values are joined by lines (similar to altitude lines on a topographical map). The vertical axis shows variations in temperature on any day of the year, the horizontal axis shows the annual variations at a certain time of the day.

Hadley cell, Illumination zone, ITCZ (Intertropical Convergence Zone), Trade wind

→ Diurnal and seasonal climate, Thermoisopleth diagram

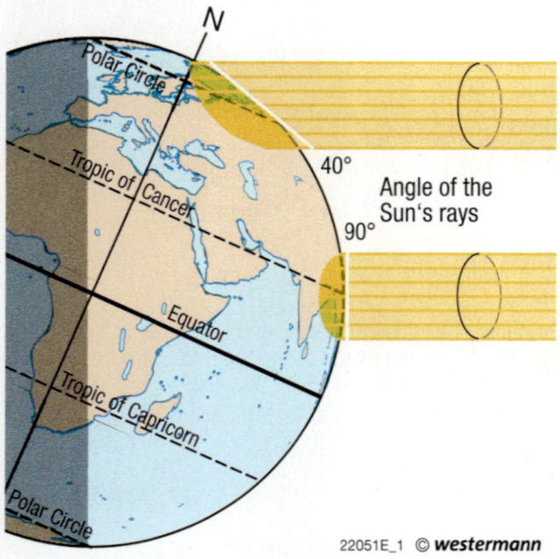

M1 Angle of the Sun's rays on June 21

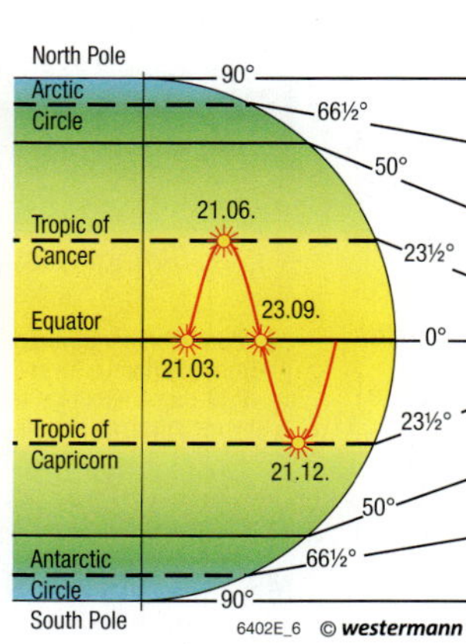

Angle of irradiation on 21.03. und 23.09.	Day length in hours 21.06.	21.03. 23.09.	21.12.	Noon altitude of the Sun highest value	lowest value	Annual average temperature
–	24	12	0	23.5°	–	-23 °C
23.5°	24	12	0	47°	0°	-7 °C
40°	16	12	8	63.5°	16.5°	+6 °C
66.5°	13.5	12	10.5	90°	43°	+24 °C
90°	12	12	12	90°	66.5°	+26 °C
66.5°	10.5	12	13.5	90°	43°	+22 °C
40°	8	12	16	63.5°	16.5°	+4 °C
23.5°	0	12	24	47°	0°	-8 °C
–	0	12	24	23.5°	–	-33 °C

Average radiant energy Latitude	$KJ/cm^2/day$
90°	1.52
80°	1.58
70°	1.74
60°	2.09
50°	2.52
40°	2.91
30°	3.24
20°	3.48
10°	3.63
0°	3.67

M2 Radiation and temperature in relation to geographical latitude

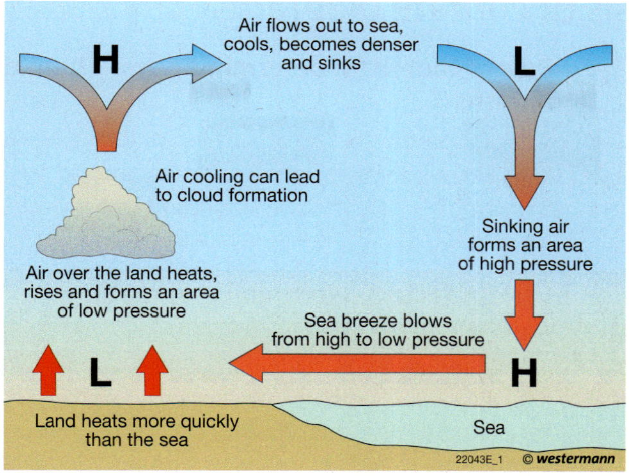

M6 Formation of wind based on the land and sea breeze system during the day

The land and sea breeze system is a localised wind system that occurs on hot days, predominantly in tropical and subtropical regions. Nevertheless it can be used to illustrate the most important principles in the formation of wind:

- Differences in temperature result in differences in pressure.
- Differences in pressure result in airflow, i.e. wind.
- The greater the differences in pressure, the stronger the wind.
- Cold air at the Earth's surface is associated with high pressure. A downward airflow prevails over an area of high pressure.
- Warm air at the Earth's surface is associated with low pressure. An upward airflow (convection) prevails over an area of low pressure.
- Winds are always named after the direction of origin (a land breeze blows from the land, a westerly from the west).

Based on: W. Latz (Ed.), Diercke Geographie, 2012

M9 Findings from the land and sea breeze system

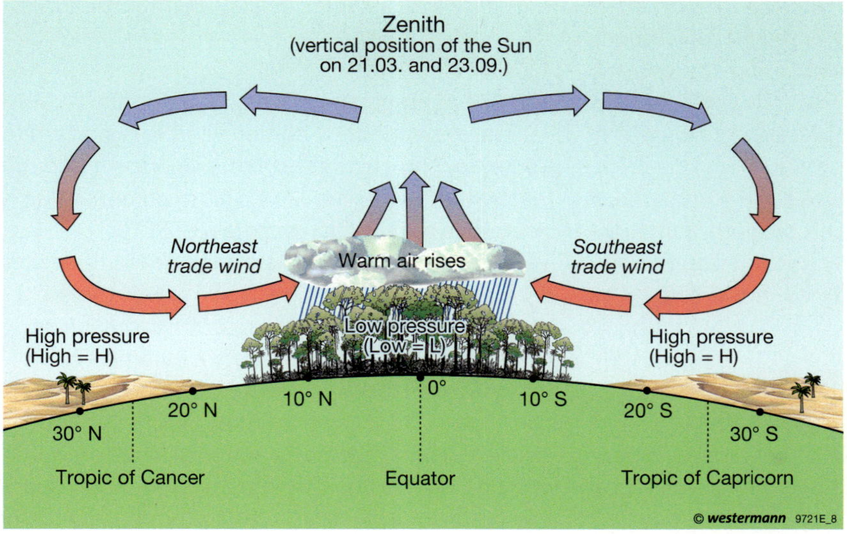

M7 Model of wind circulation in the tropics – the Hadley cell

Experiment

- Light eight tea candles and place them in a circle with a diameter of approx. 20 cm. Observe the behaviour of the flames.
- Sketch the experimental setup and your observations.
- Label your sketch using the terms high pressure and low pressure and explain the processes.
- Transfer your findings to the situation at the coast on a hot summer day and the situation at the Equator (M7).

M10 Experiment

The trade winds are formed by rising air masses near the Equator. As a result of intense solar radiation, the Earth's surface and layers of air closest to the surface that contain large quantities of water vapour heat up intensely. This process of heating causes the humid air to expand and rise (convection). The rising air masses cause the air pressure close the Earth's surface to drop. The rising air cools, resulting in cloud formation and torrential tropical precipitation.

At an altitude of about 16 to 18 km the air does not rise any further. The stream of air splits and flows off in a southerly and northerly direction. At about the level of the Tropics of Cancer and Capricorn the air is forced downwards creating belts of high pressure close to the Earth's surface, so-called subtropical/dry tropical high pressure cells. The pressure gradient between the high pressure belt in the region of the Tropics of Cancer and Capricorn and the low pressure belt close to the Equator causes compensatory currents to flow towards the Equator – the trade winds. The Earth's rotation means that the air currents do not flow directly towards the Equator, but are deflected to the right in the northern hemisphere and to the left in the southern hemisphere (Coriolis force). Thus the northeast trade wind is formed in the northern hemisphere and the southeast trade wind is formed in the southern hemisphere. The Intertropical Convergence Zone (ITCZ or ITC) refers to the region in which both trade wind streams collide. This atmospheric circulation is called the Hadley cell.

The Hadley cell not only determines the wind conditions in the tropics and dry tropics, it also determines the distribution of precipitation. In the equatorial region, the ITCZ, convection causes the formation of huge towering clouds from which it rains profusely. When the drier air masses descend in the regions of the Tropics the air warms. The clouds dissipate and the air is able to absorb large quantities of water from the surroundings. These regions therefore have very low levels of precipitation.

Because the ITCZ migrates with the Sun's zenith, the position of the northern and southern Hadley cells also shifts. This in turn impacts on the direction of the trade winds.

Based on: A. Siegmund and P. Frankenberg (Ed.): Diercke Spezial. Klimakunde, 2013

M8 The influence of the Hadley cell on the tropical climate

METHOD Thematic analysis of climate diagrams

Climate diagrams show the distribution of precipitation and the fluctuations in temperature over the course of a year. Moreover, they give information about the annual precipitation and annual average temperature. The potential evaporation (M2) is also displayed in a climate diagram.
All values in a climate diagram are average values recorded over a time period of 30 years. A climate diagram thus depicts the conditions typical of the atmosphere close to the Earth's surface at a particular location. But a climate diagram only has limited use for weather forecasting, because the term "weather" describes the momentary condition of the atmosphere close to the Earth's surface at a particular location.
In order to characterise weather and climate, climatic elements (M2) are taken into consideration. The climatic elements are recorded globally, among others, at 9,000 national monitoring stations and passed on to weather services. This information forms the basis for weather forecasts and long-term climate observation.
Climate diagrams can be used to describe annual variations in temperature and precipitation. However, climate diagrams only become meaningful when analysed and considered in the context of specific questions.

Arid, humid, Climate, Climate diagram, Climatic elements, Climatic factors, Potential evaporation (PET), Vegetation period, Evaporation, Weather

Different types of climate diagram

There are several different ways of representing a climate diagram. Many diagrams compare temperature and precipitation data on the two vertical axes in a 1 : 2 ratio (e.g. M1 or a Walter/Lieth climate diagram). 10°C on the temperature axis corresponds to 20 mm on the precipitation axis. It is assumed that at 10°C, 20 mm of water evaporates. In this way it is possible to predict whether more rain will fall than evaporate or whether less rain will fall than can evaporate at the prevailing temperatures i.e. there is a shortage of water.

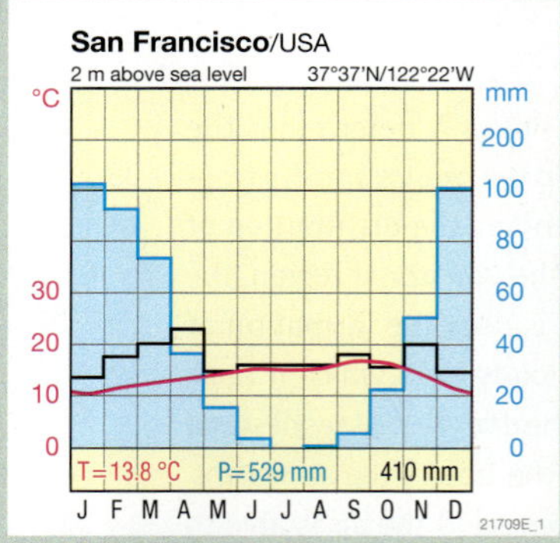

M1 Climate diagram for San Francisco, USA

Rule of thumb for determining arid / humid months

Rule of thumb: If the top of the precipitation column ends below the temperature curve it is an arid (dry) month, if the precipitation column ends above the temperature curve it is a humid (wet) month.

Various factors determine evaporation

In this rule of thumb it is assumed that evaporation is determined only by temperature. In reality the rate of evaporation is also dependent on other factors. For example, the evaporation rate of a body of water is greater than an area of land; the evaporation rate of a grain field is greater than bare agricultural land and the evaporation rate of pastureland is greater than that of a forest. Furthermore, the factors wind, humidity and sunshine all play an important role.

Potential evaporation

The rate of evaporation that actually occurs is called "actual evaporation". Additionally, it is important to know how much water could evaporate if enough water were always available. This maximum value is referred to as potential evaporation (PET).
Very few climate stations are equipped with devices for measuring evaporation, which is why climate researchers try to calculate the rate of evaporation using a combination of different climate parameters.
When precipitation is correlated with PET, a month is considered arid if the precipitation value (in mm) is smaller than the potential evaporation. If precipitation and PET are equal or if the precipitation value is greater than the PET, the month is considered humid.

1. Using two examples, present how climatic factors influence climatic elements (M5).
2. Explain the meaning of the term potential evaporation.
3. Based on the analysis of the climate diagram for Belém produce a checklist or outlines for analysing climate diagrams (M2/ M3/ M6).
4. Evaluate the climate diagram for Athens using your checklist and the information on this page (M4).
5. Ⓢ Analyse another climate diagram from this book. Name possible reasons for the characteristics of several climatic elements (book, atlas).
6. Ⓢ Using an atlas, classify the climate of the climate stations mentioned on this page.

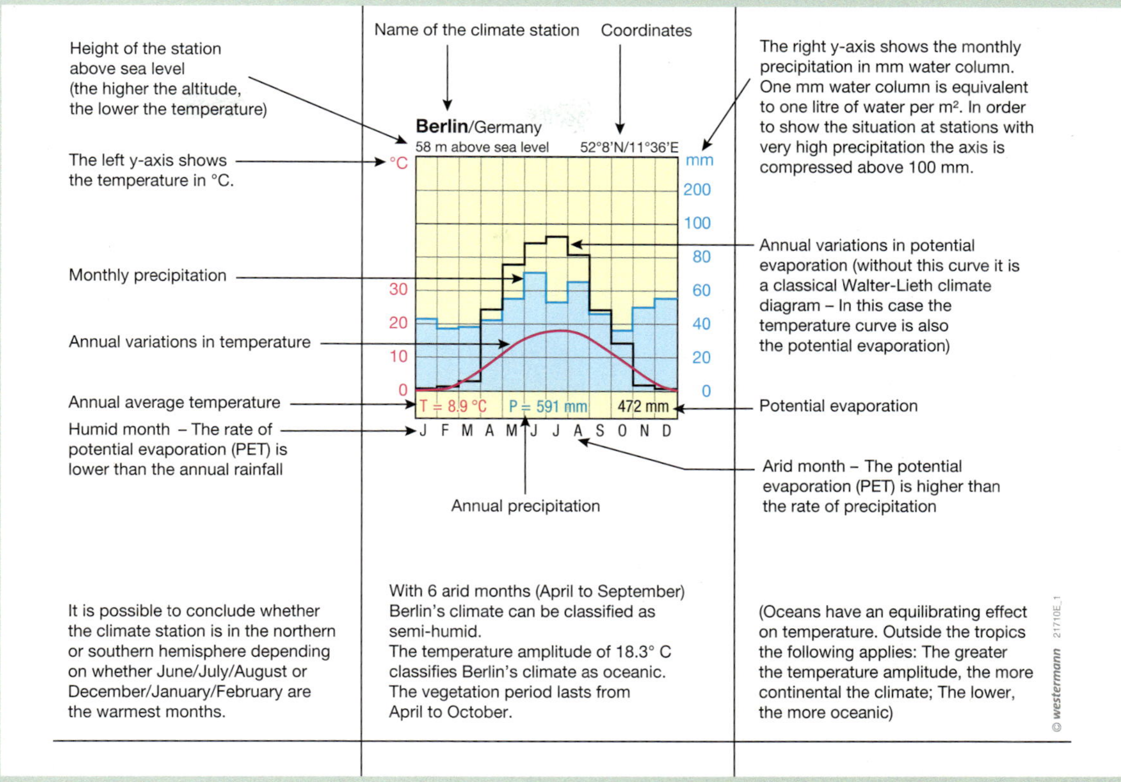

M2 Elements of a climate diagram

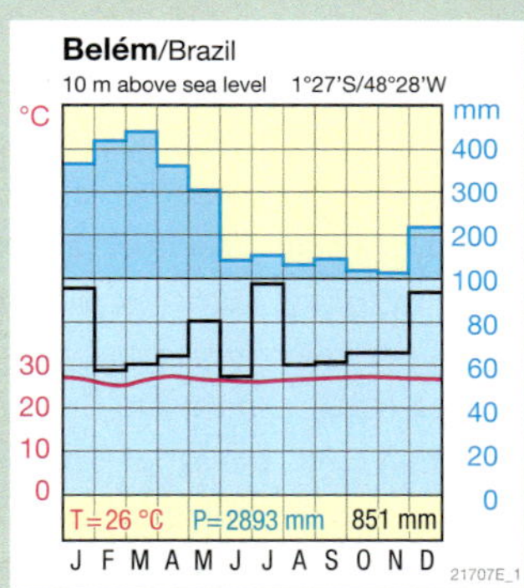

M3 Climate diagram ▷ Belém, Brazil

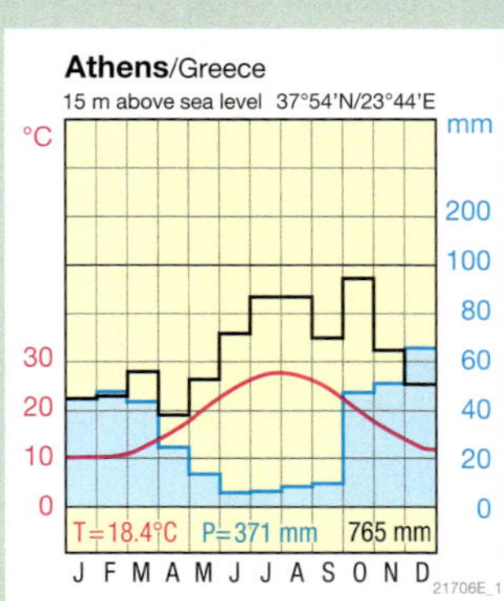

M4 Climate diagram ▷ Athens, Greece

Climatic elements	
• Temperature • Precipitation - Quantity • Air pressure • Wind - Direction - Speed	• Humidity • Clouds - Cloud type - Cloud coverage • Surface temperature • Ground temperature
Climatic factors	
• Geographical latitude • Proximity to the ocean • Altitude • Exposure • Wind • Sun • Ground cover	Large-scale Small-scale

21677E_1

M5 Climatic elements and factors ▷

Analysis example

Belém is located in the northeast of Brazil at the mouth of the Amazonas and Tocantins in a flat, partially swampy area (reference to a suitable atlas map) 10 m above sea level (coordinates: 1°27'S / 48°28'W).
In Belém it is hot all year round. The annual average temperature in Belém is 26°C. The monthly average temperatures are almost constant. February is the coldest month at 24.5°C, the warmest month is April at 26.7°C.
The annual precipitation in Belém is very high. It is 2,893 mm. The precipitation values are high all year round, however they fluctuate considerably. Precipitation from January to May is always over 300 mm per month with a maximum of 440 mm in March. The monthly precipitation is lower from June to December, lying between 110 and 220 mm, with a minimum of 110 mm in November. The potential evaporation fluctuates. With a maximum of 95 mm in July and a minimum of 54 mm in June it is lower than the level of precipitation. The climate of Belém is therefore humid.
With high temperatures and year-round precipitation, Belém's climate can clearly be classified as humid tropical.
This results in both favourable and unfavourable factors for agricultural land use. The year-round hot humid climate allows for year-round plant growth. There is neither a lack of heat or moisture.
On the other hand, the potentially heavy precipitation (e.g. 440 mm in March – in comparison the wettest month in Essen, Germany is December with 90mm) means that the possible threat of flooding and soil erosion due to torrential rain must be taken into consideration.
Another speculation that can be made on the basis of the climate diagram – but for which it is important to find further proof – is that the soils could be leached and nutrient deficient due to the heavy precipitation.

M6 Analysis of a climate diagram based on an examination that dealt with possible land use in the Belém region.

Heat and precipitation in excess – the permanently humid tropics

Tropical soils and agroforestry

In many regions of the permanently humid tropics the human impact on the tropical rainforest is very high. The people who live there need agricultural land, pasture and firewood. However, soon after clearing, apparently fertile areas of tropical forest can no longer feed the inhabitants and new areas have to be cleared.
In addition to small-scale farming, large areas of the rainforest are used for the cultivation of cash crops and livestock grazing. It is therefore almost impossible to protect the rainforest completely from deforestation. Soil erosion and extreme deterioration of soil quality are the results of this clearing. Are there options to farm the rainforest without destroying it?

1. Name the characteristics of fertile soil.
   Explain the significance of cation exchange capacity in this context (M3/ M4).
2. Present the reasons for drop in yield faced by agricultural farming in the inner tropics in a causal diagram (M3 – M8).
3. Agroforestry (ecofarming) is a form of land use that attempts to imitate the nutrient cycle and structure of the rainforest ecosystem. M5 and M8 schematically show various agricultural crops that can be cultivated based on the principles of agroforestry.
   A Produce a causal diagram to illustrate the agroforestry methods shown in M5.
   B Describe the system of agroforestry presented in M5 and explain how drop in yield can be avoided by using this system.
4. The principles of agroforestry are not only used in small-scale farming, more and more often they are applied in market-oriented cultivation e.g. of cocoa (M8).
   Describe the required growth conditions of cocoa trees and assess the suitability of agroforestry systems for the cultivation of cocoa.
5. Assess the forms of cultivation presented in M8 in the context of sustainability.

Agroforestry, Cation exchange capacity (CEC), Clay minerals, Humic substances, Humus, Soil fertility

M1 The 1:1 clay mineral kaolinite under the microscope

M2 The 2:1 clay mineral montmorillonite under the microscope

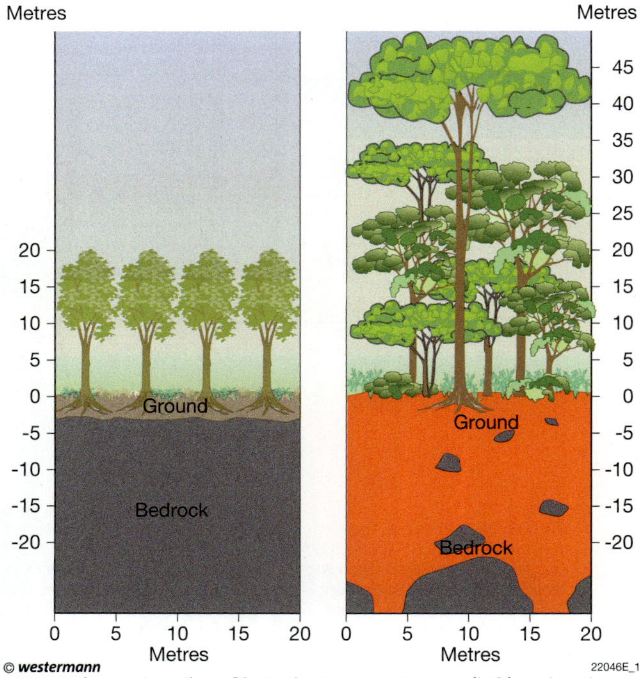

M3 Schematic soil profile in the temperate zone (left) and in the tropics (right)

Soil fertility refers to the ability of the soil to supply plants with nutrients. If there is only a thin layer of soil these nutrients can come directly from the layer of rock under the soil, i.e. from the bedrock. When this is broken down (weathered), the minerals released can be used directly by the plants. Another factor important in soil fertility is the soil's ability to bind nutrients, protect these from leaching and release them again when they are needed. This ability is determined by the cation exchange capacity (CEC). A good soil with a high CEC is like a sponge that binds the cations, preserves them from leaching and releases them when needed by the plant.
Clay minerals play an important role in the CEC. The weathering process of bedrock not only produces minerals that the plant can absorb directly as nutrients, but also forms new minerals – clay minerals. These are particularly good at binding water and mineral and organic nutrients (adsorption) and therefore increase the CEC of the soil. Clay minerals are divided into 1:1, 2:1 and 3:1 clay minerals, the ratio indicating the layer structure of each mineral. With more layers, 2:1 and 3:1 clay minerals are able to accumulate cations and water between these layers as well as on their outer surface (M2). In terms of soil fertility, they are therefore more valuable than 1:1 clay minerals (M1).
Another soil component important for soil fertility is humus. Humus is the name for the dead plant and animal components of the soil. The organic components of the soil are modified just like the rocky material. They can be used directly by plants in the form of nutrients. In addition, humus produces humic substances.
These substances have properties similar to those of clay minerals. They have a very large inner surface and have the ability to bind plant nutrients and release them when required. High humus content therefore increases the CEC. Furthermore, humic substances can store water.

Based on: D. Engelmann, F. Scholz, Diercke Spezial, 2009

M4 Soil fertility

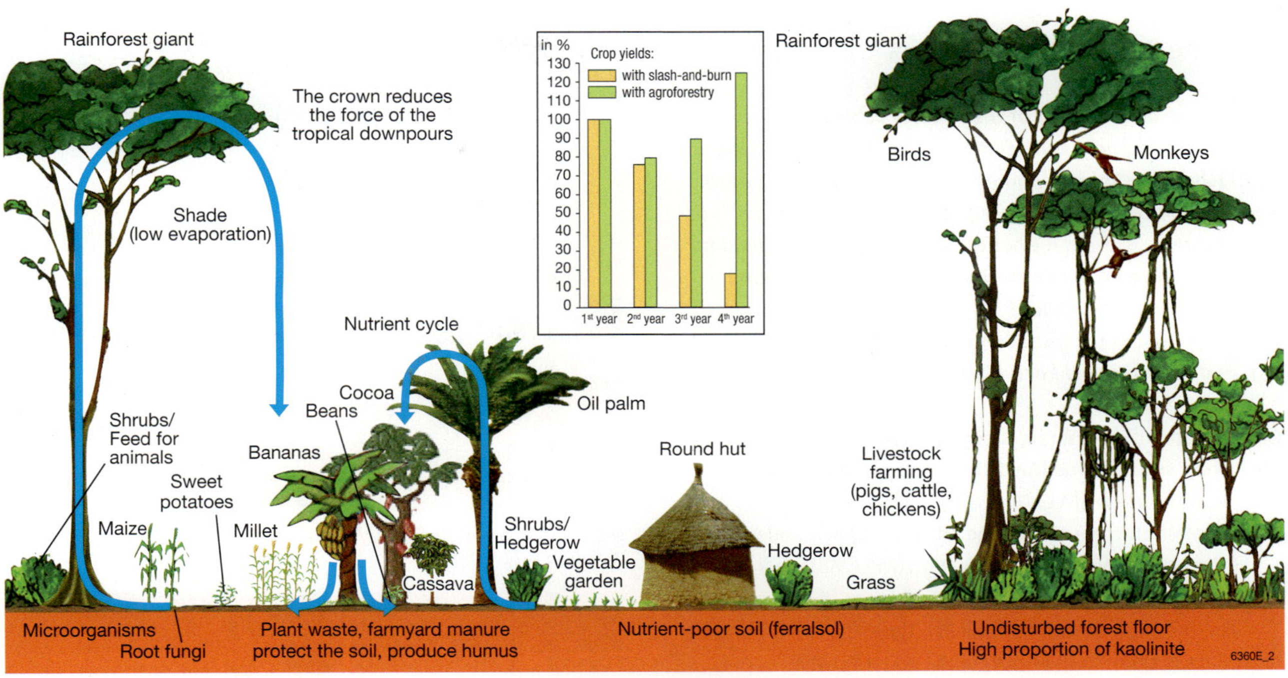

M5 Soil conservation by means of sustainable agriculture in the tropical rainforest

The cocoa tree is a tall, thin undergrowth tree that grows in the shade of larger tropical trees and doesn't tolerate direct sunlight. It can grow to 15 m in height but is pruned to a height of two to four metres in plantations.

The tree's taproots penetrate the soil to a depth of about one metre. The tree has large, smooth, sword-like leaves and is green all year round. The flowers are located directly on the approximately 20 cm thick trunk and on larger branches. They grow out of older wood and produce fruit the whole year over. The cocoa tree develops its first flowers when it is two or three years of age. The tree is most productive – up to 100,000 flowers per year – at the age of ten to twelve years.

For an optimum yield the tree requires an annual average temperature of about 25°C. It is very sensitive to extreme fluctuations in temperature. The temperature should be between 16°C and 35°C and shouldn't drop too far at night.

Precipitation should lie between 1,250 and 2,000 mm per year and be evenly distributed throughout the year. The monthly precipitation shouldn't be less than 100 mm. If the monthly rainfall is less than 60 mm for more than two consecutive months the cocoa fruits will start to wilt having a negative effect on the harvest. The growing soil should be wet, deep and rich in organic matter.

Based on: www.theobroma-cacao.de, 06.11.2013

M6 The cocoa tree

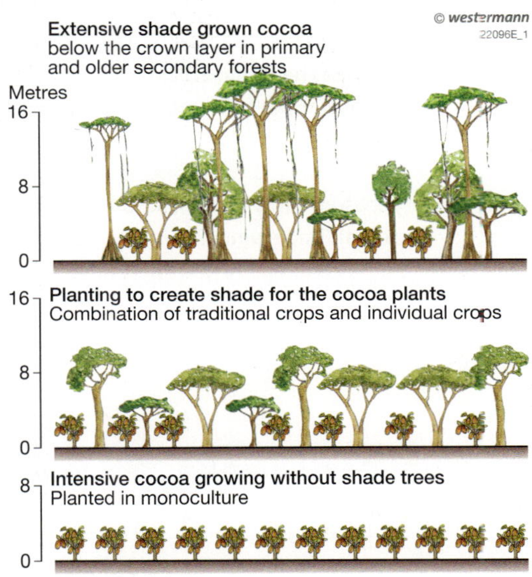

M8 Cocoa agroforestry systems

The most commonly occurring soil type in the permanently humid tropics is ferralsol. As the name implies, iron (Fe) and aluminium (Al) are the main components. These are left over after other, more easily dissolvable. minerals such as calcite and silicon have been extensively broken down by means of intensive chemical weathering. Aluminium is poisonous to plant roots and soil organisms. The iron oxide colours the soil a reddish-yellow.

The predominant clay minerals are 1:1 clay minerals such as kaolinite. The soil therefore has almost no capacity to store nutrients – either naturally existing or artificially applied in the form of fertiliser.

Even if plant material constantly falls to the ground in an intact rainforest, the humus layer remains thin because – due to the climatic conditions – the organic matter is decomposed very quickly (about five to ten times faster than in the temperate latitudes). The mycorrhizae, indispensable in the recycling process of nutrients (see pp. 14/15), take time to regenerate after the vegetation has been burnt or the soil has been ploughed.

M7 Ferralsols – typical soils in the permanently humid tropics

M9 Ferralsol soil profile, Brazil

Rainy and dry seasons – the semi-humid tropics

From the Equator to the Tropics

Once we leave the tropical rainforests of the Equator behind, we encounter grassy landscapes stretching polewards that account for almost a quarter of land area worldwide. The Spanish named these areas "sabana", which means grassy plain and from which the present-day name "savanna" is derived. As we cross these landscapes we notice large differences, depending on the time of year but also on the distance we are from the Equator. How can these differences be explained?
Why is it that in some regions the grass can grow up to five metres high (elephant grass) and in others only knee-high?

Ⓞ **1. A** Locate Abéché, Birao, Bangui and Bobo Dioulasso on the map and describe the distribution of precipitation for each location (M1/ M4).
B Explain the reasons for the alternating wet and dry seasons and their different duration (M5/ M8).
2. Produce an overview of the different types of savanna (climate and vegetation). Assign each type to a corresponding climate diagram and photograph (M1–M4, M7).
3. Based on the example of the Baobab tree explain how plants have adapted to the alternating wet and dry seasons (M7, internet).
4. Describe the global distribution of the different types of savanna (M6).
Ⓢ **5.** Analyse the climate diagram for Kisangani and explain the causes for the two precipitation peaks (M1/ M5/ M8).

Dry season, Rainy season, Savanna (wet, dry, thornbush savanna), Trade wind, Zenith

M2 Savanna with umbrella thorntrees (acacia)

Wet savanna	Dry savanna	Thornbush savanna
> 1,000 mm precipitation	c. 500 – 1,000 mm precipitation	< 500 mm preciptitation
Rain green tropical forests that shed their foliage in the dry season	Grassland (60 cm to 1.5 m in height)	Grasses max. knee-high, still unbroken in contrast to the grass cover of the semi-desert
Shrubs and low trees that keep their foliage all year round	Sparse tree population	Isolated trees
Broadleaf, 2 – 5 m elephant grass, gallery forests along rivers due to the year-round water supply, some evergreen		Thornbushes and thorntrees (acacia)

M3 Types of savanna

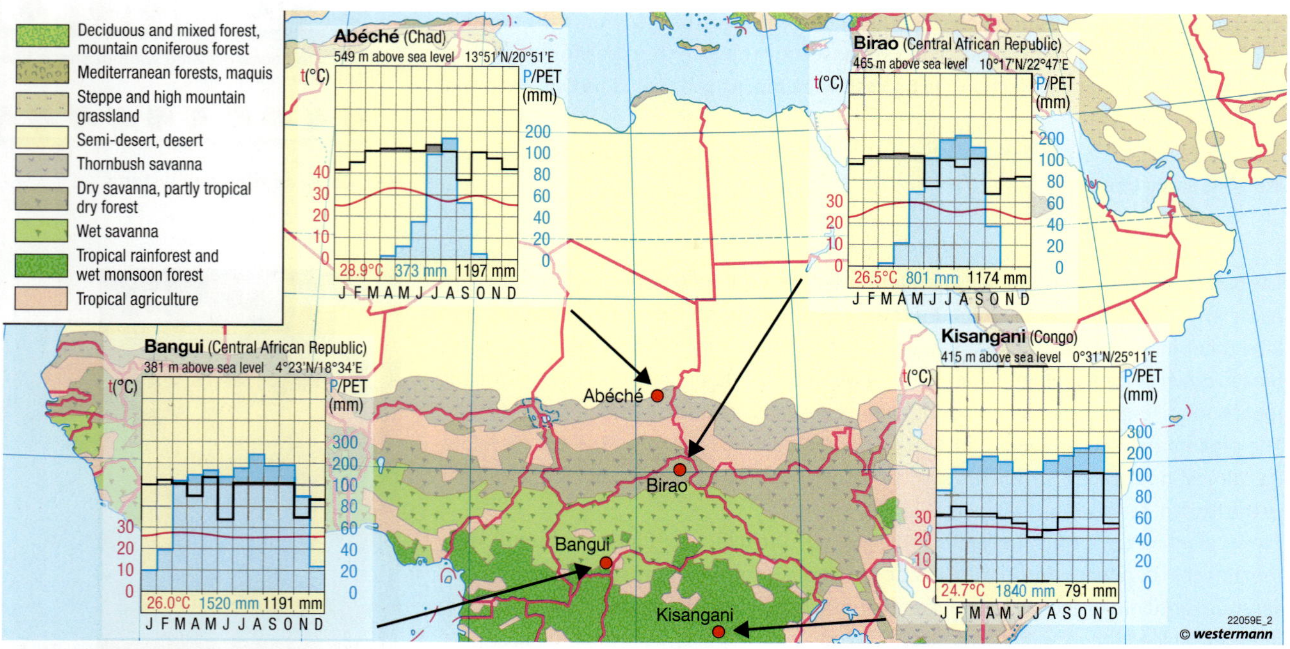

M1 Climate and vegetation zones of the savannas. In areas subject to intensive agricultural use (tropical farming) the original vegetation is to a large extent no longer intact.

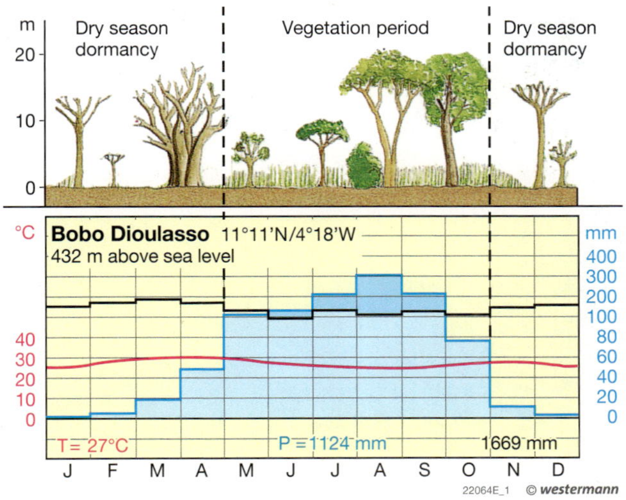

M4 Savanna: Climate und vegetation

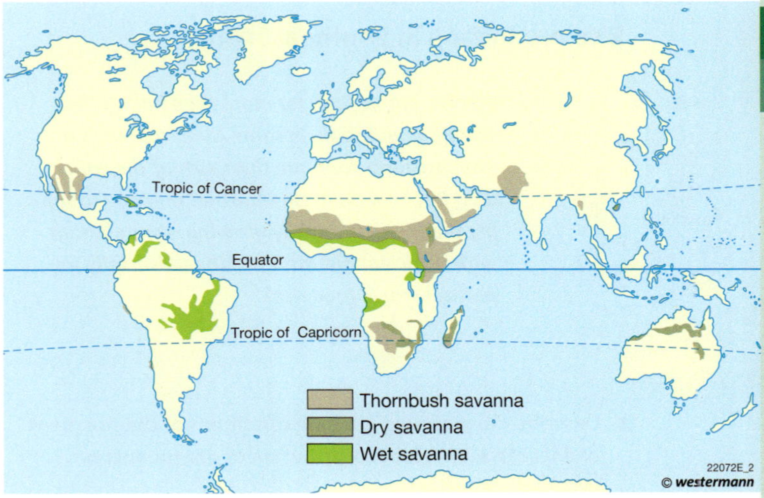

M6 Distribution of the savannas

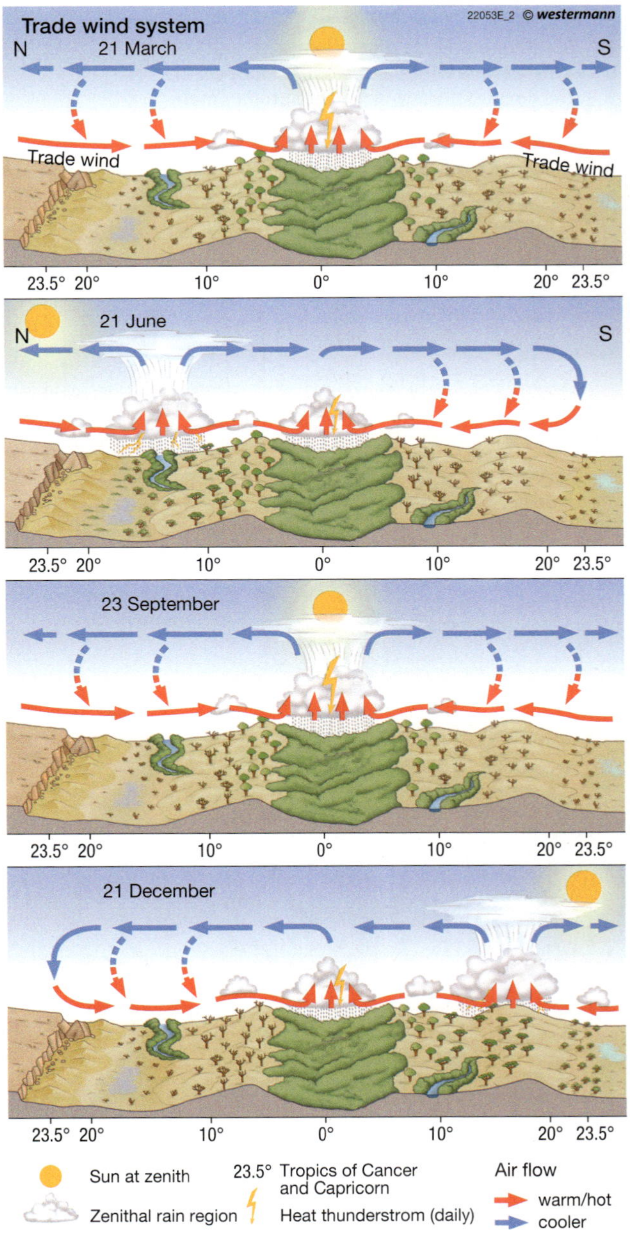

M5 Trade wind system (see also Hadley cell, p. 17)

M7 A Baobab tree in the African savanna (Tanzania). During the rainy season, the trunk can store up to 120,000 litres of water. The trunk can grow several centimetres in circumference during this process. Small feather-shaped (pinnate) leaves that are shed at the end of the rainy season reduce the evaporation surface.

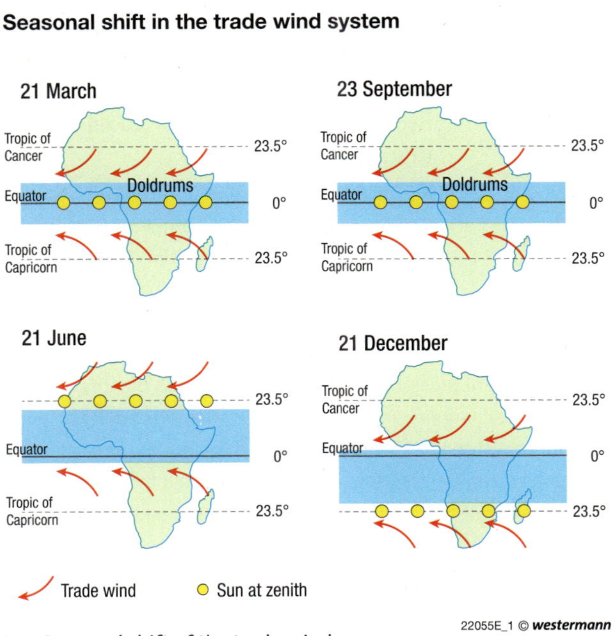

M8 Seasonal shift of the trade wind system

Rainy and dry seasons – the semi-humid tropics

Rice production in Southeast Asia

Rice is a staple food for more than half of the world's population. In some Asian countries rice even accounts for more than 80% of the entire diet. But how do these countries manage to supply their – often rapidly growing – populations with sufficient quantities of the universally important staple food rice?

1. Describe the preferred growing conditions of the rice plant and research the global distribution of rice production (M1–M6, M9, atlas).
2. ⓞ **A** Compare the rice production of individual countries in 2002 and 2012 (M6).
   B Present M6 as a pie chart and formulate the core message of your illustration.
3. Thanks to the Green Revolution the growing population in Southeast Asia had sufficient rice. Explain (M6–M10).
4. Demonstrate how the alluvial plain of the Mekong developed into a globally significant region for rice production (M3/ M4/ M5/ M8/ M9).
5. ⓢ In the Khmer Empire (9th – 14th cent., present-day Cambodia) cities and architectural structures (e.g. the temple complex Angkor Wat) unthinkable for Europe of the time were built. The capital Angkor Thom, with a population of over 1 million at its peak, is considered to have been the largest city of the Middle Ages.
   Research (M8/ M9) how it was possible to feed so many people and why the agricultural methods used at that time are rarely used today.

Green Revolution, High Yielding Variety (HYV), Monsoon

Rice has been cultivated in Asian countries such as China and Thailand for over 5,000 years. The rice plant Oryza sativa comprises many thousands of varieties, most of which belong to the two major subspecies Oryza sativa indica and Oryza sativa japonica.

Rice is a plant with high temperature requirements. The subspecies indica germinates at temperatures of 18°C and above and japonica at 10° - 12°C and above. Both subspecies require at least 20°C to grow, ideal temperatures are up to 30° – 32°C. During the growth period of 120 – 180 days rice needs at least 800 mm of precipitation, 1,200 – 1,500 mm is most favourable. The ecologically significant characteristic of rice is its ability to germinate under water and absorb oxygen through organs above ground. This means that flooded floodplains can be used for cultivation. Rice is one of few cultivated plants that does not need to be planted in rotation. This allows for several harvests each year.

Based on: Diercke Weltwirtschaftsatlas Rohstoffe, Agrarprodukte, 1981; Geographie heute, 187.

M1 Growth conditions of the rice plant

M2 Rice terraces in the village of Batad (Luzon/Philippines)

M3 Summer monsoon circulation (schema): Whereas the winter monsoon is identical to the northeast trade wind, the summer monsoon changes direction on crossing the Equator as a result of the Coriolis force

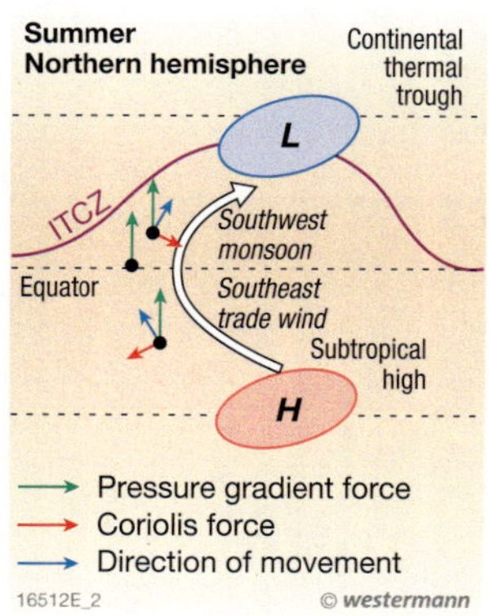

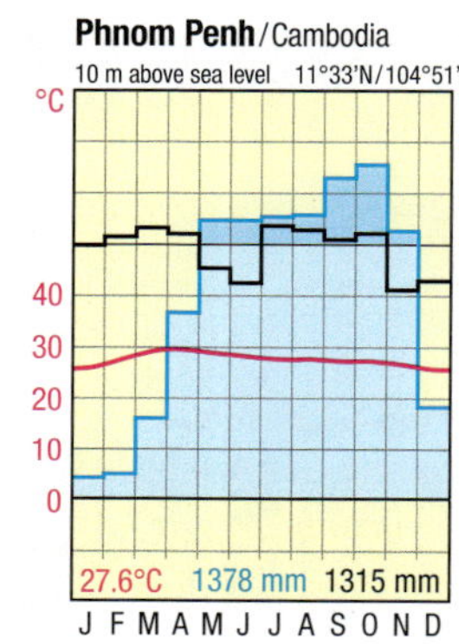

M5 Climate diagram for Phnom Penh

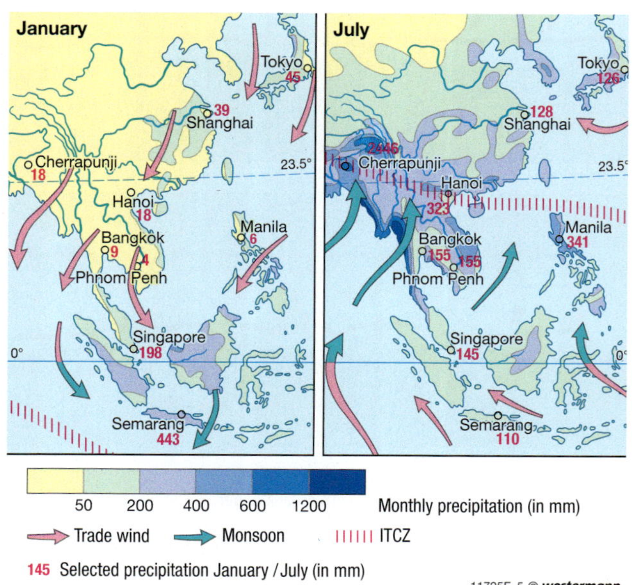

M4 Twice-yearly changes in wind direction determine the climate of Southeast Asia

Countries	Harvest vol. in mill. t 2002	Harvest vol. in mill. t 2012
China (PRC)	174.5	204.3
India	107.7	152.6
Indonesia	51.5	69.0
Vietnam	34.5	43.6
Thailand	27.9	37.8
Bangladesh	37.6	34.2
Myanmar	21.8	33.0
Philippines	13.3	18.0
Brazil	10.5	11.4
Pakistan	6.7	9.4
Cambodia	3.8	9.3
U.S.A.	9.6	9.0
Japan	11.1	8.5
Egypt	6.1	6.5
Korea, Republic	6.7	6.4
Nepal	4.1	5.1
Nigeria	2.9	4.8
Madagascar	2.6	4.0
World	571.4	718.3

Source: FAOSTAT 2013

M6 (Leading) rice producing countries

	1967	2012	Growth (in %)
Cambodia	1,216	3,000	146.7
Indonesia	1,759	5,139	192.1
Laos	845	3,736	342.1
Malaysia	2,017	4,000	98.3
Philippines	1,380	3,845	178.6
Thailand	1,750	3,000	71.4
Vietnam	1,916	5,631	193.9
For comparison			
Australia	7,407	8,910	20.3

Source: FAOSTAT 2013

M7 Rice harvest in the Southeast Asian countries (in kg/ha)

M9 Rice production in Southeast Asia

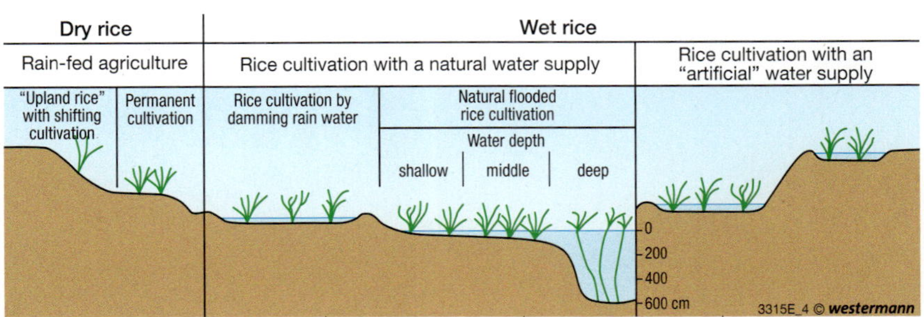

Dry rice (also upland rice)	Wet rice	Wet rice	Wet rice
Cultivated without natural or artificial accumulation of water. During the growing period it needs at least 200 mm of rain per month. Dry rice is widespread in the permanently humid tropics, the yields are however significantly lower than with wet rice cultivation.	Flooded in the main growing season by damming rainwater (with small retaining walls around the fields)	Flooded with flood water from rivers in the lowlands	Artificial irrigation (storage ponds, canals, wells)

M8 Methods of rice production

Until well into the 1960s, many Asian countries feared that the supply of their staple food rice was no longer guaranteed for the rapidly growing population.

But the introduction of High Yielding Varieties (HYV) during the Green Revolution proved to be revolutionary. By crossbreeding, the International Rice Research Institute (Philippines) succeeded in combining positive properties of traditional varieties in one rice variety (HYV-IR8).

By using this high yielding variety, modified levels of fertiliser, crop protection measures, adequate and controlled irrigation as well as careful seedbed preparation, a boost in yield was achieved.

Based on: K. Vorlaufer, Südostasien, 2009

M10 The Green Revolution

Year-round aridity – deserts and semi-deserts

Las Vegas – unlimited urban growth in the desert?

"Welcome to Fabulous Las Vegas" is how the city of superlatives greets its visitors – here to pursue the promise of experiencing the world's grandest hotels, most exclusive shows and highest concentration of casinos. The nightly sea of lights and spectacular fountains in front of the hotels seduce the visitors into quickly forgetting that they are in the middle of a desert. In 1910, Las Vegas had a population of only 1,500 – today the population is about 400 times greater. Where does the city in the middle of the desert get its water supply? And more importantly, can it continue to grow in the future? Or has Las Vegas reached its limits?

1. Describe the location of Las Vegas and characterise its climate (atlas, M1/ M12).
2. a) Demonstrate the spatial development of the metropolitan area and the population trends of the city (M1/ M5/ M9).
   b) Investigate the development of visitor numbers (M11).
3. a) Investigate the development and type of water consumption (M6/ M7/ M10).
   b) Explain the city's water supply and the consequences of water catchment (M1/ M3/ M7/ M10).
4. ⓞ Quo vadis, Las Vegas?
   A Discuss issues and show possible perspectives for the future (M7/ M9/ M10).
   B Produce a poster that illustrates the water supply issues in Las Vegas.
5. ⓢ Compare the population and visitor numbers (M9/ M11).

Artesian well

M2 In the middle of the desert

Artesian well

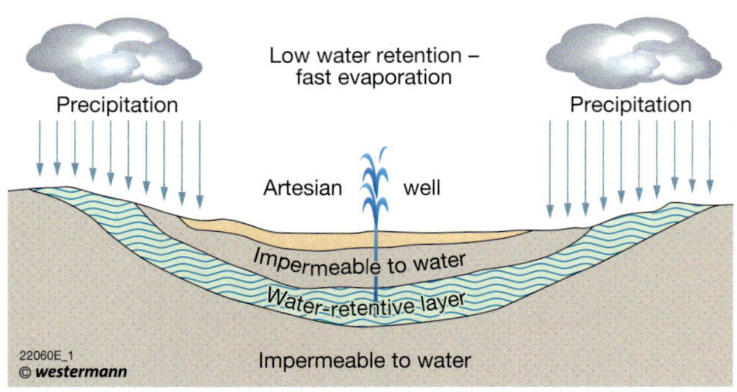

When, in 1829, a caravan of travellers ran out of water about 160 km north of the present-day city, their scouts finally found water after several days of riding. They discovered an artesian well – the Las Vegas Springs. An artesian well forms when positive pressure forces groundwater to the surface.

M3 Water from the desert

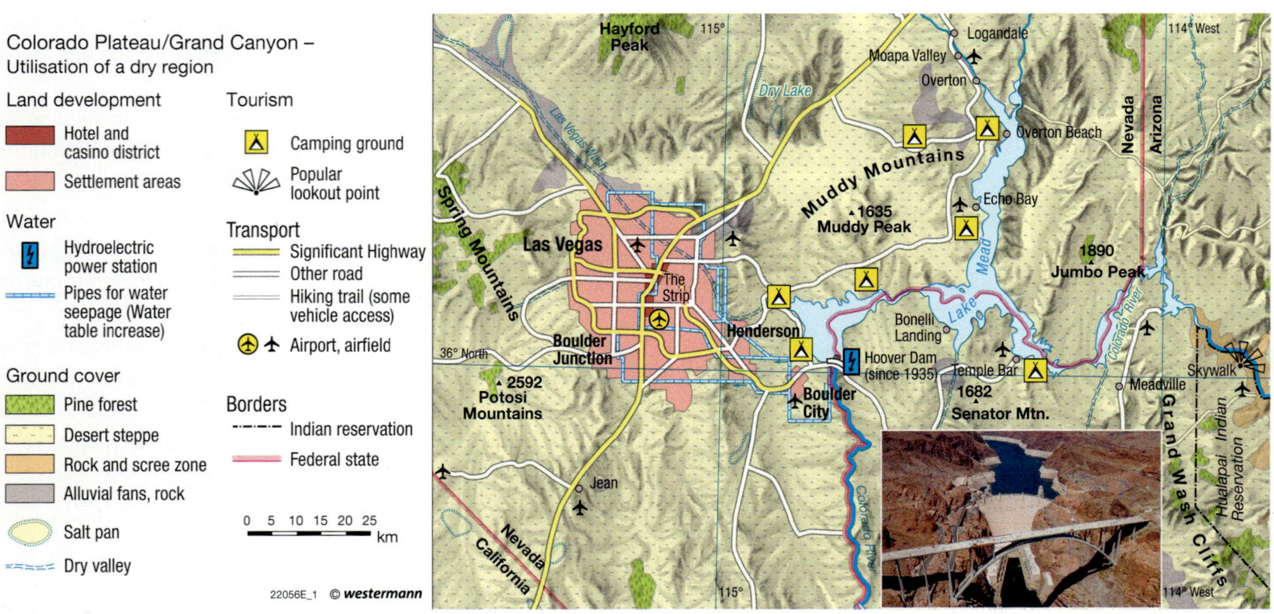

M1 Colorado Plateau and Las Vegas 2010

M4 Las Vegas – Hotel complex

M8 Las Vegas

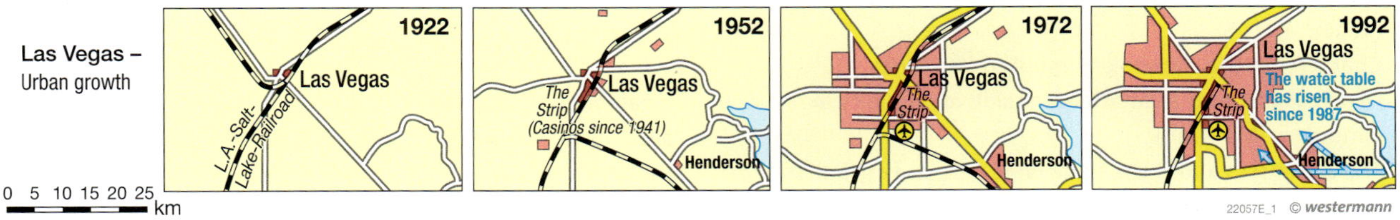

M5 Las Vegas – Urban growth

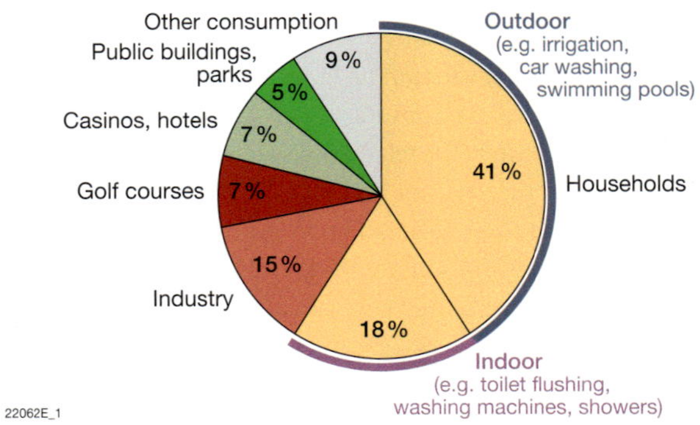

M6 Water consumption based on sectors

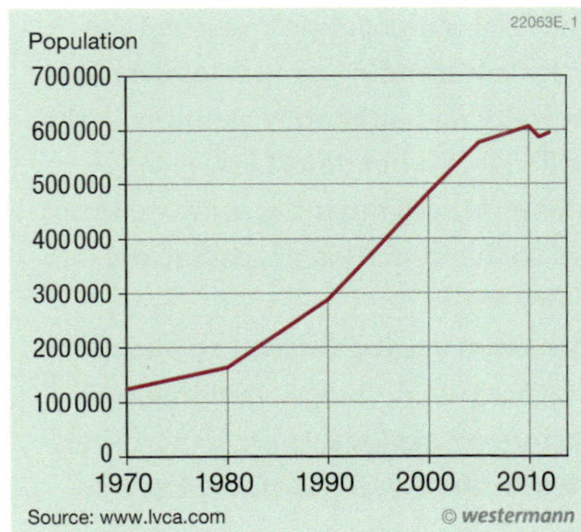

M9 Las Vegas: Population trend

Visitor numbers	
1970	6,787,650
1980	11,941,524
1990	20,954,452
2000	35,849,691
2005	38,566,717
2010	37,335,436
2011	38,928,708
2012	39,727,022

M11 Las Vegas: Visitor trend

Lake Mead, a reservoir formed by the Hoover Dam, is fed by the Colorado River and supplies Las Vegas with drinking water and electric power. Above-average rainfall in the 1970s and 1980s resulted in the reservoir reaching capacity. In recent years, the water level has dropped by 30 metres.

This is made visible by white stripes on the shores of the lake: crusts of mineral salts. Water shortage in the Colorado River is the reason for the low water levels in Lake Mead. Dry periods, less snowfall in the Rocky Mountains and higher temperatures have changed the river. The two intake pipes, or straws, which have been used up till now, are currently only a few metres below the water's surface. A third intake – Lake Mead Intake No. 3 – has been under construction since 2007. This will withdraw water from the bottom of the lake, meaning that as long as the Colorado River flows into Lake Mead, water can flow to Las Vegas through this pipeline. The "Third Straw" is scheduled for final completion in 2015.

Based on: PG 3/2006, PG 1/2009, www.ingenieur.de, 1.12.2013

M7 Water availability

Year	River water supply	Groundwater withdrawal
1922	0 m³ (no river water supply)	21 mill. m³ (groundwater level at -5 to -20m)
1952	20 mill. m³ (Lake Mead/ Colorado River)	42 mill. m³ (groundwater level at -26m)
1972	80 mill. m³ (Lake Mead/ Colorado River)	85 mill. m³ (groundwater level at -40m)
1992	338 mill. m³ (Lake Mead/ Colorado River)	84 mill. m³ (groundwater level at -51m)
2010	526 mill. m³ (Lake Mead/ Colorado River)	82 mill. m³ (groundwater level at -45m)

M10 River water supply and groundwater withdrawal

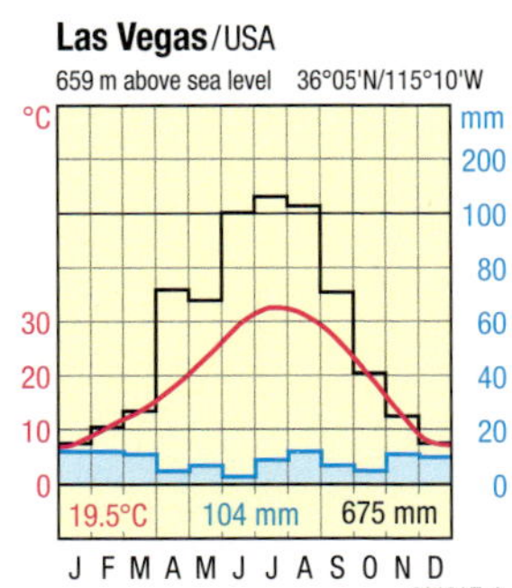

M12 Climate diagram for Las Vegas

Year-round aridity – deserts and semi-deserts

Causes of desert formation

Deserts cover about 15 – 20 percent of the Earth's land surface. Deserts can have rocky, stony, salty or sandy subsoil and sparse vegetation or no vegetation at all. All deserts have one thing in common: an extreme water shortage. Everything that lives in the desert – animals, plants and human beings – have to develop strategies to deal with this water shortage. What causes the formation of deserts? Why are deserts so dry?

1. Produce a table and enter the names of the deserts mapped in M9 and their location e.g. continent, countries (atlas).
2. Characterise the climate of Timbuktu (M1/ M2).
3. Give reasons for the variations in relative humidity in Timbuktu (M1 – M4).
4. The air masses above the trade wind deserts (e.g. Sahara) are so dry that they absorb a large proportion of the moisture from their surroundings. Using the terms "absolute humidity", "condensation", "dew-point", "relative humidity", "evaporation" and "cloud dissipation", describe the processes that lead to the descending air masses over the Sahara being extremely dry and ultimately resulting in the formation of the world's largest hot desert (M3 – M5).

Ⓞ 5. **A** Based on your findings in assignment 4, compare the processes that lead to the formation of coastal and rain shadow deserts (M6/ M7).
B Assign the deserts named in assignment 1 to one of the three desert types (trade wind, coastal, rain shadow).

Ⓢ 6. Calculate the temperature of previously humid (100% relative humidity), 20°C warm air after it has crossed an approx. 2,000 m high mountain range and explain what consequences this has for the windward and leeward regions of the mountain range (M6).

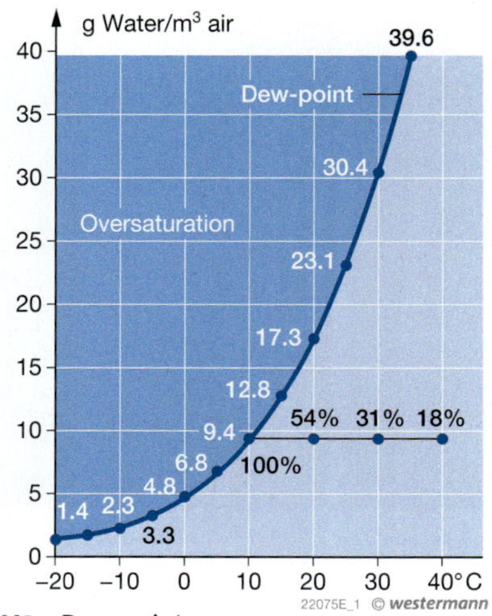

M3 Dew-point curve

The amount of water vapour in the air is measured in g/m³ and is called absolute humidity. The maximum amount of water vapour that can be absorbed by the air (maximum humidity) depends on the temperature and can be derived from the dew-point curve (M3). The dew-point is the temperature at which water vapour condenses. The relationship between absolute and maximum humidity is referred to as relative humidity and is expressed as a percentage. When dew-point is reached the relative humidity is 100% – the air cannot absorb any more water vapour. The lower the relative humidity, the more water the air can still absorb.
In the world's hot deserts the relative humidity is extremely low. This is partially due to the very hot temperatures that prevail. Because of the lack of cloud cover, solar radiation is often more intense than at the Equator. More importantly, the air masses above deserts are particularly dry because they have already lost their moisture in other regions.

M4 Water in the atmosphere – relative und absolute humidity

Absolute humidity, Coastal desert, Dew-point, Rain shadow desert, Relative humidity, Trade wind desert

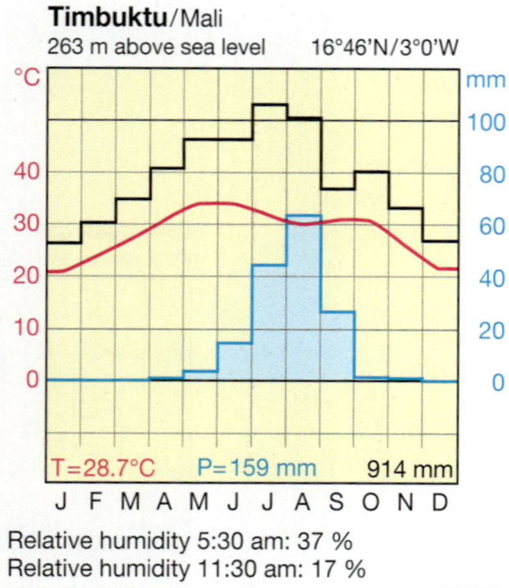

M1 Climate diagram and variations in relative humidity during the day (July), Timbuktu, Mali

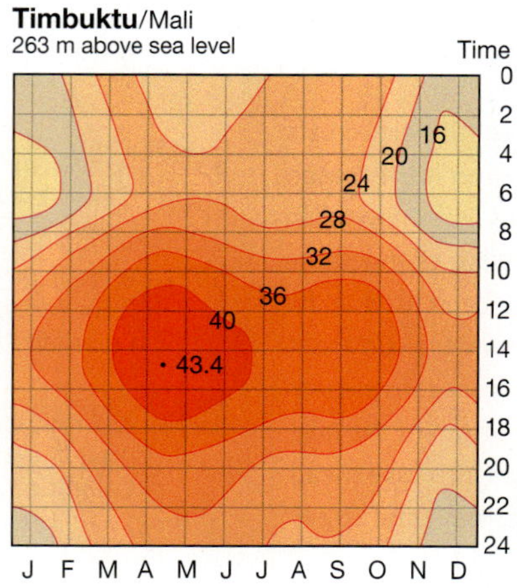

M2 Thermoisopleth diagram for Timbuktu, Mali

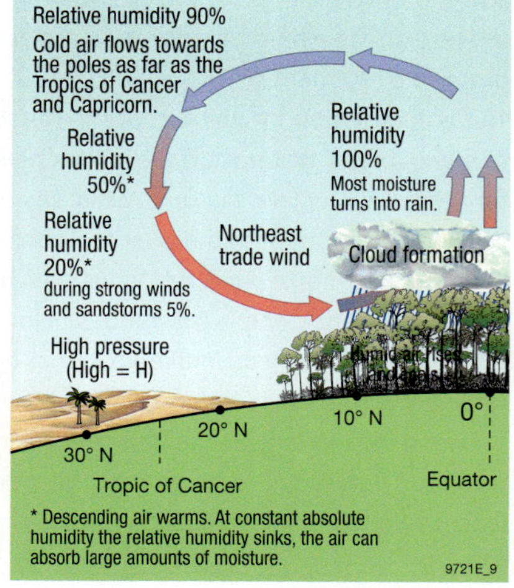

M5 Formation of trade wind deserts

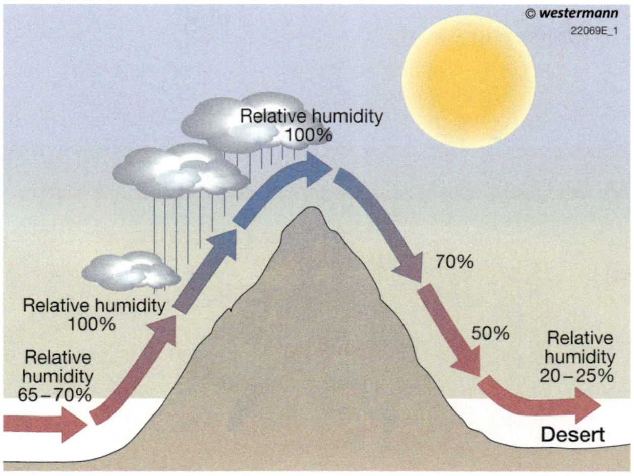

M6 The formation of rain shadow deserts

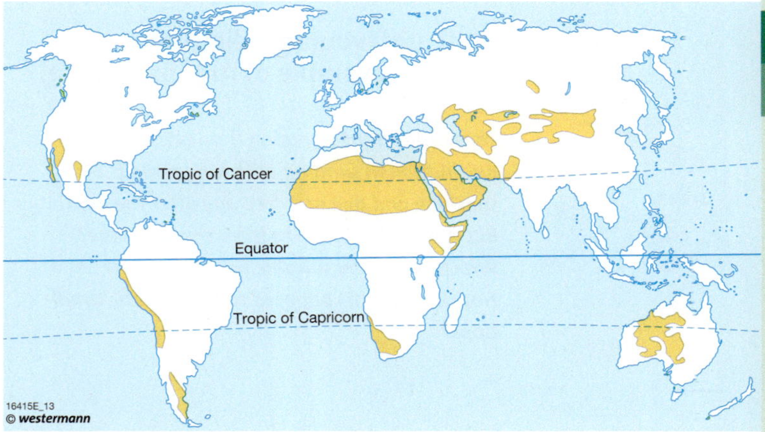

M9 Distribution of deserts and semi-deserts

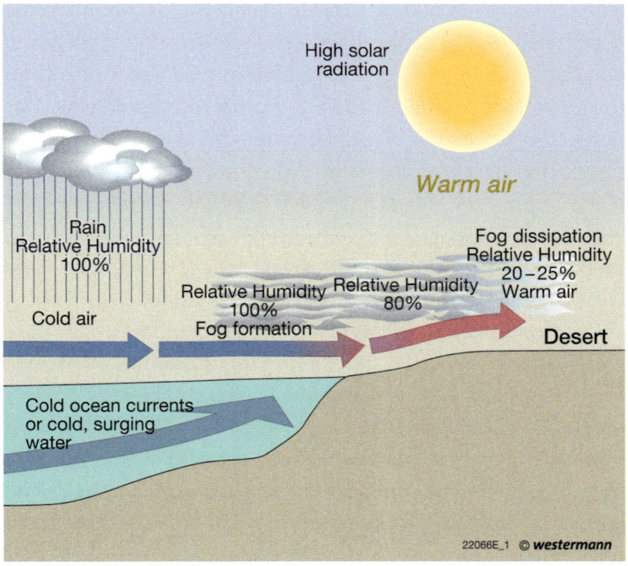

M7 The formation of rain shadow deserts

Everybody who has shivered after going swimming is familiar with the cooling effect of evaporating water. However, when water condenses – transitions from water vapour to liquid water – it releases heat. These processes play an important role in the formation of deserts.

When air rises and cools the water held in the air condenses. This process of condensation releases energy in the form of heat. A rule of thumb is that ascending, dry air cools at a rate of about 1°C per 100 m in altitude (dry adiabatic ascent). On the other hand, humid air only cools at a rate of approx. 0.6°C per 100 m (saturated adiabatic ascent) because it receives the thermal energy from the process of condensation.

Conversely, dry, descending air heats at a rate of approx. 1°C per 100 m in altitude (dry adiabatic descent). Humid air heats more slowly when descending because part of the thermal energy is extracted when the moisture evaporates. Therefore, in the case of saturated adiabatic descent the air is only warmed at a rate of approx. 0.6°C per 100 m.

M10 The influence of condensation and evaporation on temperature

M8 The Atacama Desert, Chile – The driest desert on Earth

Abundance and shortage – the winter-humid subtropics

Growth despite unfavourable distribution of precipitation – the example of Argolis (Greece)

The Mediterranean region is an attractive tourist destination in summer with long periods of sunshine and one rain-free day after the next. But how is it possible to farm the land here? Which plants are suitable for cultivation? How have plants adapted to the climatic conditions?

1. Locate the Argolis region and describe its climate (M1). Derive favourable and unfavourable factors for agricultural use.
2. The Argolis region is intensively farmed today. Investigate how this is achieved and characterise the present-day cultivation structure (M2–M6/ M8).
3. Ⓢ The karstic spring in Kiveri is a blessing for the agricultural sector in the Argolis region, but also a particular feature of the natural environment. Consider what other possible solutions there are to obtain water for irrigation.
4. The plants of the Mediterranean region have adapted to the climatic conditions. Describe different adaptation mechanisms of some typical Mediterranean plants (M5).

Groundwater, Karst

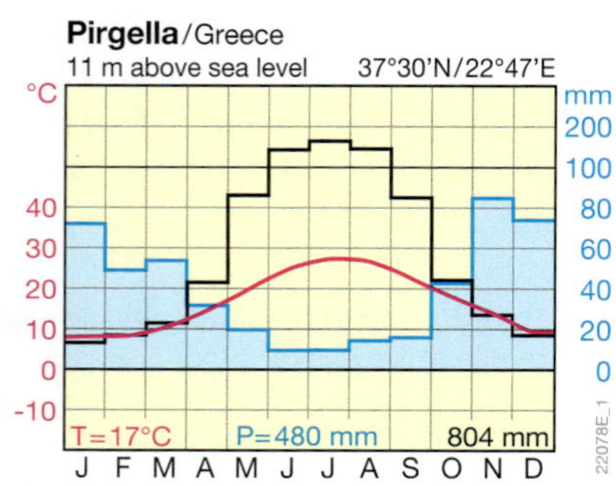

M1 Climate diagram Pirgella (Greece) ▷

In the 1930s, only a small area of land in the basin surrounding the Greek townships of Argos and Nauplia was irrigated by means of groundwater pumps. At that time, farmers produced primarily for their own consumption because the otherwise fertile plain was prone to water shortage. Yet, water literally flows just around the corner. The high plains near Tripoli, where there is always abundant precipitation, have almost no natural drainage. The water finds its way through karstic rock and comes to light several hundred metres deeper either on land or at sea. Whereas springs on land have always been utilised, for centuries farmers could only stand by and watch as the precious resource flowed into the sea.

A German engineer harnessed one such spring near the township of Kiveri – a spring that pours out water at a constant rate of 13m³ per second. The facility was inaugurated in 1972. A dam that separates freshwater from seawater encloses the spring that lies only 25 m from the steep coast. The water is then channelled from several reservoirs via an open canal system as far as the other side of the Gulf of Nauplia.

Based on: www.argolis.de.htm, 28.8.2013 and Diercke Handbuch, 2006

M3 Freshwater spring and reservoirs in the Gulf of Nauplia

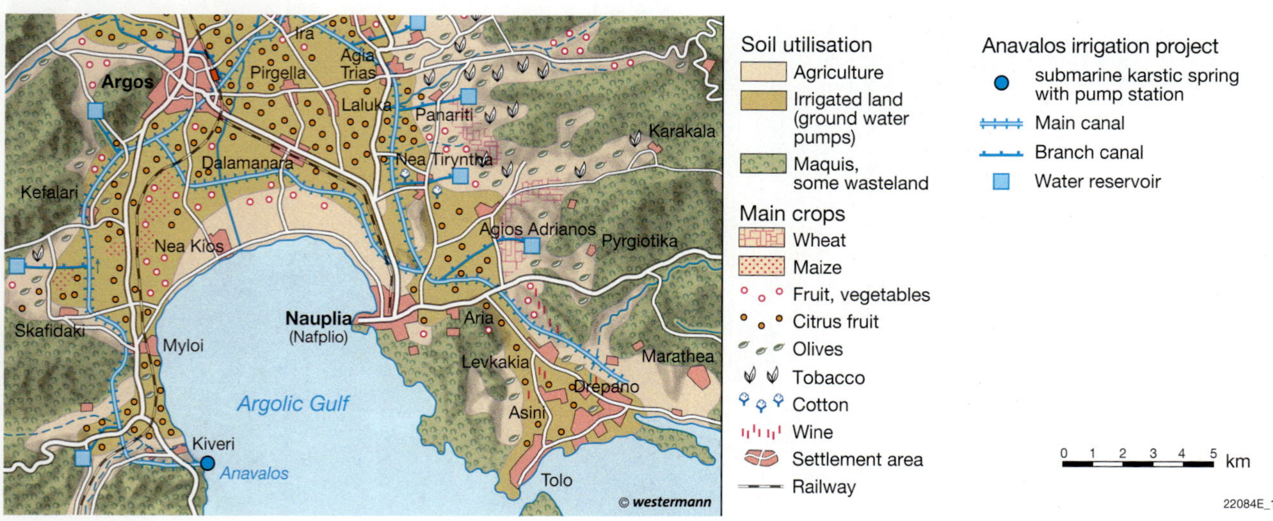

M2 Argolis (Greece)

The olive tree (Olea europaea) is native to the eastern Mediterranean region and the Middle East. It has been cultivated as one of the most economically significant plants since the third millennium BC. In ancient times, the olive branch was considered a symbol of victory and peace. The gnarled, often bizarre tree can grow to a very great age. 800 – 1,000-year-old trees are not uncommon. The olive tree thrives at moderate temperatures of 15 – 22°C, in frost-free conditions and with 500 – 700 mm precipitation.

The tree develops an extensive root system of up to 6 m in depth and 12 m in width, enabling the tree to survive on only 200 mm rain and less favourable soil conditions. The evergreen foliage is adapted to the summer aridity.

Based on: M. Geiger: Apfelsinenbaum und Ölbaum, In: PG 11/ 1990

M4 The olive tree

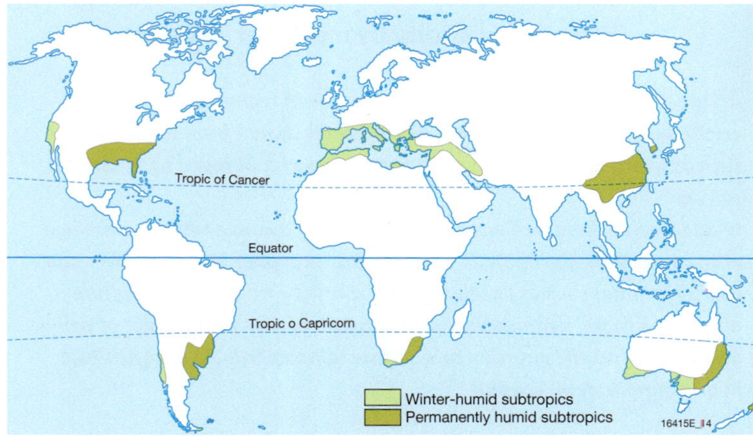

M7 Distribution of the winter-humid and permanently humid subtropics

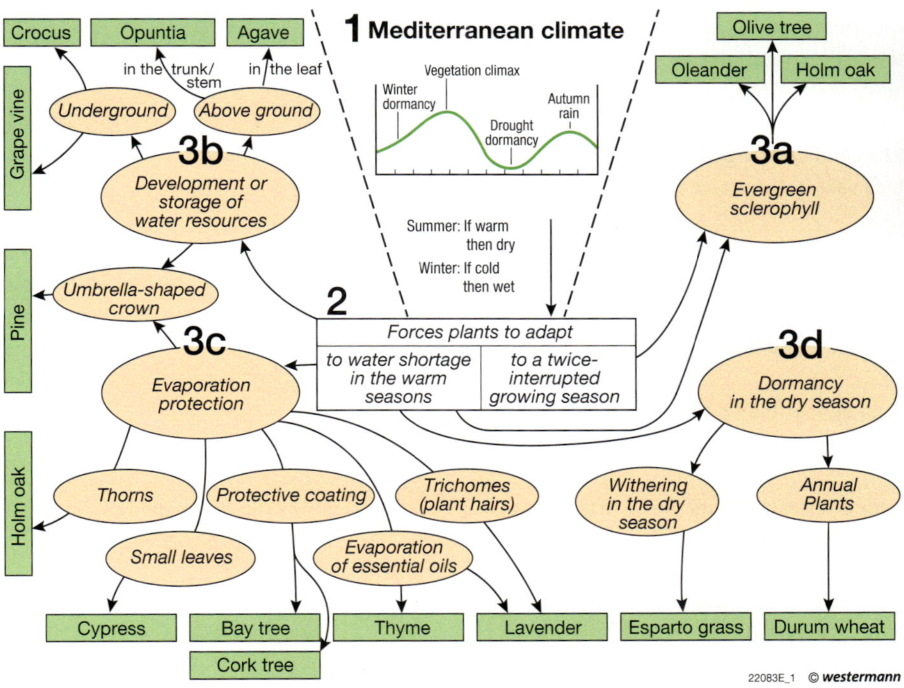

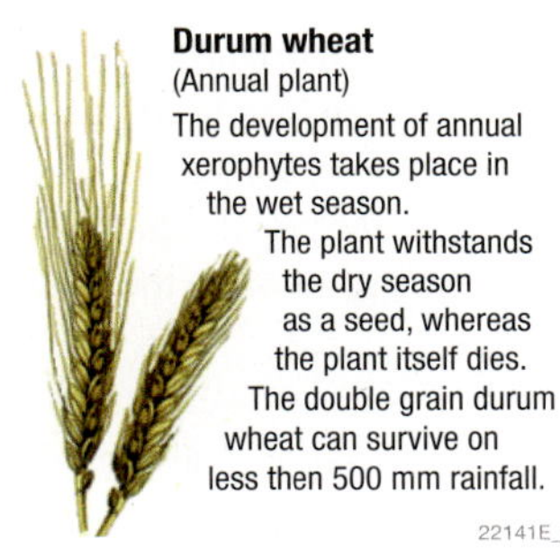

M5 Adaptation of plants to the Mediterranean climate

M6 The orange tree

The orange tree (Citrus sinensis) originated in southern China, where it has been cultivated since the 2nd century BC. The German word for orange, "Apfelsine", is derived from the Dutch word "appelsien", which means "apple from China".

The orange tree requires high quantities of sunshine, a warm climate with annual average temperatures of 17 – 20°C, and regularly distributed, abundant precipitation of at least 1,200 mm per annum. The evergreen tree produces new large, laurel-like, shiny, leathery leaves three times a year. The spring shoots are the strongest with the most blossoms and produce the most prolific harvest in autumn/winter. The white flowers are very fragrant. Fruit develops out of these in the form of large berries with a wax-covered peel containing essential oils. The fruit pulp is made up of spongy juice sacs stored in wedges divided by membranous walls.

Based on: M. Geiger: Apfelsinenbaum und Ölbaum, In: PG 11/ 1990

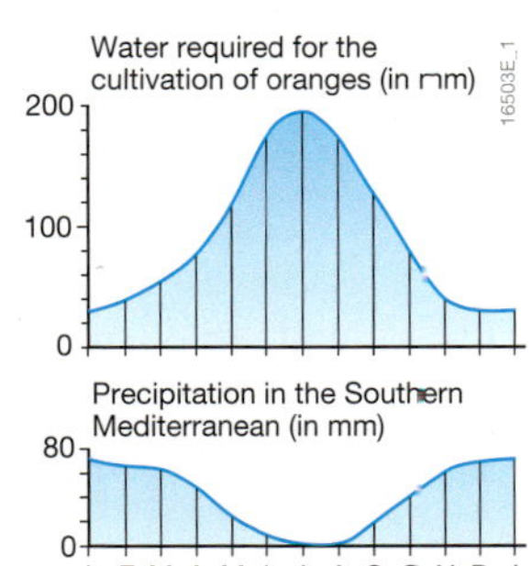

M8 Water required for the cultivation of oranges in the Mediterranean region

Abundance and shortage – the winter-humid subtropics

Overcoming aridity by means of irrigation

Irrigation makes agricultural activity possible in regions that would otherwise be too dry. The water needed is usually drawn from surface water such as rivers, lakes and reservoirs or groundwater pumped to the Earth's surface.
In addition to the risk of water resources being overexploited – combined with conflicts over usage, e.g. between the local population, farmers and tourist facilities such as hotels and golf courses – irrigation agriculture is also linked to soil degradation (deterioration of soils to the point of being unusable). Thus the question arises: Is sustainable irrigation agriculture in arid regions at all possible?

1. Describe and compare the irrigation methods presented on this page (M1/ M4/ M5/ M7/ M8).
2. (O) **A** Explain the processes that lead to soil degradation through salinisation (M3/ M6, experiment).
   B Produce a causal diagram that shows how irrigation can lead to salinisation and soil degradation (M3/ M6, Experiment).
3. Both the installation of drainage ditches and drip irrigation systems represent solutions with which salinisation can be avoided.
   Describe each technology and explain how salinisation is avoided (M1/ M5/ M10).
4. Evaluate these technologies in light of water conservation and their implementation in both small-scale and industrial farming.
5. Discuss which irrigation methods would be feasible for the Argolis region in light of water stress and the risk of salinisation (p. 30/31).
6. (S) Examine whether the cultivation of saline-tolerant plants could be a solution to the problem (M9, internet).

Irrigation cultivation, Soil degradation, Soil salinisation

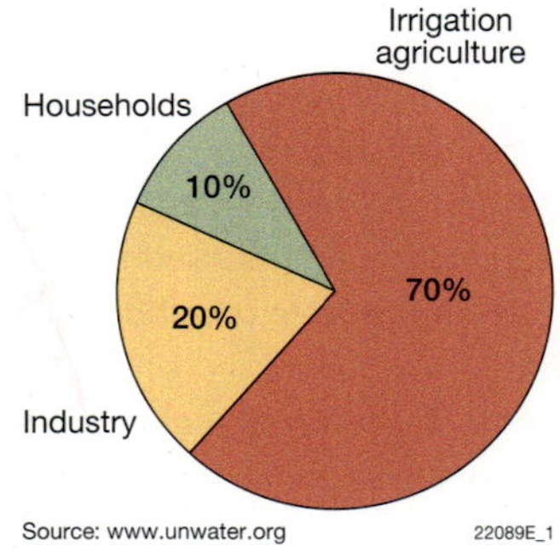

M2 Distribution of global freshwater consumption across households, industry and irrigation agriculture

Soil salinisation in the Grand Valley, Colorado, USA

Soil salinisation is a huge problem. The salinisation of soils results in plant growth being severely restricted. Many plants cannot grow on saline soils making the soil unusable for agricultural purposes. It is estimated that today almost 10 – 15% of irrigated areas in arid regions are already affected by salinisation. In some countries the figures are significantly higher, in Australia it is 20 percent, in Egypt it is 30 percent and in Iraq it is as high as 50 percent. The global trend is continuing to rise.

Based on: C. O. Stockle. Environmental impact of irrigation: a review, www.swwrc.wsu.edu, 25.8.2013

M3 Soil salinisation - A problem of global dimensions

	Surface irrigation	Sprinkling	Drip irrigation possibly combined with foil tunnels
Technology	Areas are flooded by means of ditches and channels.	Areas require sprinkling (e.g. with sprinkler systems)	Small amounts of water are applied directly to the roots of the plant by means of hoses fitted with small jets.
Evaporation loss	high	high	low
Seepage loss	moderate	low	low
Water use efficiency	40 – 50 percent	60 – 70 percent	80 – 90 percent
Installation costs	low	high	very high
Demands on soil and subsoil	Heavy, non-sandy soils, levelling may be necessary	All soils, no or very minimal slope	All soils, every slope
Further aspects	Labour-intensive, not capital-intensive	Capital-intensive (installation and operation), Fertiliser and pesticides can be administered in the water, sensitive to wind	Capital-intensive (primarily installation) Labour-intensive (maintenance, water supply must be precisely regulated) Fertiliser and pesticides can be mixed with the water.
Suitable crops	Water-tolerant varieties (e.g. rice) or varieties that only require temporary irrigation (e.g. citrus fruit)	Primarily annual plants (e.g. alfalfa, vegetables, cereals)	Primarily permanent crops, also vegetables

M1 Comparison of different irrigation methods

M4 Sprinkling

M7 Furrow irrigation

M5 Drip irrigation

M8 Cultivation under plastic sheeting (combined with drip irrigation)

Experiment on soil salinisation

- You need about 500 ml of soil, 3 Tbsp of salt and water.
- Put the soil and salt in the bowl and combine well. Then wet the mixture thoroughly and place the bowl in a warm, dry place for a few days.
- Sketch or photograph the results and note your observations.
- Explain the consequences of your observations for agriculture in arid regions.

Saline-tolerant plants	Moderately saline-tolerant plants	Plants with low saline-tolerance
Date tree	Wheat	Pea
Barley	Tomato	Bean
Sugar beet	Oats	Orange
Cotton	Alfalfa	Apple
Asparagus	Rice	Plum
Spinach	Corn	Almond
	Olive	
	Grapevine	

M9 Plants with varying saline-tolerance

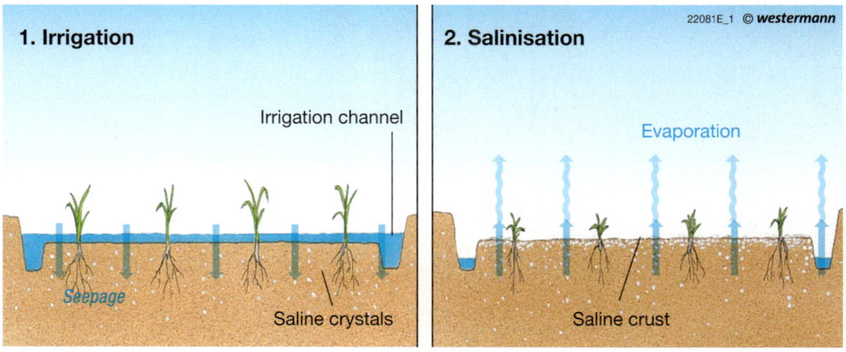

M6 Soil degradation due to irrigation in arid regions

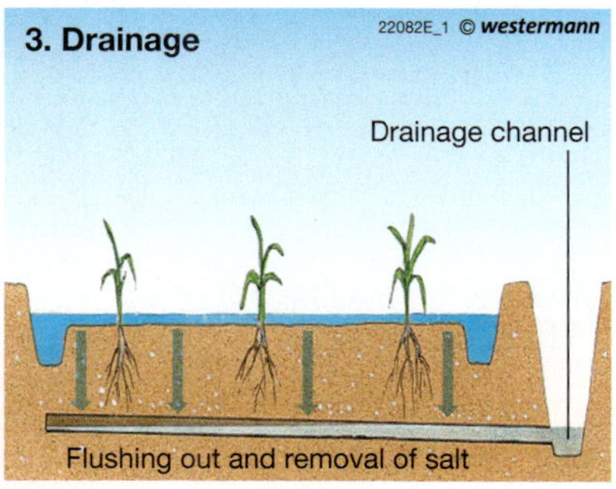

M10 Drainage as a measure of avoiding soil salinisation

Cold winters, hot summers – the dry middle latitudes

Nomadic pasture farming in Mongolia

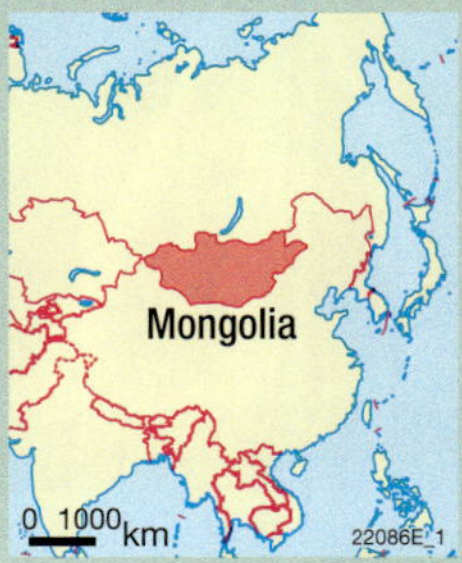

Steppes, grasslands outside the tropics, traditionally provided a habitat for huge livestock herds. In North America in the 1870s, the Plains Indians followed the vast herds of grazing buffalo the largest of which are said to have had over 60 million animals. Animals were also hunted and grazed in other steppe regions such as in Mongolia. Livestock farming in the steppes was traditionally carried out by nomads, people who don't live a sedentary lifestyle.
What has happened to this habitat today?
Are there still livestock herders who migrate with their animals to different pasture areas?
What are the characteristics of both natural and cultural regions?

Ⓞ **1. A** Locate the global distribution of steppes on a map and describe the different types of steppe (M8, → definition, atlas).
B Produce a table to list the distribution of steppes according to country and type of steppe (M8, → definition, atlas).
2. Show the distribution of vegetation zones in Mongolia taking the climatic conditions into account (M1/ M10, → definition, atlas).
3. a) Describe the traditional agricultural method of the Mongolian nomads (M1/ M4/ M5).
b) Show the present-day significance of the nomadic agricultural method (M3/ M9).
4. Discuss the extent and consequences of extreme weather conditions for the Mongolian nomads (M6/ M7/ M9).
Ⓢ **5.** Explain the reasons for the increase in goat numbers in light of cashmere production and present the related problems (M6/ M9).

Nomadism
→ Steppe regions

M2 Nomad in Mongolia

	Mongolia	Germany
Population (2012)	2,796,000	81,890,000
Population density	2 inhabitants per km²	229 inh. per km²
Capital city (2010/2012)	Ulan Bator, 1,144,954 inh.	Berlin, 3,460,725 inh.
Land area (in km²)	1,564,100	357,121
Mongolia		
Pastureland (in ha)	112,970,500 (72% of the total land area)	
Agricultural land (in ha)	315,300 (0.4% of the total land area)	
Total households	742,300	
Nomadic households	160,265 (22% of all households)	
Nomadic households with cars	44,600 (28% of all nomadic households)	

Sources: Fischer Weltalmanach, 2014; K. Zoritza among others, Landwirtschaft und Ernährungssicherung in der Mongolei, In: GR 12/2012.

M3 Facts about Mongolia 2010, 2012

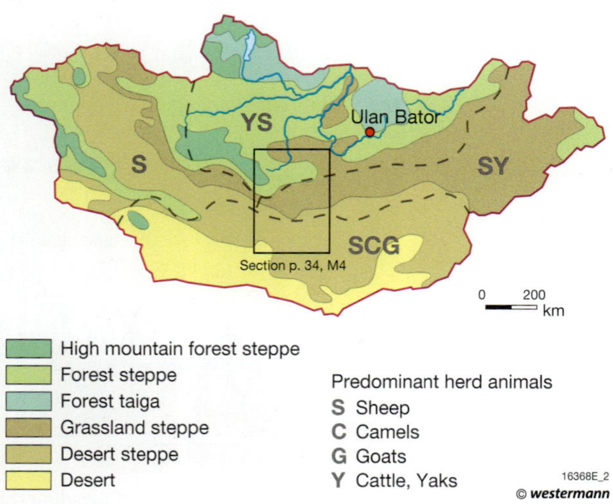

M1 Mongolia – Vegetation zones und pastureland

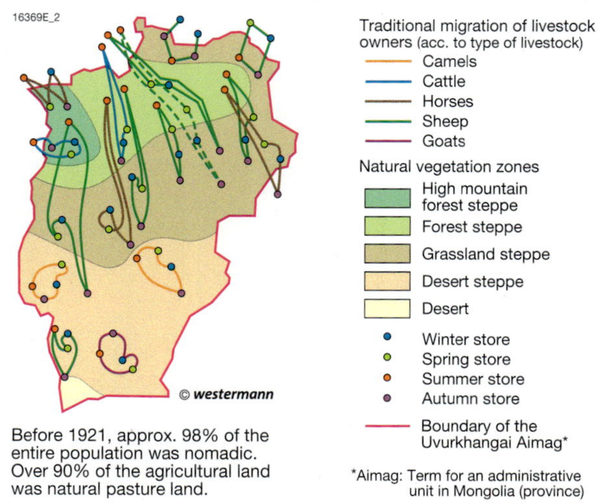

Before 1921, approx. 98% of the entire population was nomadic. Over 90% of the agricultural land was natural pasture land.

*Aimag: Term for an administrative unit in Mongolia (province)

M4 Traditional nomadic migration

M5 On the Mongolian steppe

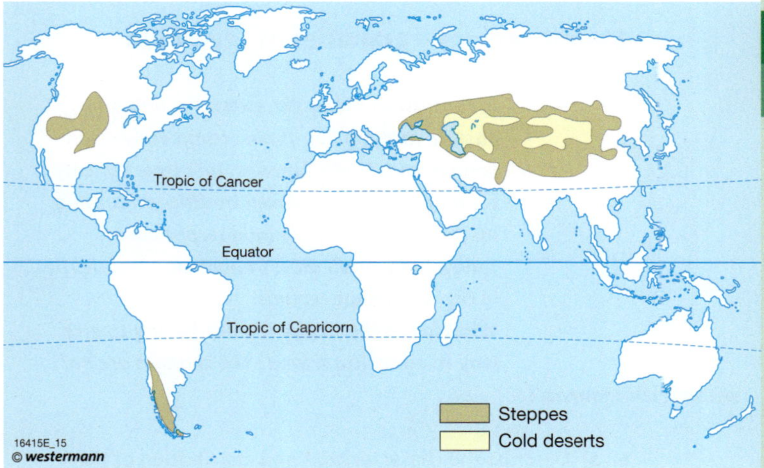

M8 Distribution of steppes and cold deserts

One problem is the commonly occurring Dzud (very cold period, high snowfall) in winter. In the winter of 2009/2010 the temperature dropped below -40 degrees Celsius in 19 of the 21 provinces. Due to the deep snow, animals are not able to find food on the winter pastures. In the winter of 2009/2010 alone, over 40,000 nomadic households lost over half of their livestock. Livestock numbers dropped from 44 million (2009) to just under 33 million (2010).

However, an increase in livestock numbers occurred due to the rise in cashmere goats. The wool is sold to Chinese textile companies.

Unlike sheep, cashmere goats, originally from Tibet, graze by ripping the grass out from the roots.

Based on: K. Zoritza, GR 12/2012; www.radio-utopie.de

M6 Dzud – An extremely cold winter

→ Steppe regions

Steppes are the grasslands of the middle latitudes. Depending on the availability of water, different types of steppe have formed:

Long-grass steppe or **wet steppe** with up to two-metre-high grasses (e.g. Pampas, Argentina)

Mixed-grass steppe is a transitional region between long-grass and short-grass steppe (e.g. the prairies of the Great Plains, U.S.A.)

Short-grass steppe or **dry steppe** (e.g. Mongolia) with up to 40 cm high tufts of grass. As well as grasses, herbs and low-growing shrubs can be found.

In addition to the three types of steppe in the vast central grasslands of North America and Eurasia, **forest steppe** has formed beyond the wet steppes and **desert steppe** has formed beyond the dry steppes.

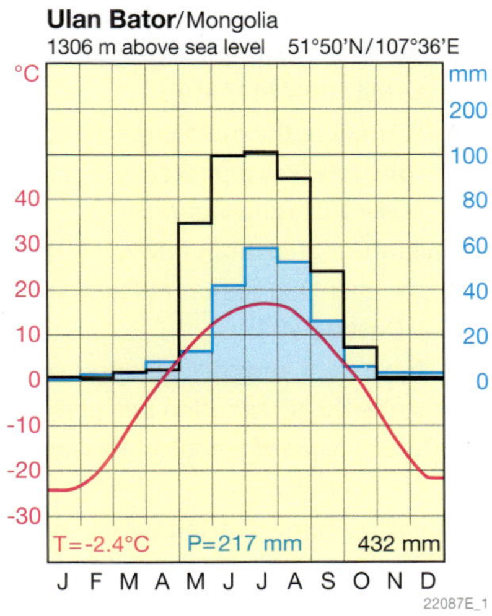

M10 Climate diagram for Ulan Bator

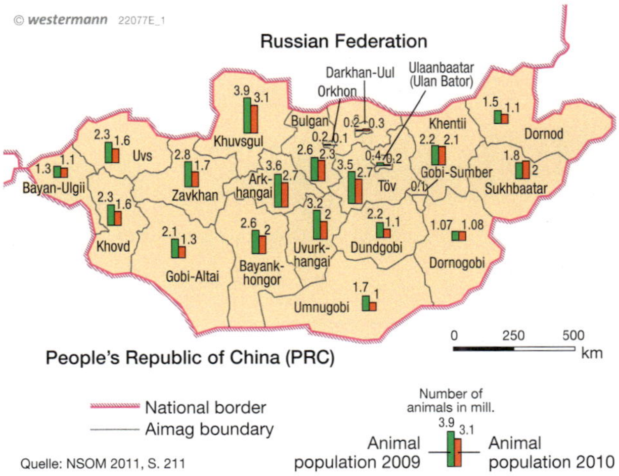

M7 Livestock numbers in Mongolia before and after the extremely cold dzud of 2009/2010

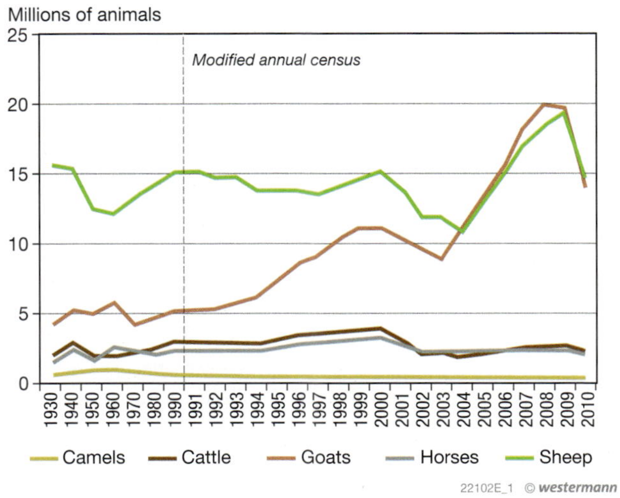

M9 Development of livestock numbers 1930 – 2010

The humid middle latitudes

Steppe regions – granaries of the World

In contrast to Mongolia, extensive nomadic pasture farming has almost completely disappeared from North America. Intensive agricultural farming has replaced traditional farming methods in the more humid steppe zones. Today, the world's most important granaries lie in the wet steppe regions. What led to the steppe regions having to give way to agricultural land? The changes are not without consequence!

1. a) Describe the characteristics and distribution of the different types of steppe in the USA (M1/ M2/ M4).
b) Explain the relationship between the respective climatic conditions and the characteristics of the vegetation (M1 – M4).

2. Explain why the humid steppe regions are used intensively for agricultural purposes (M1/ M6) and investigate present-day land use in the USA (atlas).

Ⓞ **3.** Show the risk potential and the effects of intensive land use based on the example of the Great Plains (M1/ M4/ M5/ M7/ M9).
A Formulate the main implications and summarise these concisely (key phrases).
B Make a causal diagram.

Ⓞ **4.** Find information about (M7, internet)
A the events in the Great Plains in the 1930s ("Dust Bowl").
B the extent and consequences of the drought in 2012.

Ⓢ **5.** Describe the distribution and present-day usage of the different types of steppe in Ukraine (M1, atlas).

Black earth, Prairie, Precipitation variability, Steppe
→ Agronomic dry boundary, Soil erosion

M2 Short-grass steppe in Colorado (USA)

→ Soil erosion
The removal of soil through water and wind that exceeds the natural process of erosion. It is triggered and promoted by soil cultivation.

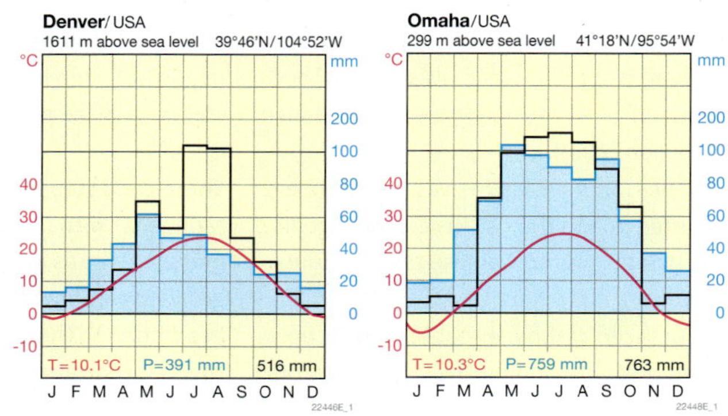

M3 Climate diagrams for Denver and Omaha

→ Agronomic dry boundary
Boundary of rain-fed agriculture. In the prairies the dry boundary is at an annual precipitation rate of approx. 500 mm.

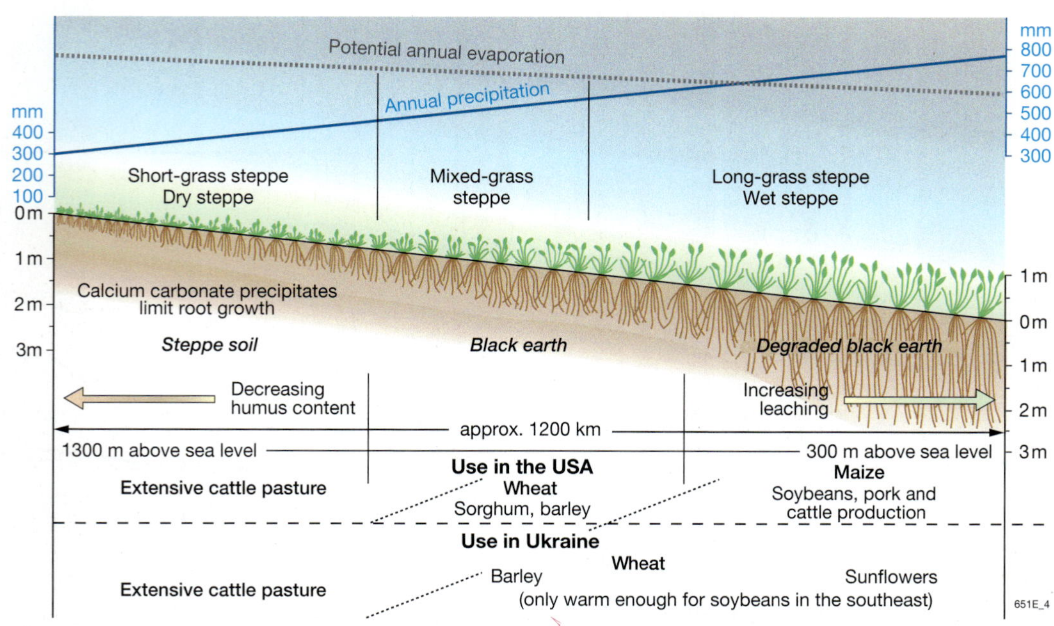

M1 Schematic profile of dry steppe to wet steppe

M4 High-grass and short-grass steppe on the Great Plains. Prairie: Name for the steppes of North America

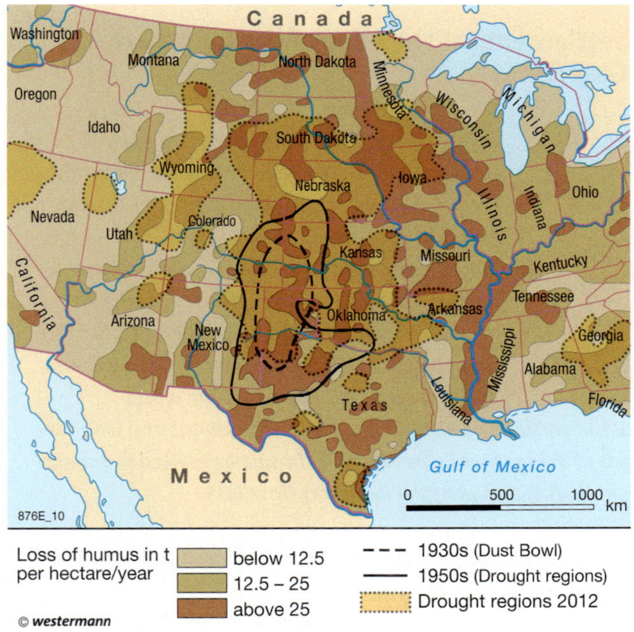

M7 Soil erosion on the Great Plains

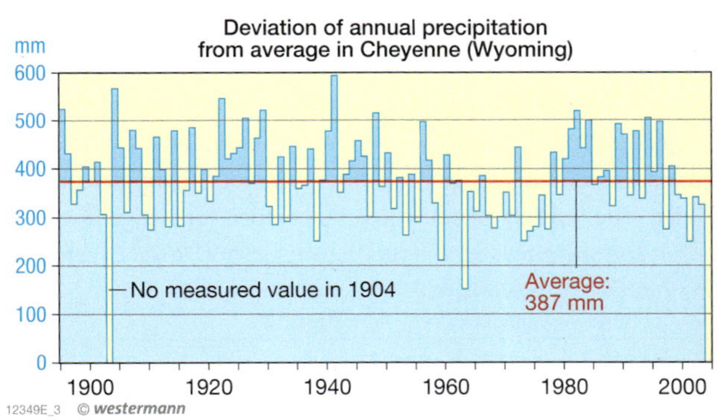

M5 Annual precipitation variability in the steppes

M8 Wheat farm in the USA

M6 Soil type black earth

Black earths (Russian Chernozem) are very humus-rich topsoils that can be up to one metre thick. They have their origin in the high production of biomass at the time of their formation (up to about 10,000 years ago). Plants were carried deep into the soil by numerous soil organisms but the unfavourable continental climate (cold winters, dry summers) prevented them from mineralising completely.

A high exchange capacity and a continual nutrient supply from loess bedrock make black earth one of the most fertile soils in the world.

Based on: W. Latz (Ed.), Diercke Geographie, 2010

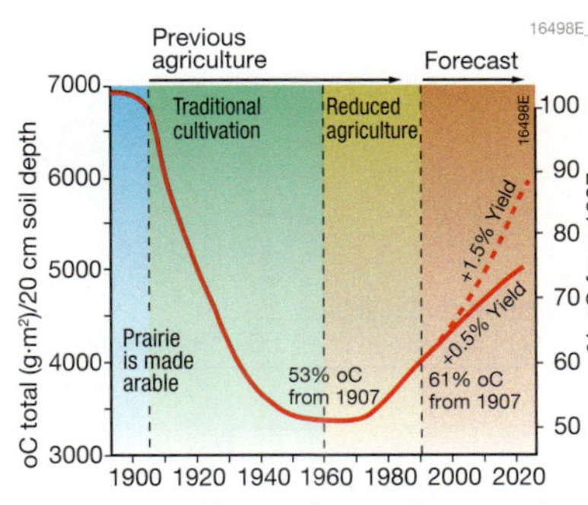

M9 Consequences of agriculture on the humus-content of the steppe ecosystem using the Great Plains as an example (oC = organic carbon content)

Long winters, extreme cold – the polar and subpolar zones

Settlements at the edge of the ice desert – Greenland

Severe climatic conditions prevail in the polar and subpolar zones. The poleward regions belong to the polar ice deserts and are permanently covered with ice.
The extreme living conditions make permanent settlement in the polar ice deserts impossible.
In contrast, the "treeless mountain tract" of the subpolar zone, the tundra, belongs to the inhabited and economic regions of the world.
But what are the particular climatic features of this habitat and what are the limitations of living there? One example: People have lived in Saqqaq (West Greenland) at the edge of the ice desert since 1755.

1. Locate Saqqaq and describe your first impressions based on the photo (M6/ M7).
2. **a)** Explain the climatic conditions and light conditions in Saqqaq and in Greenland (M1 – M3, M1/ M2 on p. 16).
   b) Name the consequences for the vegetation, everyday life and economic activities of the people living there (M5).
3. Present the infrastructure and economic situation in Saqqaq (M5/ M7/ M9).
4. ⓞ Saqqaq, a settlement with a future?
   Since the mid-1990s, the population of Saqqaq has been about 200. But what future does such a small settlement have in the long-term?
   A Discuss the future perspectives from the point of view of the inhabitants (e.g. youth, elderly, supermarket owner, fishers).
   B Should the government provide subsidies to sustain the settlement? Collect arguments for and against.
5. Ⓢ Produce a table and enter the Greenlandic place names mentioned in M2 and the total number of polar nights and polar days at each location.

Polar day, Polar night,
Polar zone, Subpolar zone

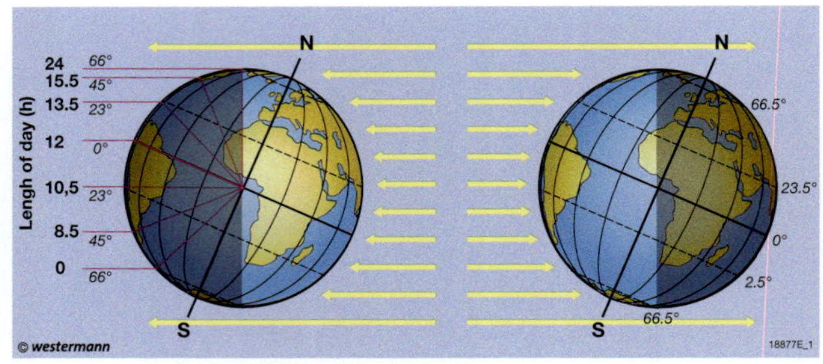

M3 Illumination of the Earth at summer and winter solstice

February/March:
Average temperature: c. -20°C
Minimum temperature: < - 40°C
Summer months:
Average temperature: c. 5°–8°C
Maximum temperature: 20°–25°C
Based on: R. Brodmann, M. Willeke, PG 12/ 2011

◁ **M4** Climate data from Saqqaq

In summer, a mail boat from Ilulissat calls on Saqqaq once a week and supply ships transporting essential goods call at irregular intervals. In winter, Saqqaq can only be reached by helicopter. Roads and footpaths are unsurfaced.
The residents move around on foot or more commonly by quad bike. A central generator supplies all the houses with electricity. The water supply is accessible at several central extraction points around the village. The residents transport the water to their houses in canisters. Most men work as fishers or hunters. Besides fishing, seals and reindeer are hunted for self-subsistence, food for sledge dogs and for fur. The state-owned fish processing factory "Royal Greenland" employs five male workers.
Based on: R. Brodmannn, M. Willeke, PG 12/ 2011

M5 Saqqaq – infrastructure and economy.

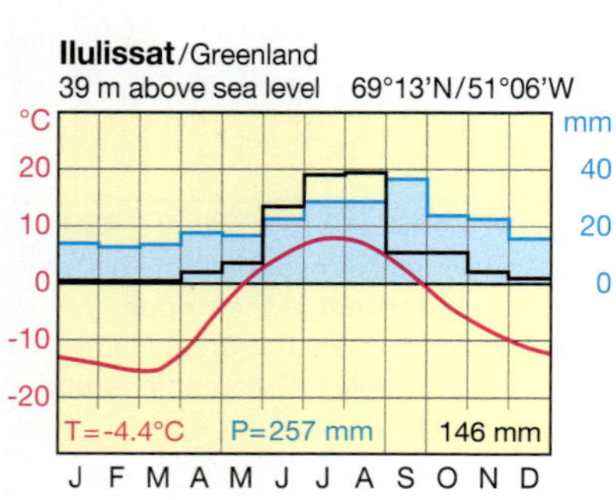

M1 Climate of Ilulissat (Greenland)

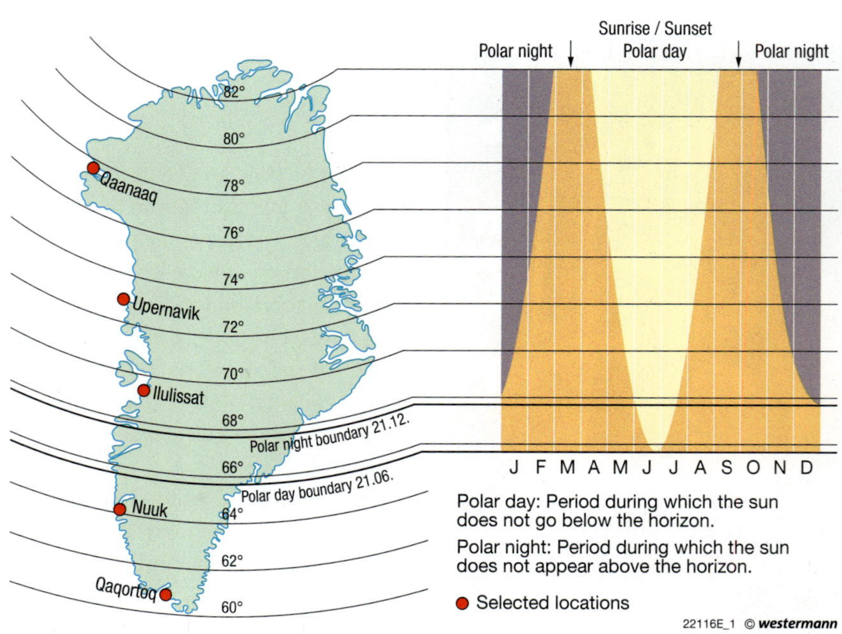

M2 Greenland – light conditions

M6 Greenland

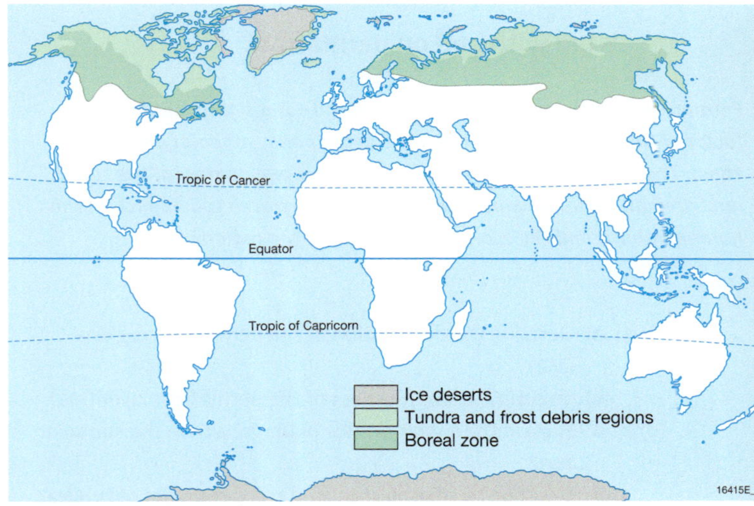

M8 Distribution of the subpolar und polar zones

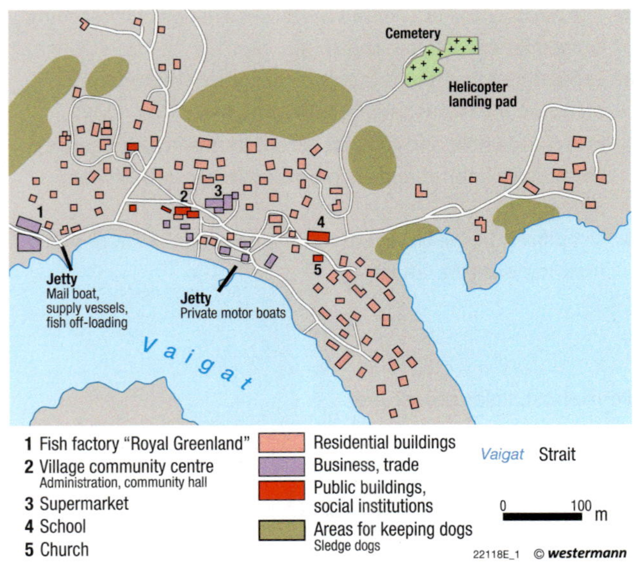

M9 Saqqaq – building utilisation

M7 Saqqaq – view of the settlement from the north

Long winters, extreme cold – the polar and subpolar zones

Living with and on permafrost

Permafrost – deep, year-round frozen subsoil – is a significant limiting factor of the natural-spatial potential in the polar, subpolar and boreal zones. The permafrost regions make up almost one quarter of the total land area on Earth. How does the permanently frozen soil impact on this human habitat and the economic activities of the region?

1. a) Describe the distribution of the permafrost regions (M2).
b) Investigate the thickness of the permafrost layer (M1).

Ⓞ **2. A** Explain the conditions for plant growth in the subpolar zone.
B Describe the distribution and characteristics of tundra, taking into account the significance of terrain for plant growth (M1/ M7).

3. The devastation of tundra vegetation can have far-reaching consequences. Explain this (M5).

4. Explain building issues faced in permafrost regions and technical measures used to overcome these (M4 – M6).

5. Investigate how and to what extent permafrost influences the human habitat and economic activities in Greenland (M6, pp. 38–39).

Ⓢ **6.** List places (in the Northern Hemisphere) that lie in permafrost regions (M2, atlas).

Permafrost, Polar zone, Tundra

M3 Overground pipeline on permafrost

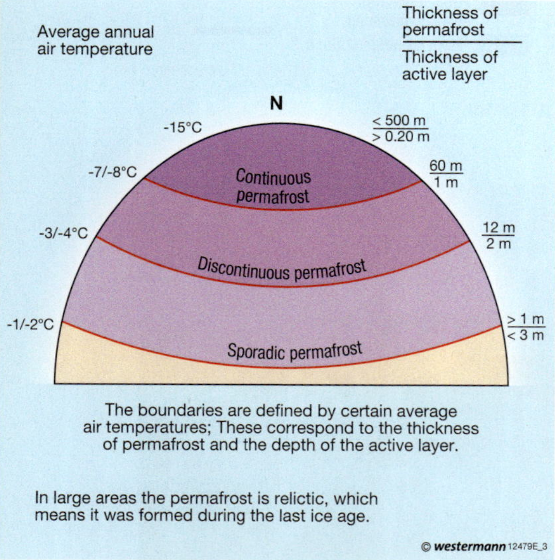

M1 Diagram of permafrost zones

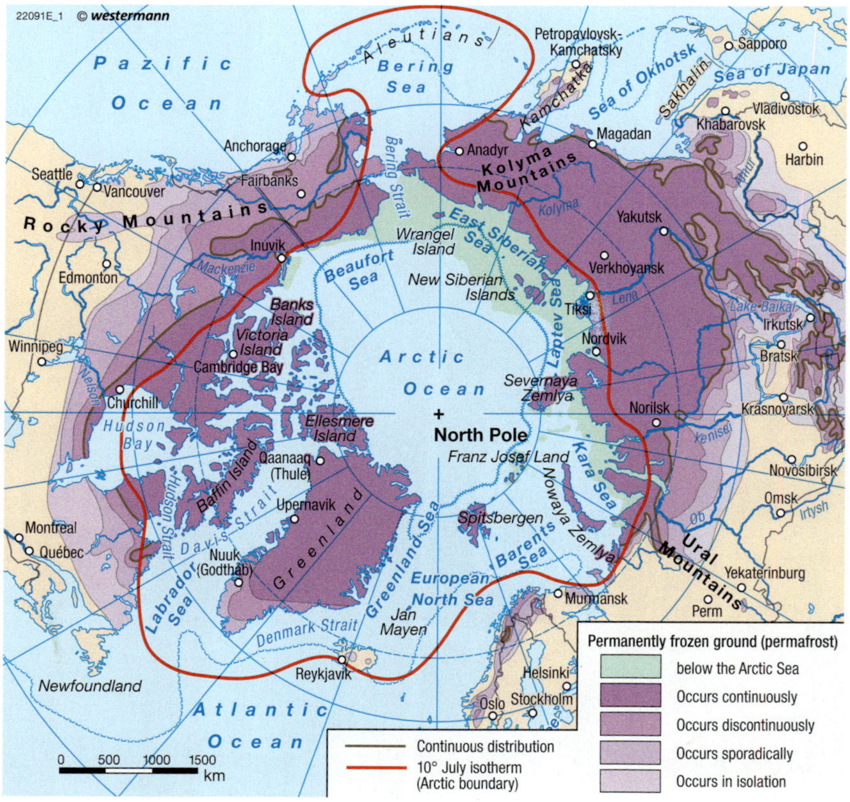

M2 Arctic region: Ice cover and permafrost

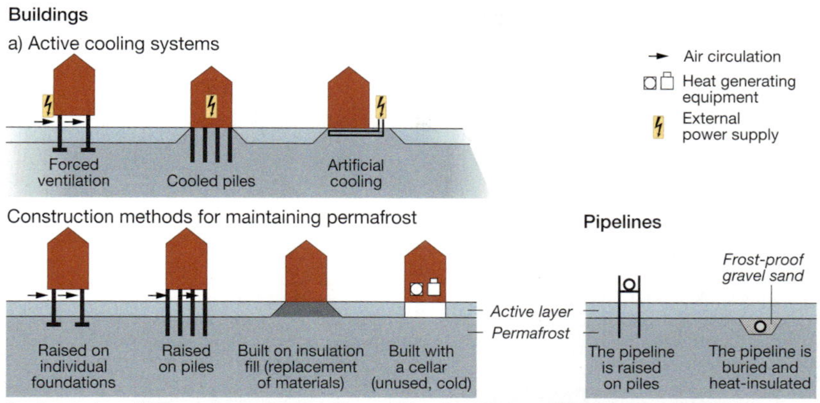

Cooling systems influence the underground temperature in order to stabilise or reduce the thickness of the active layer. Active cooling systems require an external power supply, whereas passive cooling systems do not need additional power.

22100E_1

M4 Building methods on permafrost

M6 Permafrost and usage ▷

All construction projects in permafrost regions are confronted with the problems of building on frozen subsoil.

- Permafrost is very sensitive to thermal change (i.e. change caused by heat).
- Permafrost is relatively impermeable to moisture. Water build-up can occur on frozen soil.
- Permafrost reacts differently depending on ice-content and substrate. Frozen rock, scree and sand usually contain little ice, whereas frozen fine-grained materials such as clay and peat contain a lot of ice and are susceptible to frost processes.

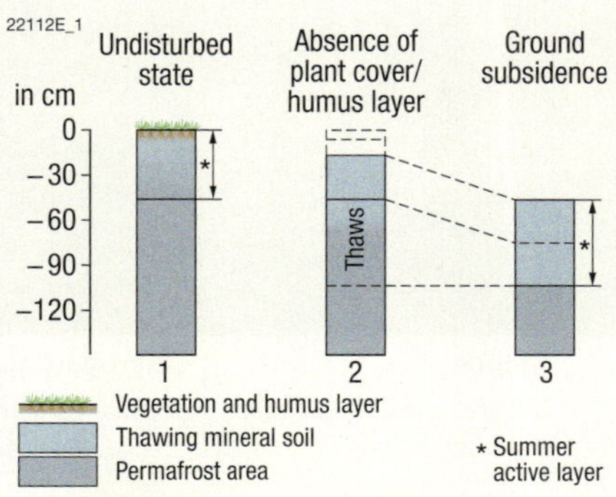

Example: Soil under moss tundra

1. Vegetation and humus layer are intact: The permafrost area comprises 50 % ground ice (supposition). Thickness of the active layer: 45 cm.
2. Humus layer/plant cover (15 cm) is removed: An additional 60 cm layer of permafrost thaws.
3. The thickness of the active layer increases by a total of 60 cm, the ground level subsides.

This subsidence can become even greater:

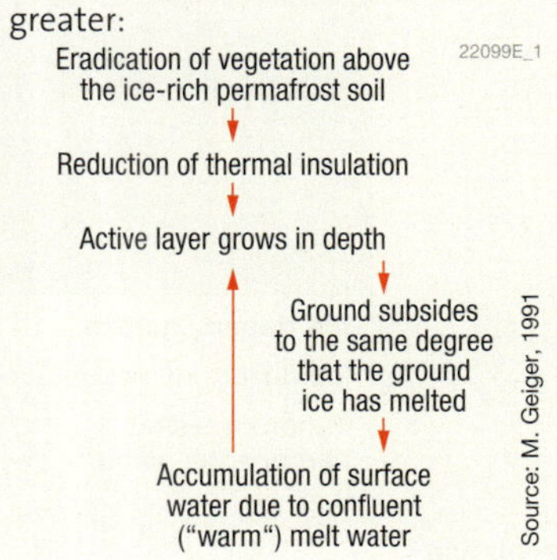

M5 The consequences of tundra vegetation damage

The tundra has no trees. The vegetation cover is characterised by dwarf shrubs, dwarf flowering plants and creeping ground cover plants. Tundra occurs north of the Arctic Circle. The growing season in the subpolar regions is too short (< 100 days with > 4°C) for tree growth, the summer is too cold and the virtually snow-free winters are too frosty.

Most tundra areas are covered with dwarf birches, dwarf willows, cotton grass, mosses and lichens. The plants grow very slowly, the regenerative capacity of the vegetation cover is very low and requires a long time. Damage to the vegetation is often irreversible.

The tracks made by a heavy vehicle remain visible for decades in the tundra. The distribution of tundra vegetation is dependent on the terrain. Whereas valleys, coastal regions and (high) plateaus often have vegetation, slopes and mountains are often only covered with glacial debris. Glacial debris is rocky debris that results from intensive frost weathering processes.

Based on: Westermann Lexikon d. Geographie „Tundra"

M7 Tundra

Landscape zones of the Earth

Climate zones	**Polar zone**	**Subpolar zone**	**Middle latitudes (temperate zone)**		**Subtropi**

Climate diagrams

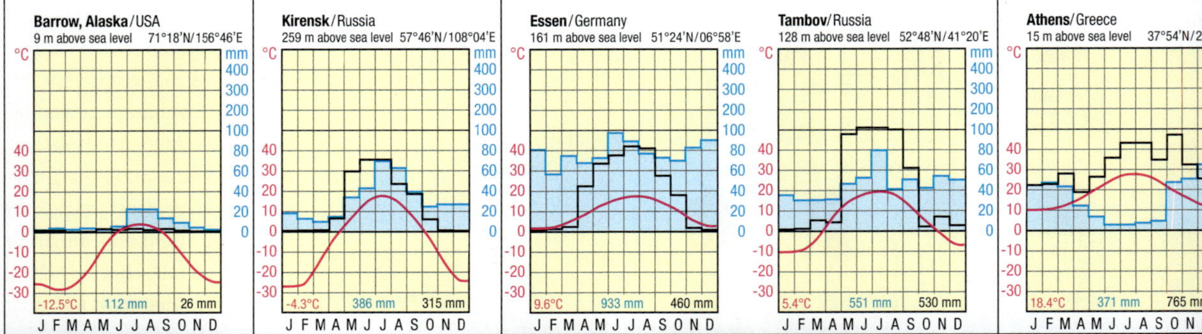

Vegetation zones	**Ice desert, tundra** Ice desert: no vegetation Tundra: grasses herbaceous plants	**Northern (boreal) Coniferous (Taiga)** Larches, spruce, moors	**Deciduous and mixed forests**	**Steppe** tall grass, can grow taller than head-height	**Sclerophyllous shrub** (Mediterranean vegeta

Seasons			**Distinct seasons**		
	More than 8 months winter	6 – 8 months winter	Mild winter Warm summer	Cold winter Hot summer	Significant differences in temperature between summer and winter
Average annual temperature	Below –10 °C	–10 °C to 0 °C	0 °C to 12 °C		12 °C to 24 °C
Precipitation	Less than 300 mm	Less than 600 mm	More than 600 mm	Less than 600 mm (however in some regions up to 900 mm)	400 – 1000 mm, dry in summer, rain in winter (winter rain
Possible growth	Less than 30 days	30 – 180 days	More than 180 days	Less than 180 days	More than 150 days

Soil properties

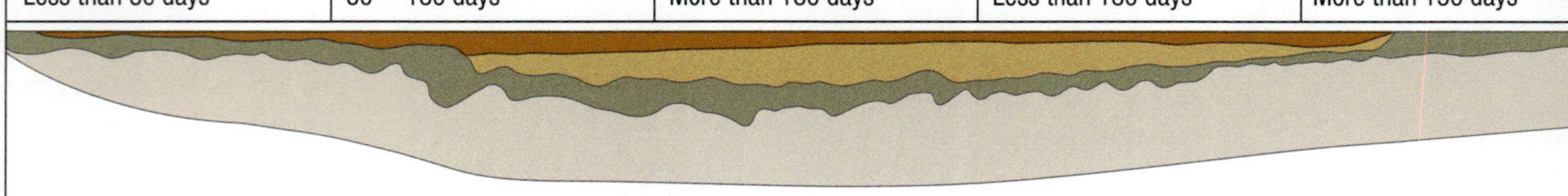

- Unweathered rock
- Depth of weathering
- Formation of minerals with two-layer clay minerals, low cation exchange capacity
- Accumulation of manganese and iron oxides (laterite), soils have very low cation exchange capacity
- Formation of minerals, soils have high cation exchange capacity
- Humus

Restrictions for agricultural use	Permafrost, low primary production	Short vegetation periods, limited availability of nutrients	Short vegetation periods, limited availability of nutrients	Short vegetation periods, water shortage	Water shortage, limited availability of nutrients
Crops, livestock farming	Reindeer (nomadism)	Frequent Forestry	Cold boundary Rye, potatoes, corn, wheat, cattle, pigs	Wheat, sugar beet, sunflowers, sheep, cattle	Wine, olives, fruit, citrus fruit, rice, sheep, goats

		Tropics			
		Semi-humid			Permanently humid
Climate diagrams	**Keetmanshoop**/Namibia, 1067 m above sea level, 26°32'S/18°07'E, 20.7°C, 146 mm, 493 mm	**Abéché**/Chad, 549 m above sea level, 13°51'N/20°51'E, 28.9°C, 372 mm, 1142 mm	**Bamako**/Mali, 380 m above sea level, 12°32'N/7°57'E, 27.8°C, 878 mm, 1354 mm	**Bouaké**/Ivory Coast, 376 m above sea level, 7°44'N/5°04'E, 25.7°C, 1101 mm, 1200 mm	**Pontianak**/Indonesia, 3 m above sea level, 0°1'N/109°24'E, 26.4°C, 3298 mm, 1403 mm

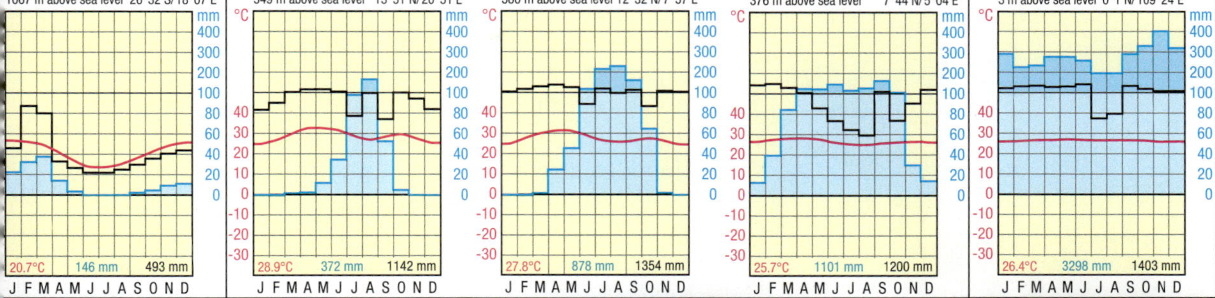

Climate zones

Vegetation zones	**esert and semi-desert** o vegetation or lolated dwarf shrubs, ʼasses	**Savannas** Thornbush savanna: Knee-high grass, shrubs, isolated trees	Dry savanna: Chest-high grass, trees	Wet savanna: Head-high grass, clusters of trees, forests	**Tropical rainforest**
Seasons	gnificant differences temperature between ımmer and winter	**No seasons**			
Average annual temperature		Above 24 °C			
Precipitation	ess than 250 mm, ery short rainy season, ften drought	250 – 500 mm, up to 4 months rainy season, often drought	500 – 1000 mm, 4 – 6 months rainy season	Over 1000 mm, 6 – 10 months rainy seasons	Over 1500 mm, wet year-round
Possible growth	ess than 60 days	Less than 180 days	More than 180 days	More than 300 days	Year-round
Soil properties					
Restrictions for agricultural use	/ater shortage, mited availability of utrients	Water shortage	Water shortage, limited availability of nutrients	Limited availability of nutrients	Limited availability of nutrients
Crops, livestock farming	asis cultures, otton, rice, amels, sheep	Rice, cotton, sheep, goats, cattle, camels (nomadism) (Dry boundary of rain-fed agriculture)	Millet, maize, peanuts, cotton, cattle, sheep	Cassava, maize, peanuts, cotton, cattle (not Africa)	Cassava, bananas, cocoa, coffee, sugar cane

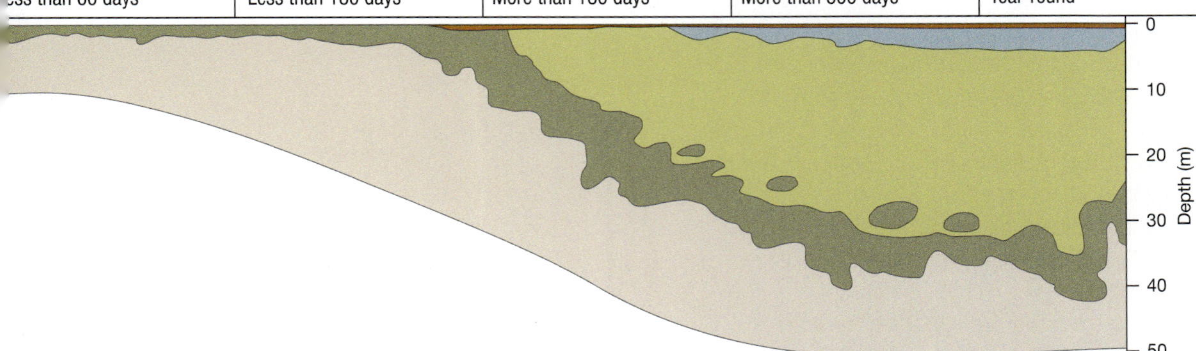

16080E_7

The essentials in brief

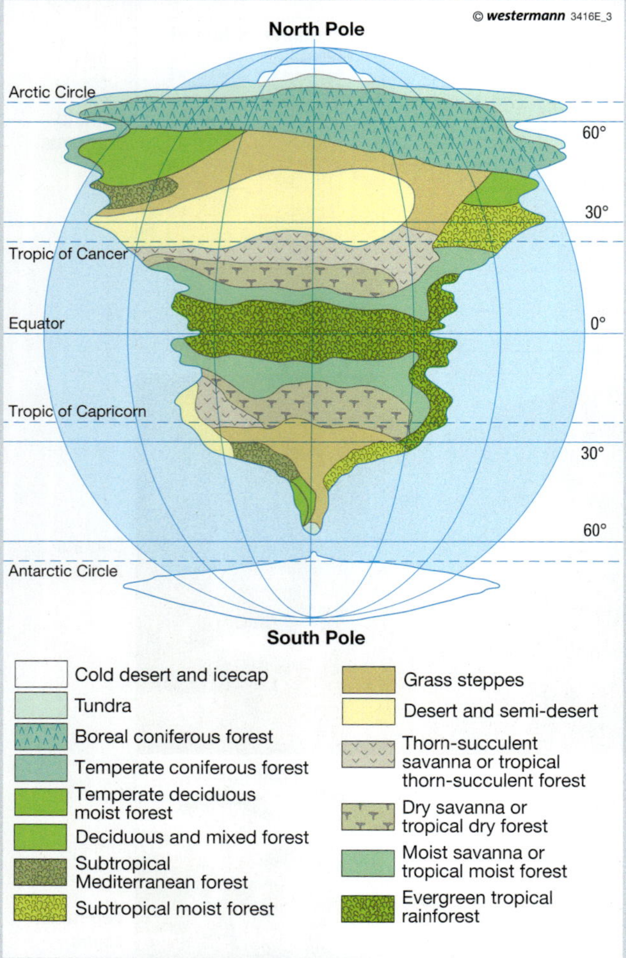

M1 The vegetation of the "archetypal continent"

Diagram M1 shows the typical distribution of the landscape zones on an archetypal continent. The archetypal continent is a theoretical construct in which all the continents of the world have been "pushed together". In so doing, special care was taken to ensure that the latitude of each region was not altered. The archetypal continent shows the typical distribution of vegetation on Earth. It becomes evident that there is a natural progression from the Equator (the permanently humid tropics) towards the poles (the cold zone). At the same time it becomes clear that this structure is sometimes irregular and that some zones stretch further north or south, whereas others are missing completely on the western side of the continent.

The reasons for this distribution have been illustrated in this chapter. For example the radiation patterns on Earth – length and intensity of solar radiation – play a significant role. This effect is counteracted e.g. by the equalising influence of the oceans or the influence of the respective terrain.

The geofactors temperature and precipitation have an all-encompassing influence on human life and activity on Earth. The natural vegetation and agricultural usability are substantially determined by these two geofactors. Whether a region is considered favourable from a human perspective and therefore suitable for habitation and economic activity (inhabitable/ecumene) or whether it is hostile (uninhabitable) depends largely on the temperature and precipitation conditions. The agronomic dry boundary (the boundary of rain-fed agriculture) and the agronomic cold boundary (the threshold of cold for agriculture) constitute natural boundaries for agricultural land use (M2).

The same applies to settlements in "uninhabitable" regions (e.g. deserts or polar regions). Geofactors also set limits here resulting in difficult living conditions and extremely sparse population. As can be seen in the example of the winter-humid subtropics (Argolis) and the desert, humans can overcome these natural limitations. Traditionally, nomads have managed to live in deserts by making use of oases or "following" the rainy season with their animals. Today, the agronomic dry boundary is overcome primarily with irrigation. Settlements, even cities, exist in deserts (e.g. Las Vegas). The utilisation of (dammed) river water and groundwater makes this possible. At the same time this chapter shows that overcoming the dry boundary is at a high cost. The over-exploitation of rivers and groundwater aquifers as well as the salinisation of soils are the obvious consequences.

Cold is also a factor that limits the inhabitable regions. Life on permafrost requires particular adaptation and engineering achievements when building houses, roads and pipelines and for ensuring supply-lines for the inhabitants. Additionally, everyday life in winter is characterised by the cold and dark. Furthermore, the geofactor soil plays an important role in determining whether a region is ecumene. It has been shown that, despite ideal conditions with regard to temperature and precipitation, agricultural farming in tropical rainforest regions is very difficult. This is because the soils in these regions undergo weathering to a great depth and therefore have no contact to the bedrock. Extremely fast chemical weathering leaves the soils low in humus, humic substances and clay minerals and they have an extremely low cation exchange capacity. Agricultural yields drop dramatically after only a few years and fertilisation has almost no effect. Agroforestry has proved to be one option for sustainable, small-scale utilisation.

Although almost anything seems possible today with suitable technology – from cultivating vegetables in Greenland to World Cup football in the desert – climate still characterises the vegetation, human life and economic activities in the different landscape zones.

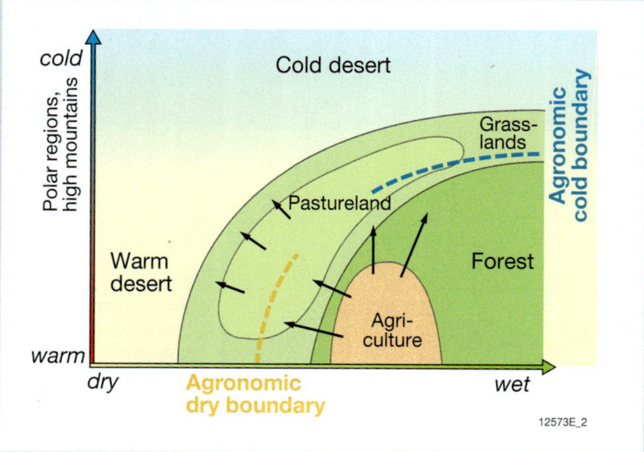

M2 The expansion of land use

Skill check

All the skills that you were able to acquire in this chapter are listed here.
You can judge for yourself if you have mastered each skill: ***securely, moderately or barely.***

Explanatory skills

Can I ...		Unsure? Look it up on these pages
1.	describe and explain the climate of the tropics and how it influences the vegetation?	16/17, 22/23
2.	explain how plants have adapted to the climatic conditions of the diverse landscape zones?	14/15, 20–23, 30/31, 36/37, 40/41
3.	explain possible solutions for agricultural land use in the humid and semi-humid tropics?	20/21, 24/25
4.	demonstrate possible solutions for overcoming aridity in semi-arid and arid regions?	30–33
5.	explain the natural geographic boundaries for the habitation of different regions and possible solutions for overcoming these?	26/27, 38–41
6.	characterise the Earth's landscape zones based on climate and vegetation?	12/13, 42/43

Methological skills

Can I ...		
7.	describe and explain landscape zones and their utilisation as habitats using physical and thematic maps?	12–43
8.	identify favourable and unfavourable factors of different habitats and formulate questions for overcoming the boundaries between inhabitable and uninhabitable regions?	12–41
9.	read climate diagrams and extract relevant information for task-based characterisation of the climate of a particular location?	18/19, 22–43
10.	analyse different kinds of presentation and working materials such as statistical data, graphs and texts to answer questions about the human habitat in various landscape zones?	14–41
11.	research Information e.g. about the "Dust Bowl" or the trade winds using suitable research methods in the library and in the internet and evaluate the information based on the respective task?	16/17, 36/37
12.	formulate key statements about tropical circulation and the trade wind system based on specific models?	16/17, 22/23
13.	present geographical information graphically, e.g. as a causal diagram or map sketch?	16/17, 20/21, 36/37

Evaluation skills

Can I ...		
14.	evaluate the adaptation of agricultural methods to climatic and soil conditions?	20/21, 24/25, 34–37
15.	assess the suitability of arid regions for settlement?	26/27
16.	evaluate problems and possible solutions for settlement at the edge of an ecumene region?	26/27, 38/39, 40/41
17.	assess different irrigation methods in ecological and economic terms?	32/33

Action skills

Can I ...		
18.	present the results of work on favourable and unfavourable habitats as well as possible solutions for overcoming the boundaries of inhabitable and uninhabitable regions using appropriate terminology and in a factually relevant way?	14–41

Examination training – sitting an examination

Fruit growing – an agricultural industry with a future? The example of kiwifruit production in New Zealand

1. Locate New Zealand on a map and describe the climatic requirements for growing fruit and the distribution of fruit production in New Zealand.
2. Taking the required growing conditions of the kiwifruit plant into account, explain the distribution and development of kiwifruit production in New Zealand.
3. Evaluate the possibilities for expanding kiwifruit cultivation in New Zealand in view of climatic and economic aspects.

These additional materials are required to solve the assignment:
M1 Atlas maps of your choice

Kiwifruit is the berry of a shrub-like climbing plant. It grows shoots of up to 8 m in length, has thick foliage and white to pink-coloured flowers.

It is cultivated in rows, similar to wine growing. Five years after the kiwifruit cutting is planted, it bears fruit for the first time. The vines can grow to a height of 2.5 m and to 50 years of age, whereby they can be harvested annually.

In New Zealand, kiwifruit plants flower in November and the fruit is ready for harvesting in May. Kiwifruit plants require a very sunny location, sheltered from the wind. The kiwifruit plant tolerates light frost, however not long periods of temperatures well below zero. With an annual volume of 1,000 – 1,500 mm the kiwifruit requires a lot of water.

Based on: www.saengerhof.de and New Zealand Kiwifruit Marketing Board (based on G. Feller, Kiwi-Anbau in Neuseeland. In: PG 7/8, 2001).

M2 Required growing conditions for kiwifruit

Sitting an examination – work steps

STEP 1

→ **Read the examination topic carefully!** The general geographical wording gives the first indication of the examination assignment (here: Fruit growing – An agricultural industry with a future?). Furthermore, the regional case study to be analysed in the examination assignment is stated in the topic (here: Kiwifruit production in New Zealand).

→ **Each part of the assignment – usually there are three – represents a guide for structuring your answer** (rough outline). The instructions also contain operators (here: locate, describe, explain, evaluate) to help you.

→ **Plan your time well:** Give yourself enough time to work through the material and do a final check at the end.

STEP 2

→ **To begin with, put the case study in its geographical context** (M1: atlas maps of choice, where applicable information on further material).

→ **Next, analyse the working material based on the assignment!** Highlight important statements and use your calculator if necessary to illustrate developments or comparisons by means of statistical data. Thoroughly analyse thematic maps and diagrams according to the question. Pay attention to possible annotations or explanatory notes (footnotes).

→ **Structure your results in a meaningful way.** In order to complete the assignment it is usually necessary to link individual statements found in the working material. A mind map can be useful.
(***Beware:*** Do not deal with individual pieces of information found in the different materials one after the other in your text!)

→ **Do not begin with your clean copy until you have thought through every part of the assignment and thoroughly evaluated the materials.**
Formulate both an introduction for the whole assignment and short transition sentences.

→ **Argue using the material at hand,** i.e. your explanations must contain concrete references. Indicate which material you are referring to (Mx) at the end of each sentence or paragraph.

→ **Particularly in part 3 of the assignment,** in which you will be asked to evaluate, give your opinion or make an assessment, make specific reference to the regional case study. Formulate a concluding paragraph.

STEP 3

→ **Finally, check the formal aspects of your text** (e.g. spelling, punctuation, page numbers).

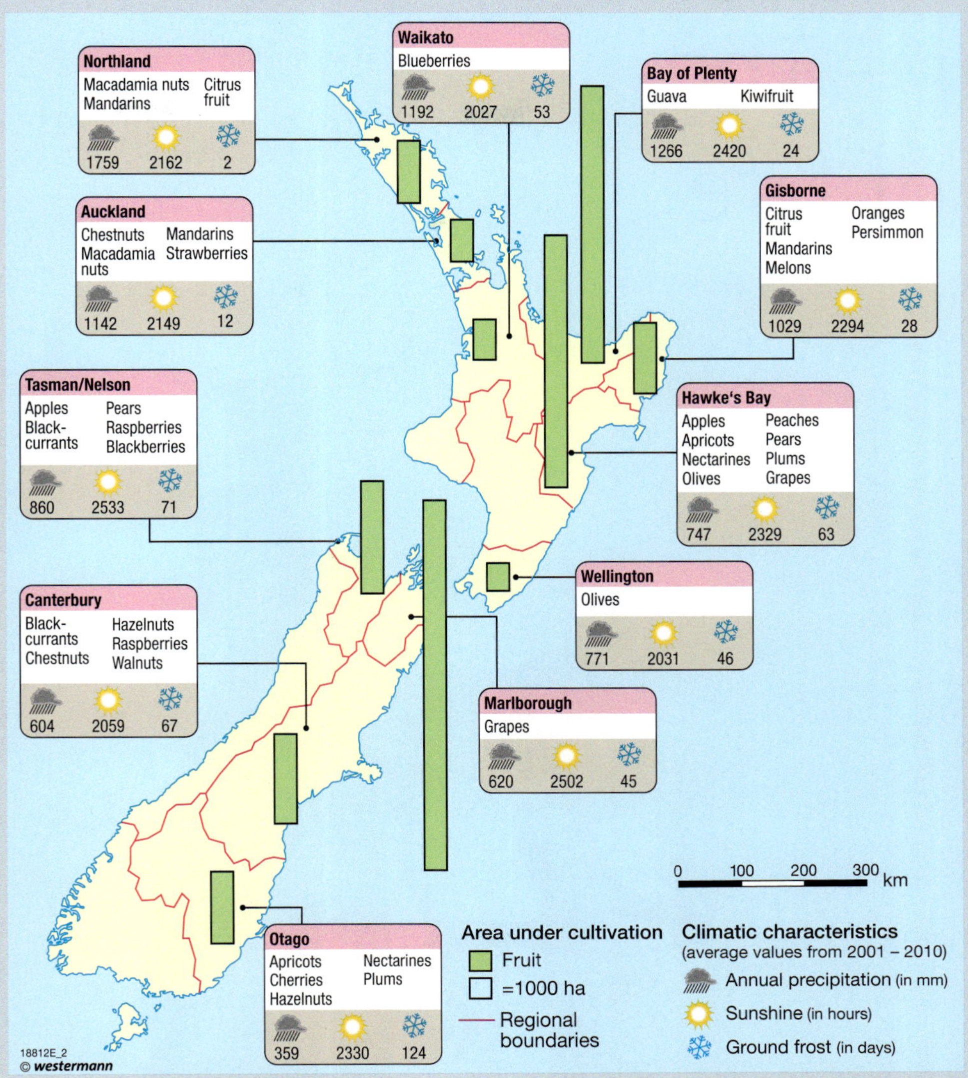

M3 Fruit growing in New Zealand

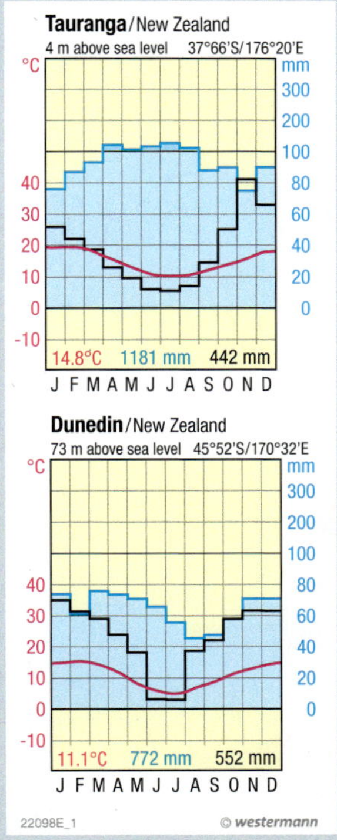

M7 Climate diagrams

1.	Italy	431,558
2.	New Zealand	420,231
3.	Chile	237,104
4.	Greece	140,400
5.	France	73,480
6.	U.S.A.	34,200

M8 Leading kiwifruit producing countries, 2011 (in t)

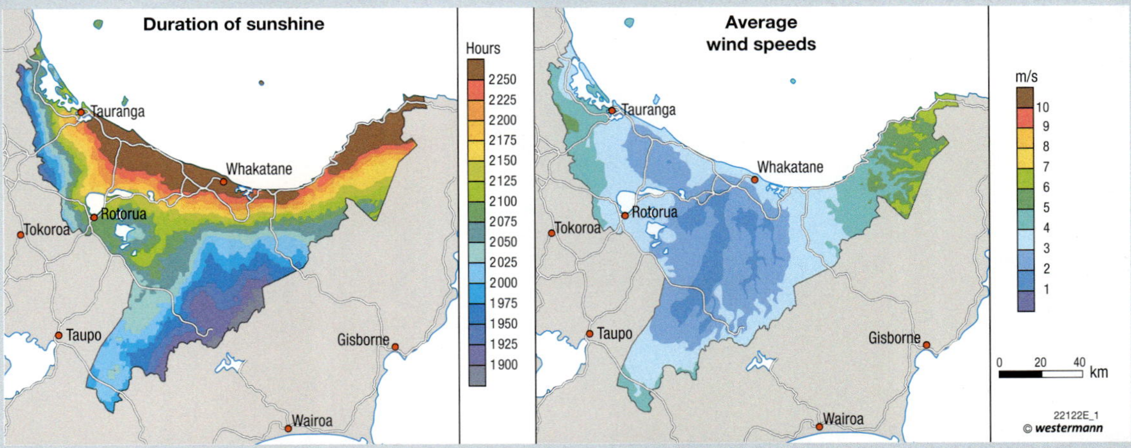

M4 Climatic conditions in the Bay of Plenty (North Island)

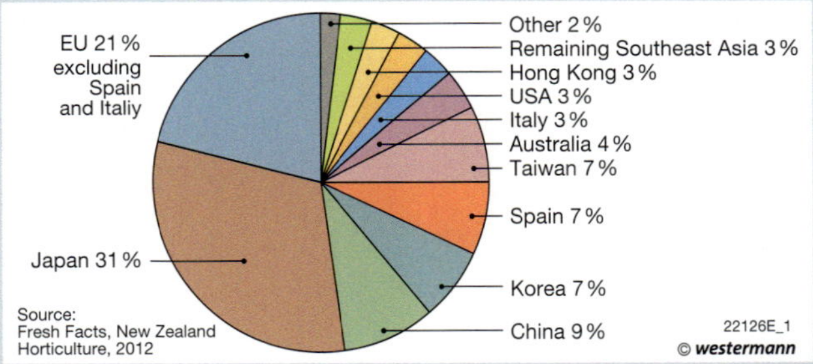

M5 Main export markets for New Zealand kiwifruit 2012

	2000	2007	2012
Kiwifruit sold (in t)	186,480	288,360	392,760
Crop yield (t/ha)	19.1	27.1	31.9
Crop area (ha)	10,234	11,967	12,502
No. of kiwifruit farmers	2,681	2,754	2,662
Packing stations	118	80	63
Cold storage	106	87	79

M6 Development of kiwifruit production in New Zealand

II Water – The Basis of Life

Hard work: Wells like this one, where women draw water in tin buckets, are often the only source of water in arid regions.

Water – between shortage and abundance

M1 Ipswich (12.01.2011, west of Brisbane)

M2 Semperoper in Dresden (05.06.2013)

Water: Life-giver, risk factor and cause of conflict

Floods inundate Bangladesh, hurricanes devastate entire regions in the USA, floods threaten parts of Germany and at the same time extreme drought is prevalent in the arid regions of the Earth. How is this possible?

For humankind, water is both a life-giver and a risk factor. On the one hand, water is the basis of life while, on the other hand, it often proves to be a natural threat. The people of Bangladesh, for example, look forward to the monsoon because it is essential for good crops. At the same time though, they dread it because it can result in severe flooding, turning the blessing into a disaster.

This chapter is dedicated to the diverse facets of water as the basis of life. The significance of water on Earth is illustrated in a general introduction to the topic. Subsequently, the effects of the absence and inappropriate use of water are examined.

A particular focus is placed on the Sahel Belt, a region that is regularly affected by drought. "Besides war and economic crises, droughts belong to the critical phenomena that cause famine" *(B. Hornetz, R. Jätzold: Savannen-, Steppen- und Wüstenzonen, 2003).*

However, catastrophic famine resulting from drought is not the only problem. Droughts influence the economy of affected countries. If economic performance drops due to drought, the buying power in the respective country also drops, so that many people are left with even less money to live on.

But water shortage is not the only threat to humans and nature. An abundance of water also exposes the vulnerability of the human habitat: "When rivers play at fate – Flooding in Bangladesh", "Red alert on the Danube – The flood situation becomes critical" or "Water rising – Dam threatens to burst" are all headlines from newspapers in recent years that demonstrate how vulnerable we are to flooding. Events such as the 100-year flood in Passau in 2013 or catastrophes caused by tropical cyclones such as Typhoon Haiyan in November of the same year bring an over-abundance of water and destroy the basis of life for hundreds of thousands of people.

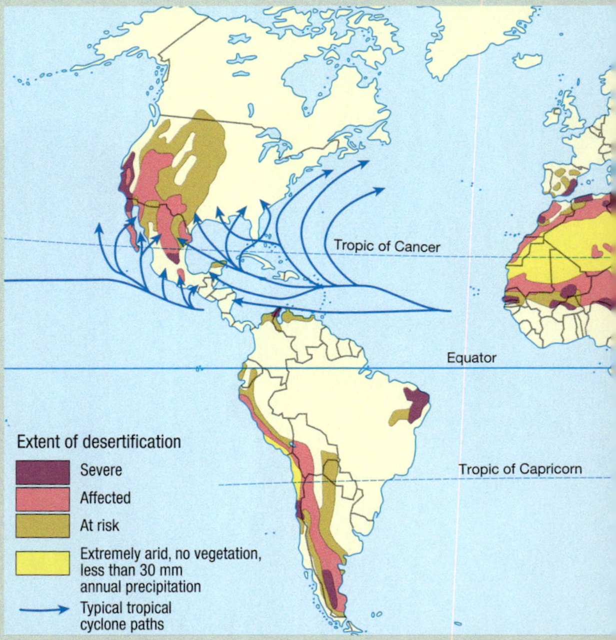

M3 Regions prone to drought and tropical cyclones

All these examples demonstrate our vulnerability in regard to water. The term vulnerability describes how susceptible, sensitive or exposed people, society and the infrastructure of the human habitat and economic regions are to disaster. We speak of ecological, social and technological vulnerability. These and further examples in this chapter will show that these aspects must always be considered in context.

M4 Lake Faguibine (80 km west of Timbuktu)

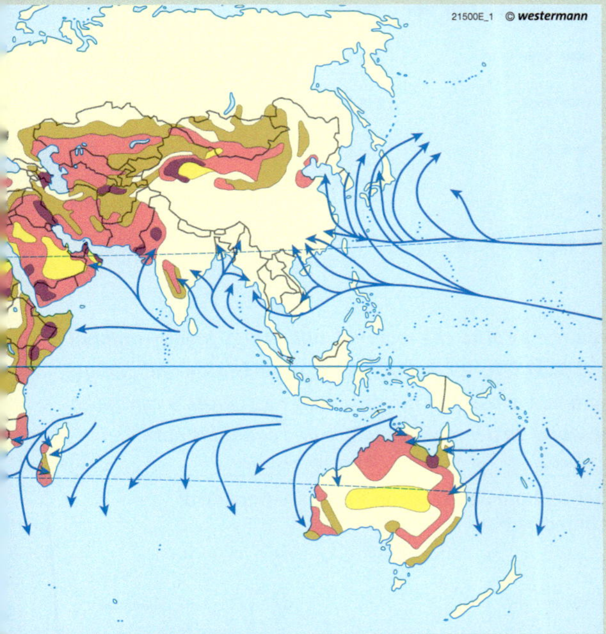

M6 Hurricane Katrina on 29.08.2005

Early Monsoon

More than 300 dead due to flooding in India
(21.06.2013, Süddeutsche Zeitung)

Severe drought at the Horn of Africa

UN fears millions will starve
(21.07.2011, Süddeutsche Zeitung)

Fatal deluge of water

Day after day of monsoon rain in July and August 2010 caused a 100-year flood in Pakistan: One fifth of the country was submerged in water, more than 1,700 people lost their lives and millions of people lost their homes. Rebuilding is difficult; the state is unable to cope.
(no date, Spiegel online)

Catastrophic drought in Ethiopia

According to UNO, 12.6 million are currently starving in war-torn Ethiopia. The drought periods have an almost cyclic character. But the disasters are not only to blame on the weather – they are partly self-inflicted.
(17.06.2003, Spiegel online)

Fatal deluge of water

100-year flood in Africa's drought belt
(21.09.2007, Spiegel online)

M5 Excerpts from articles on the internet

Ⓞ **1.** Locate the places in the images (M1/ M2/ M4/ M6) and reports (M5) on the map M3:
A Describe the geographic location of the places in the photographs. Use an atlas to help you.
B Using an atlas, find the places in the photographs and newspaper articles. State the country and continent and locate these on M3.

2. Research Lake Faguibine (M4) using Google Earth. Zoom in on different areas of the former lake. What forms of land use can you recognise? Using a ruler, determine the expanse of the former lake and compare it to Lake Constance.

3. Explain the term vulnerability based on examples around the topic of water.

Ⓢ **4.** Research further severe flood, drought and tropical cyclone disasters. Note the heading, internet link or newspaper edition and key words about each event.

Water – between shortage and abundance

Water – a commodity in great demand

Water is our life-giver: Without water we would die of thirst. But we would also have no clothing and not be able to drive a car or work at a computer. Because clean drinking water is used, contaminated or consumed in the production of almost everything. The water used is referred to as virtual water. But what are the social, economic, ecological and political dimensions of the exploitation of this precious and often rare commodity?

1. Explain the term virtual water (M2/ M7).
2. (O) **A** Describe water consumption in Germany and the distribution of our daily consumption (M4/ M5).
   B Convert M2 into a bar graph.
   C Analyse water consumption in Germany in relation to the use of virtual water. In so doing, link M4/ M5, M1/ M7 and M6.
3. Explain where conflicts could arise over water (M3/ M6).
4. Water volume in size comparison: Imagine all the water on Earth could fit into a bathtub (140 l).
   Using M7, calculate:
   a) How much freshwater is in the bathtub?
   b) What proportion of freshwater on Earth exists in the form of ice?
5. (S) Produce a mind map for water as a cause of conflict.
6. (S) Using an atlas and based on two examples for each continent, examine whether the availability of freshwater can be attributed to favourable/unfavourable climatic conditions or a low/high population density in the respective country.

Virtual water, Water availability

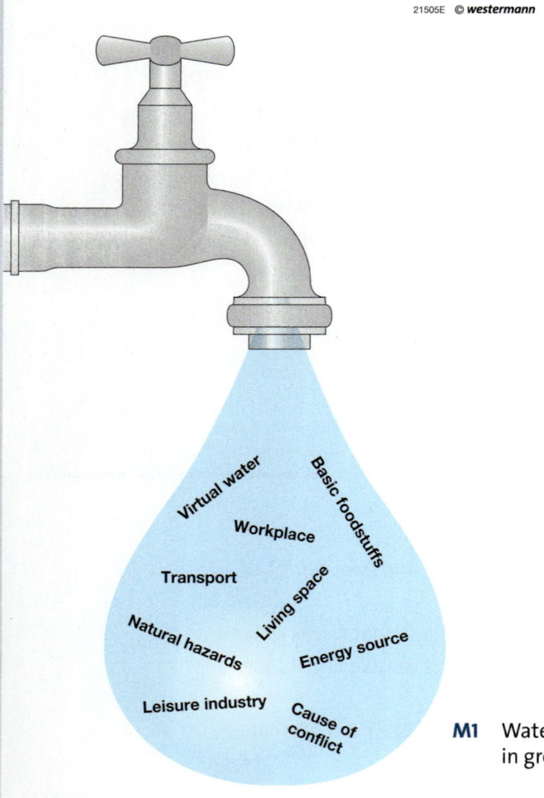

M1 Water - a commodity in great demand

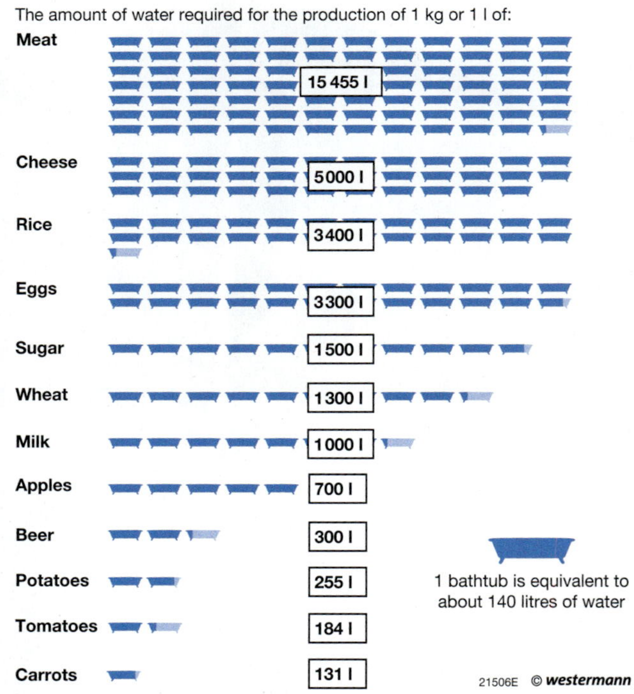

M2 Virtual water in food

„Water links the local to the regional, and brings together global questions of food security, public health, urbanisation and energy. Addressing how we use and manage water resources is central to setting the world on a more sustainable and equitable path."
Source: Ban Ki-moon; World Water Development Report 2012

M3 Quotation from Ban Ki-moon, Secretary-General of the United Nations

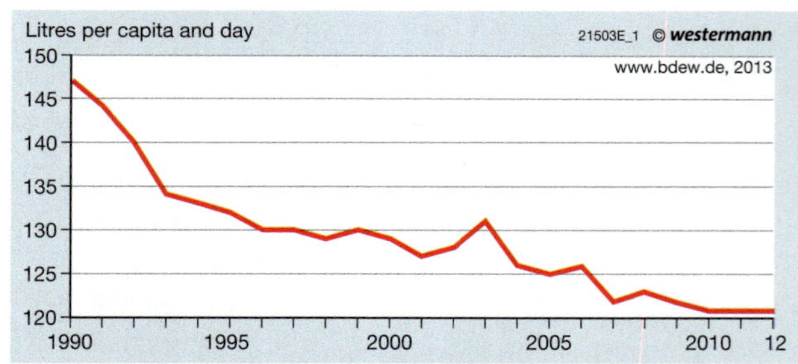

M4 Development of water consumption in litres per person and day in Germany

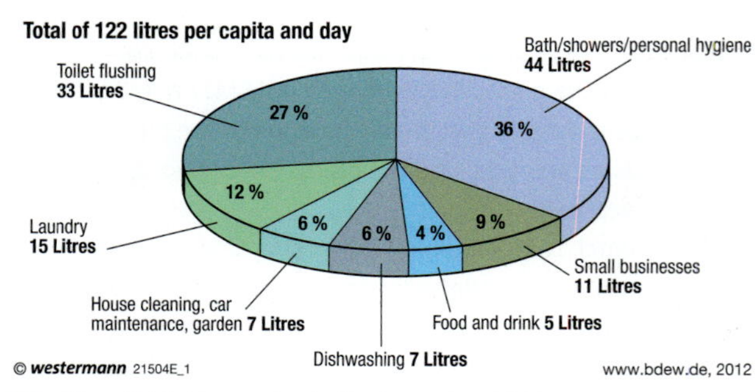

M5 Use of drinking water in the household 2012

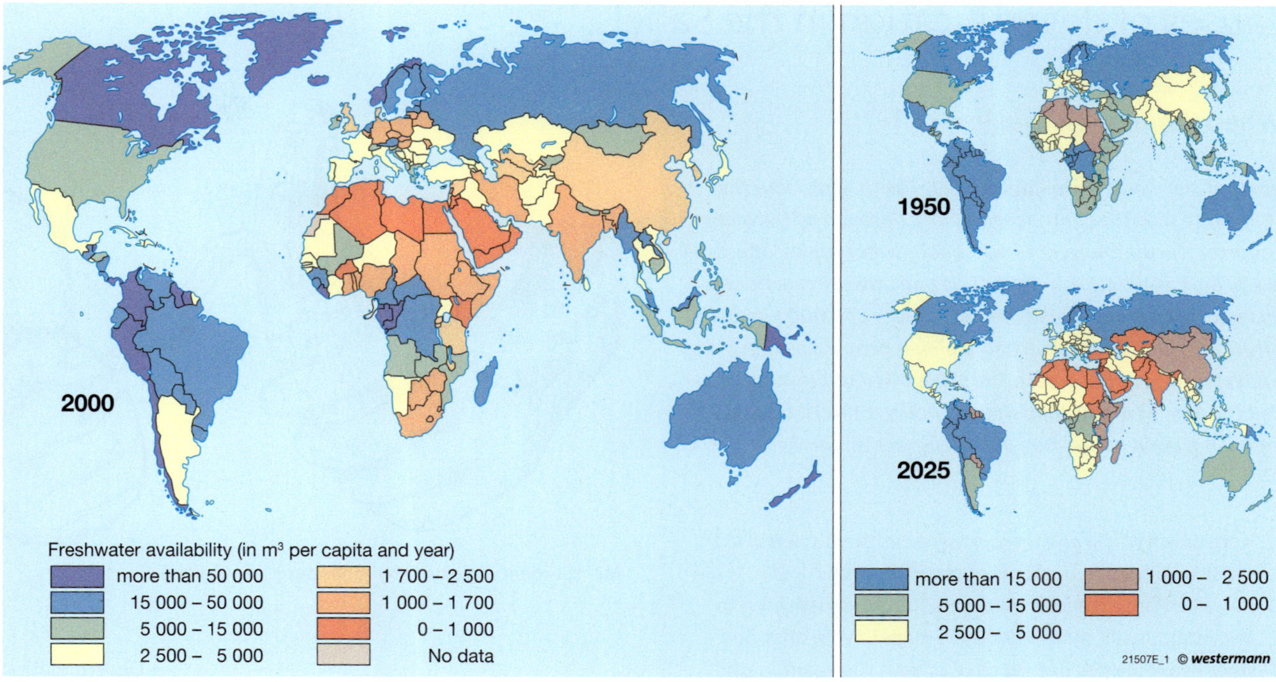

M6 Freshwater availability worldwide

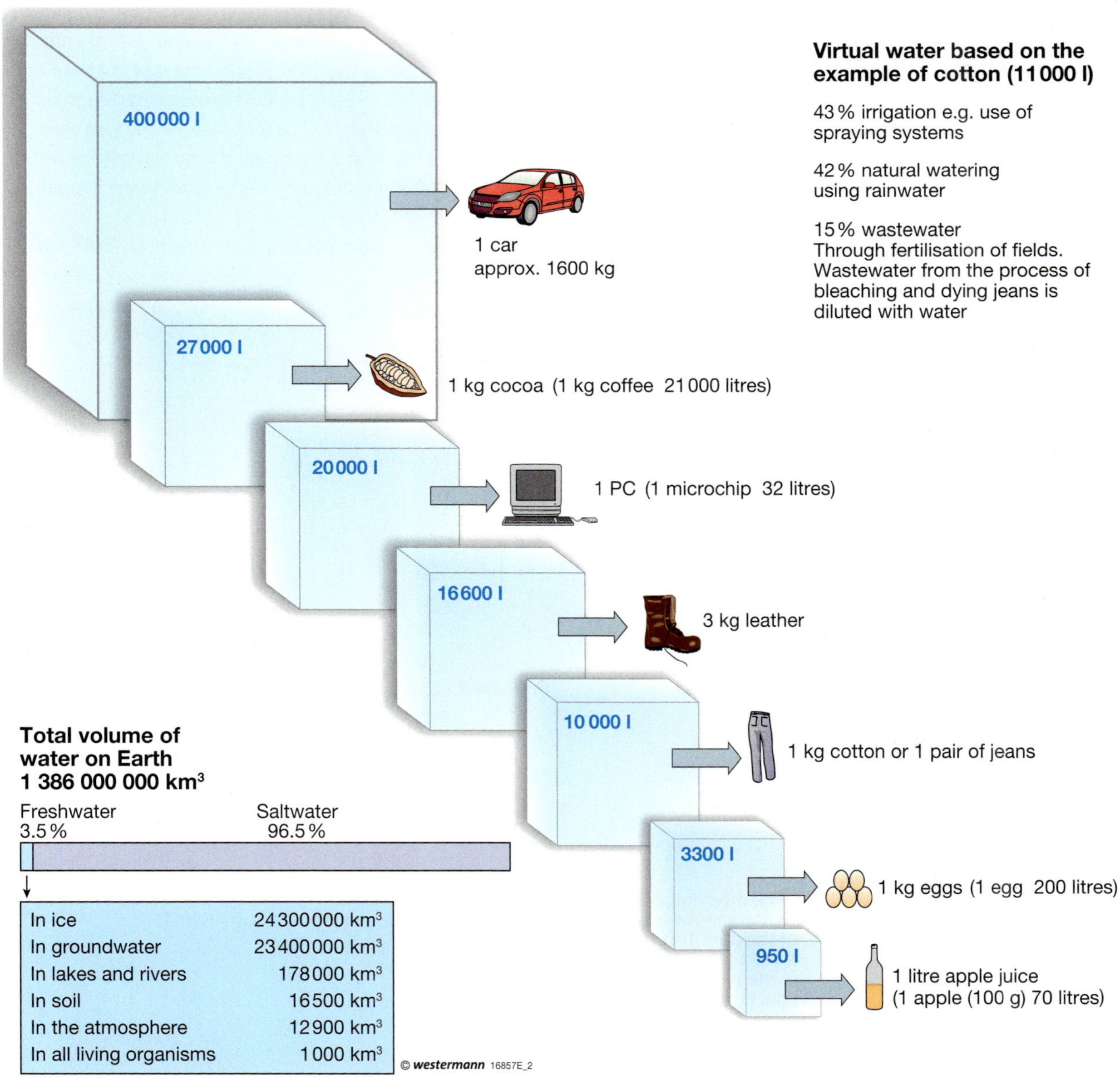

M7 Water volume worldwide and virtual water in everyday products

Causes of desertification in the Sahel

When there is less and less soil ...

About 7.5 percent of the world's population live in North Africa and the Middle East. But these countries only have about 1.2 percent of the global freshwater resources. Furthermore, the inadequate water supply affects a region marked by high population growth. This and other circumstances lead to soils being overused and a land area of 20,000 – 40,000 km² being lost to agriculture annually. More than half a billion people are affected by the subsequent expansion of the desert. The soil is often not capable of regeneration and the area of arable land shrinks. And yet this is supposed to feed an ever-growing world population. How long can this go on?

1. Note which of the environmental issues mentioned in M1 can be assigned to the topic of desertification.
2. Describe the distribution of desertification by referring to M3. Additionally, look for suitable maps in your atlas and state five countries from each of at least three continents that are affected by desertification.
3. (O) Explain issues surrounding desertification as illustrated in M4.
   - **A** Describe the illustration from top to bottom.
   - **B** Explain how migratory movements can aggravate the process of desertification at the final destination.
4. (S) Interpret the caricature in M2. Use the process of desertification as illustrated in M4 in your interpretation.

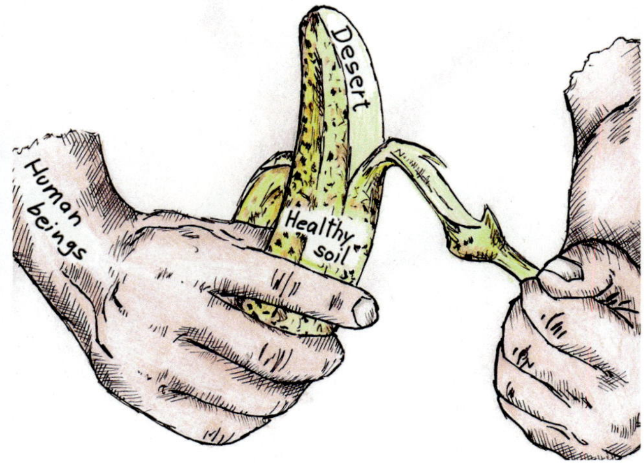

M2 Student caricature by Johannes M. & Giulia G.

→ Desertification

Desertification is the anthropogenic transformation of the landscape in arid regions, which leads to desert formation due to overexploitation of an ecosystem. Social, economic and political developments cause methods of utilisation and cultivation to change and move away from the principles of sustainability. At first glance, natural processes appear to cause desertification. However, cultivation and pastoral farming alter the type of vegetation and as a result, and more importantly, the soil and water balance. Desertification can significantly increase the natural formation of deserts.

Based on: H. Leser (Hrsg.): Wörterbuch allgemeine Geographie, 2005

Desert
→ Desertification

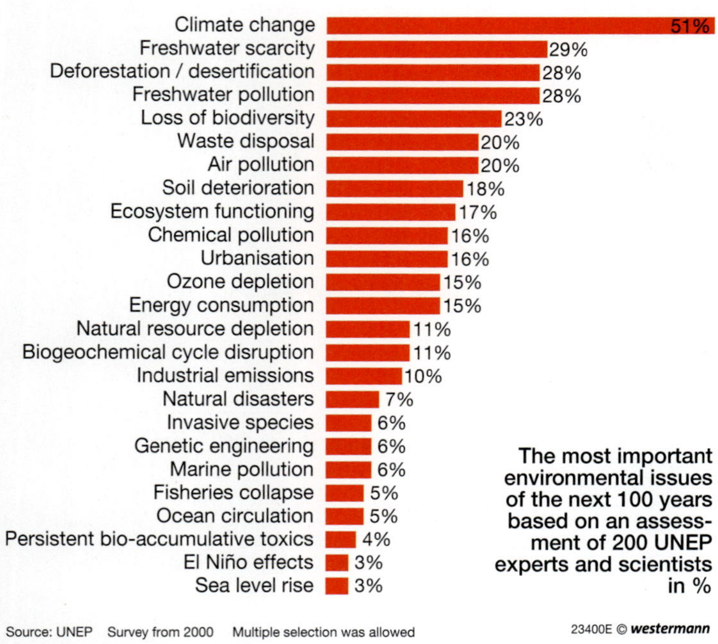

M1 Environmental issues of the 21st century

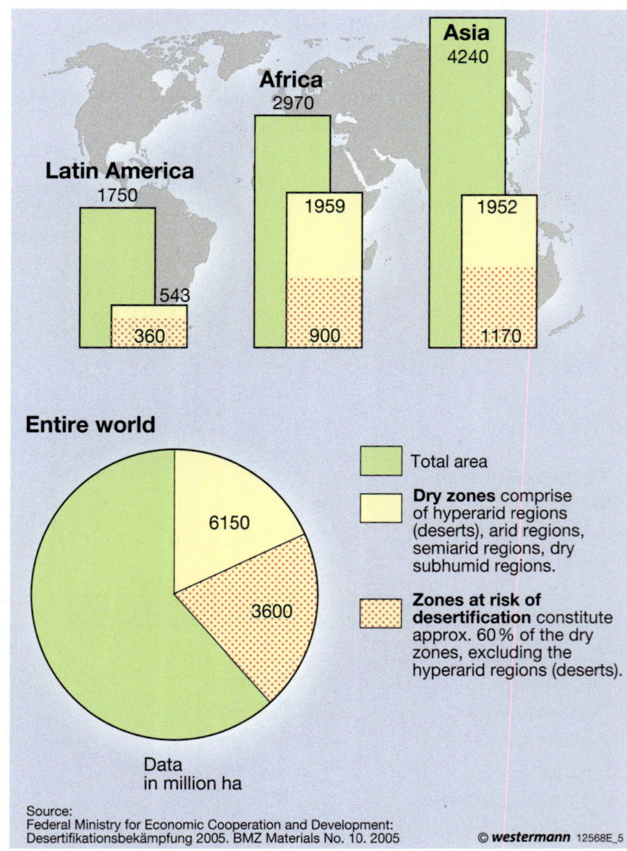

M3 Regions at risk of desertification

Desertification - an unstoppable process? – Station exercise

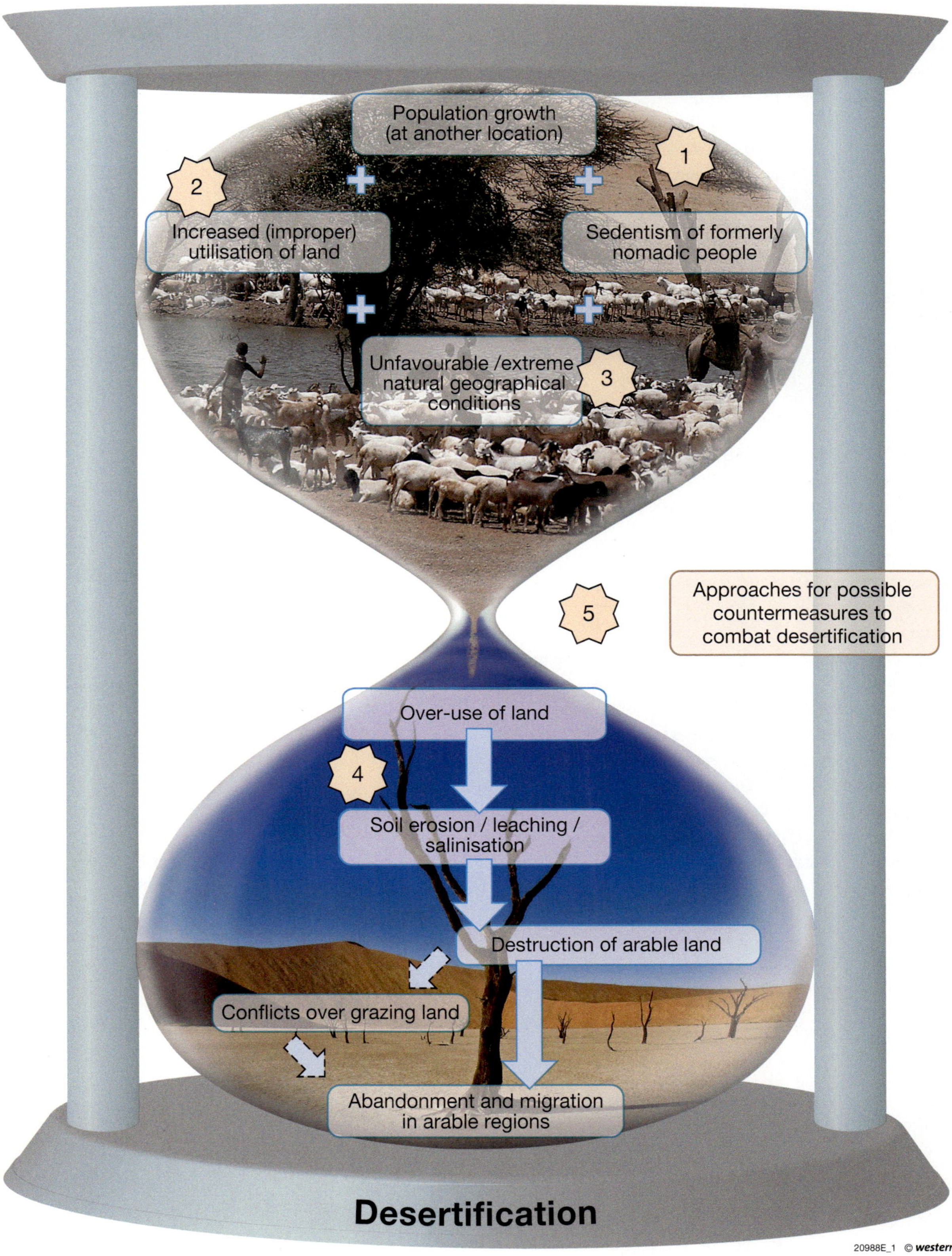

M4

Each of the topics marked with a star has a two-page spread in the following chapter. The topics can be worked through and treated as **learning stations**. The key question being: *"Desertification – an unstoppable process?"*
After you have worked through the material of one station, create a mind map (see illustration in the assignment block on each page) or expand the already existing one.

Procedure:

1. Divide into five equal groups, one for each station.
2. Work through the material according to the work assignments. For quick thinkers: Search the internet or other sources for further facts relating to your station.
3. After a set amount of time all change stations (1 to 2, 2 to 3, 3 to 4, 4 to 5, 5 to 1).

Causes of desertification in the Sahel

Population growth – catalyst for the process of desertification in the Sahel?

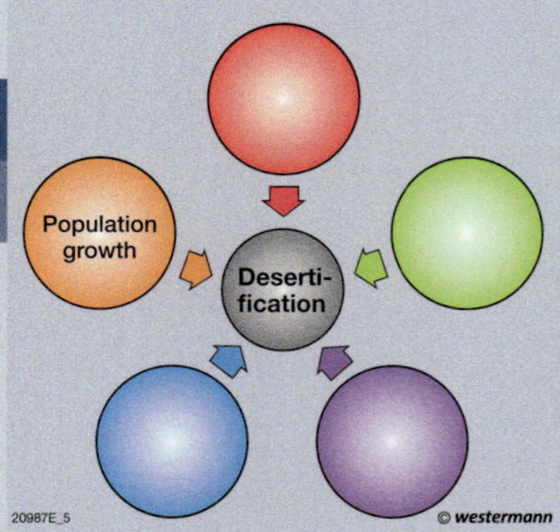

Rapid population growth in newly industrialised and developing countries can be considered a cause for the increasing deterioration of the nutritional state in these countries. What impact does a larger and denser population have on regional land use? Which regions are being newly developed? What implications does this have? How has the situation developed to the present day?

1. Explain why population growth is a significant cause of the advancing process of desertification (M1/ M4/ M7).
2. ⓞ The demographic structure in Sudan will change between 2014 and 2034.
   A Investigate the changes in the age distribution and demographics. Combine the development that is recognisable between M2 and M3 with that shown in M8.
   B Convert M8 into a diagram. Then analyse the diagram in terms of the demographic structure in the Sahel.
3. Explain agricultural land use in the region of El Fasher (M5). Discuss whether this land use is sustainable.
4. Ⓢ Discuss issues surrounding the immigration of nomads and their becoming sedentary with regard to natural population growth in Sudan (M1/ M5/ M6).
5. Expand the branch of "population growth" in your desertification mind map.

Degradation, Demography, Fertility, Mortality

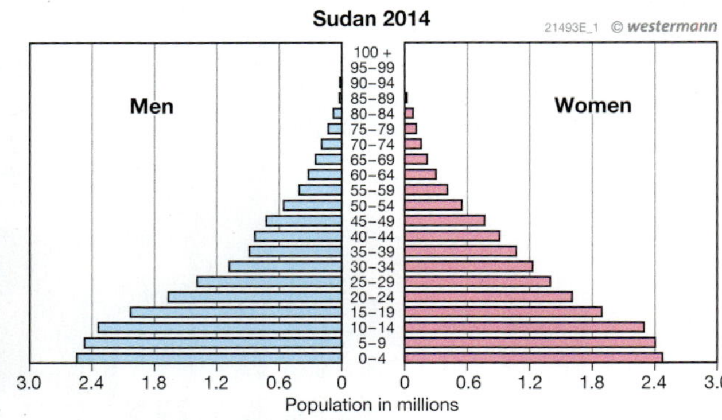

M2 Age pyramid for Sudan today

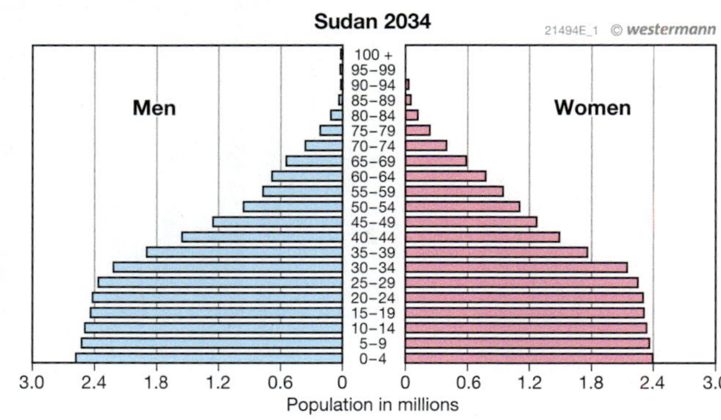

M3 Age pyramid for Sudan in the future

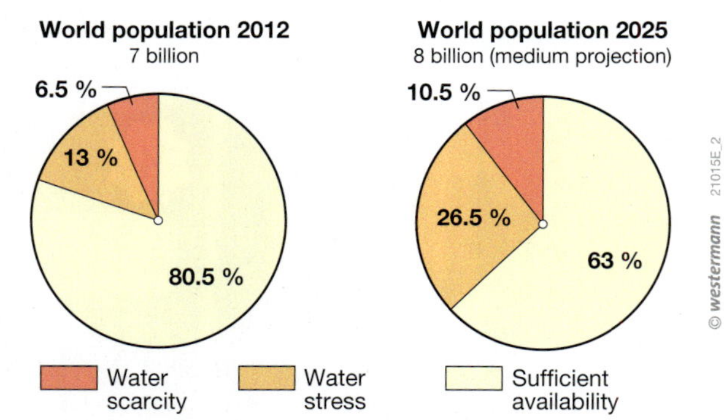

M4 Development of world population and water availability

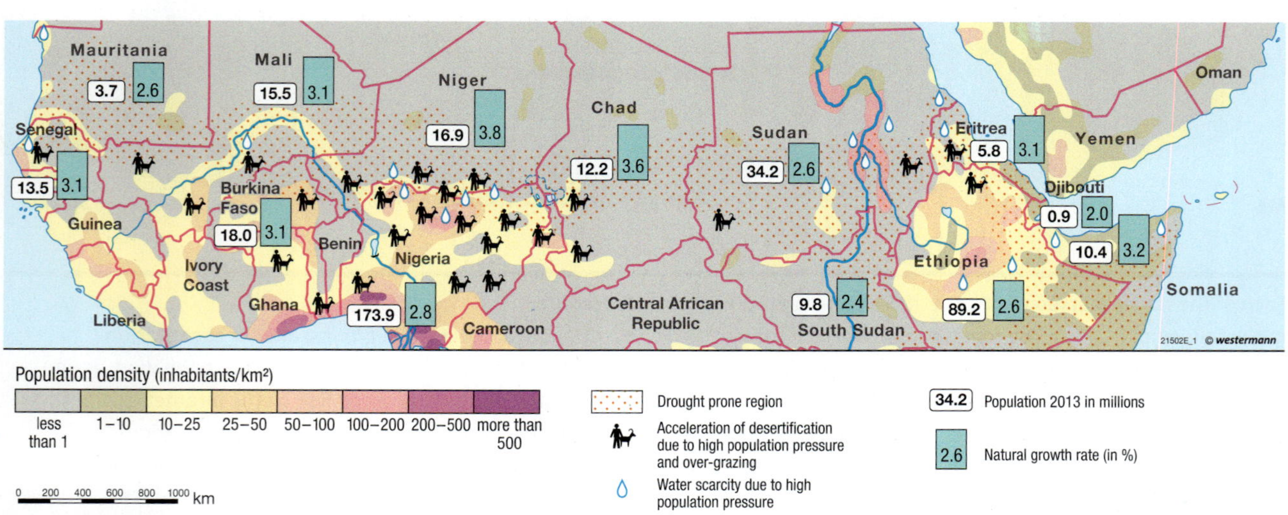

M1 Population growth and water scarcity in the Sahel

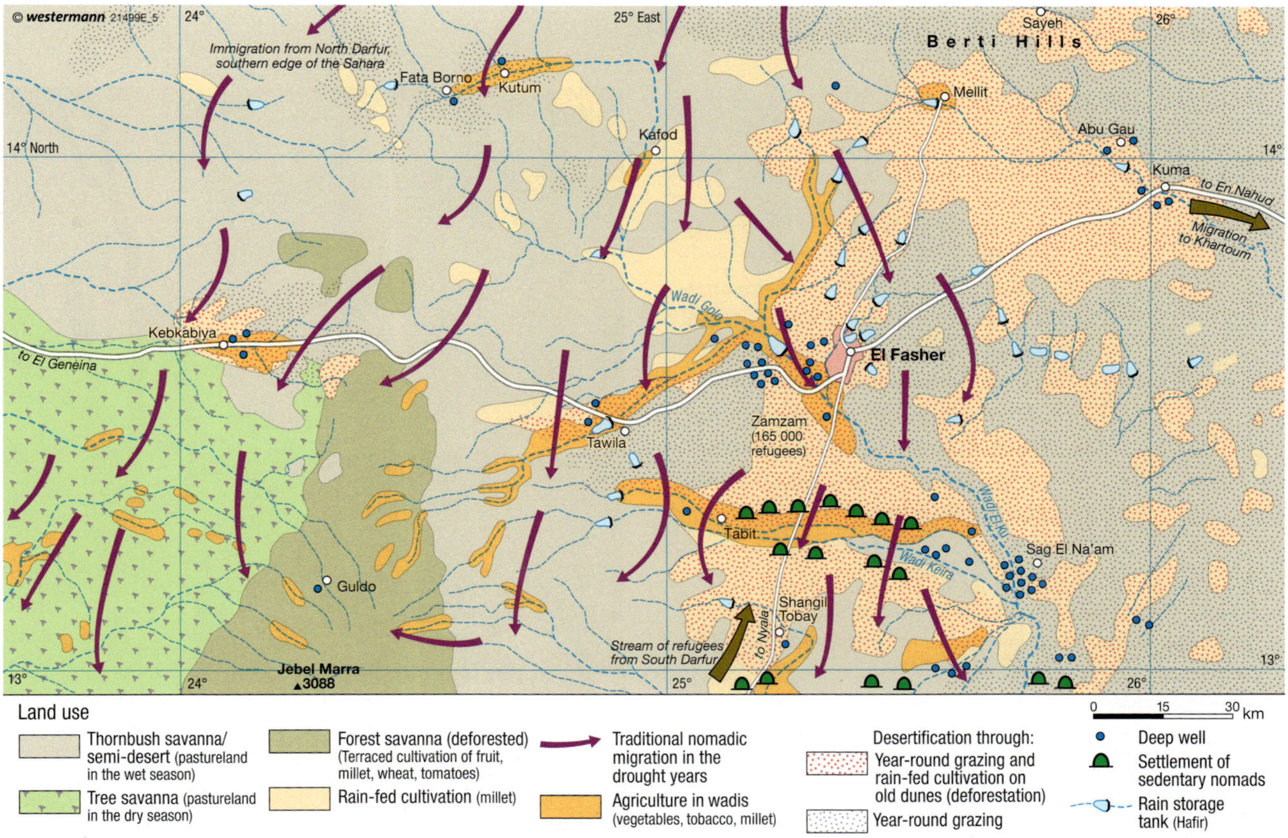

M5 Land use, migration patterns and settlements in El Fasher (Sudan)

In the traditional hierarchy of the savanna tribes, nomads occupied an elevated status and neighbouring peasants served them as slaves. Today, this allocation of roles is almost reversed.

Most nomadic tribes are marked by constant economic and social decline. It began with the development of transport infrastructures leading to the loss of their transport monopoly: camels were replaced by trucks.

Constant population growth intensified competition for the increasingly scarce resources land and water. Growing numbers of people inevitably led to higher numbers of livestock, a trend reinforced by increasing commercialisation of livestock farming ("professional pasture farming") and numerous veterinary projects.

Finally, the barren pastures could no longer supply sufficient feed for livestock. The result was overexploitation, degradation and desertification.

Based on: U. Scholz, 2011

M6 Changes in the social structure

Many parts of the thornbush savanna, in particular, are so densely populated that overexploitation and degradation occur.

On examining the population trend in individual regions, a sharp rise caused by the growth of the resident population can be observed on the one hand. On the other hand, until now arable parts of the arid regions provided important buffer zones for the population overflowing from the densely and overpopulated humid regions and those regions used by nomads. [...]

With a presently reduced, but still too high, population growth rate of over two percent, the existing buffer zones are being settled well into the extremely drought-prone regions of the agronomic dry boundary zone (see pp. 60/61).

Furthermore, as a result of the prevailing practice of "equal inheritance", farm sizes in the arid regions are continuing to shrink below the critical minimum size.

Based on: B. Hornetz, R. Jätzold, 2009

M7 Overexploitation due to high population density

Demographic indicators	1995	2005	2014	2025*
Population				
Population (in 1,000)	24,517	29,883	35,482	42,733
Growth rate (in percent)	2.6	2.6	1.8	1.7
Fertility				
Fertility rate (births per female)	6.1	5.4	3.9	2.8
Birth rate (per 1,000 inhabitants)	43	39	30	24
Births (in 1,000)	1,056	1,162	1,065	1,010
Mortality				
Life expectancy at birth (years)	52	51	63	67
Child mortality rate (per 1,000 births)	87	83	53	39
Death rate (per 1,000 inhabitants)	13	13	8	6
Deaths (in 1,000)	326	399	279	171

*Forecast

Source: US Census Bureau

M8 Demographic trend in Sudan (1995–2025)

Causes of desertification in the Sahel

Excessive and inappropriate land use in the threatened Sahel Belt

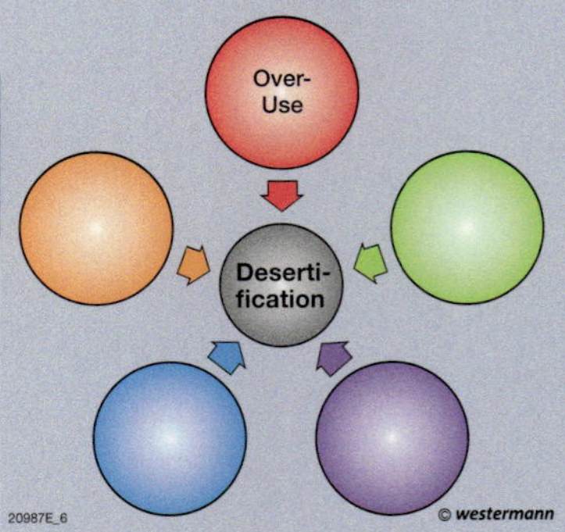

A growing population and increasing numbers of livestock have had a severe impact on the Sahel Belt. Environmental scientists and agricultural planners cannot keep up with this utilisation pressure – concepts that are both sustainable and resource-friendly are often still lacking. Is this contributing to desertification? How have livestock numbers developed?

ⓞ **1.** Analyse the utilisation pressure on the Sahel Belt and in particular Sudan. To do this, quote from different material sources and link your statements.
   A Explain the connection between the development of agriculturally used areas and the carrying capacity of the region (M2/ M4 / M5).
   B Outline the situation from the point of view of a farmer in the Sahel.
   C Find the core statement in each material source and analyse it in regard to utilisation pressure.

2. Explain the role of water in excessive and inappropriate land use in the Sahel (M6/ M7).

3. Goats are particularly "environmentally unfriendly" animals because not only do they eat the shoots and leaves of a plant, they also rip out the roots. They are also good at climbing. Use this information as a starting point for your analysis making reference to M6.

Ⓢ **4.** In recent decades, production for the global market has increased significantly, so too in the Sahel.
Using an atlas, find out which cash crops are grown in the Sahel.

5. Expand the branch of "over-use in the Sahel Belt" in your desertification mind map.

Food crop, Subsistence farming
→ Carrying capacity, Cash crop

→ Carrying capacity

The term carrying capacity generally refers to the capacity of a habitat to sustain life. We differentiate between agricultural, natural and total carrying capacity (see also p. 62).

→ Cash crop

A cash crop is an agricultural product produced for the market. Cash crops stand in contrast to crops grown for self-sufficiency (subsistence farming, food crops).

M1 Goat herds at a watering hole in the Sahel

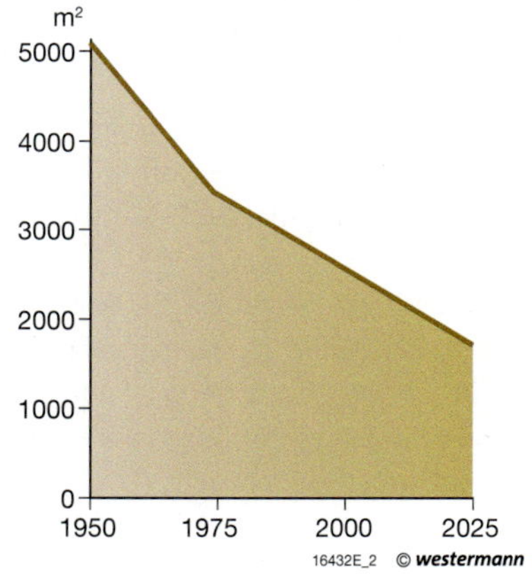

M2 Agricultural land area worldwide per capita

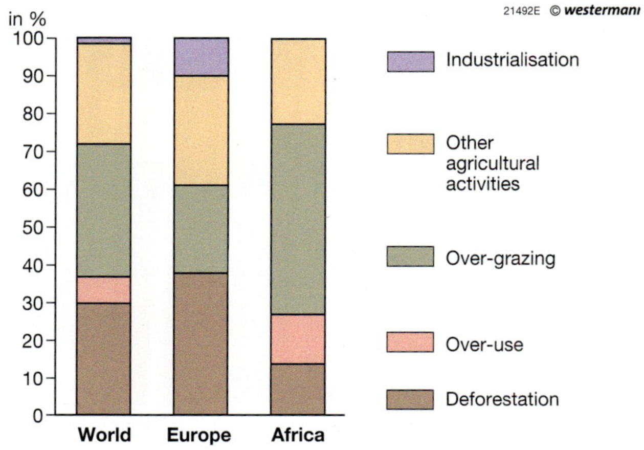

M3 Anthropogenic causes of soil degradation

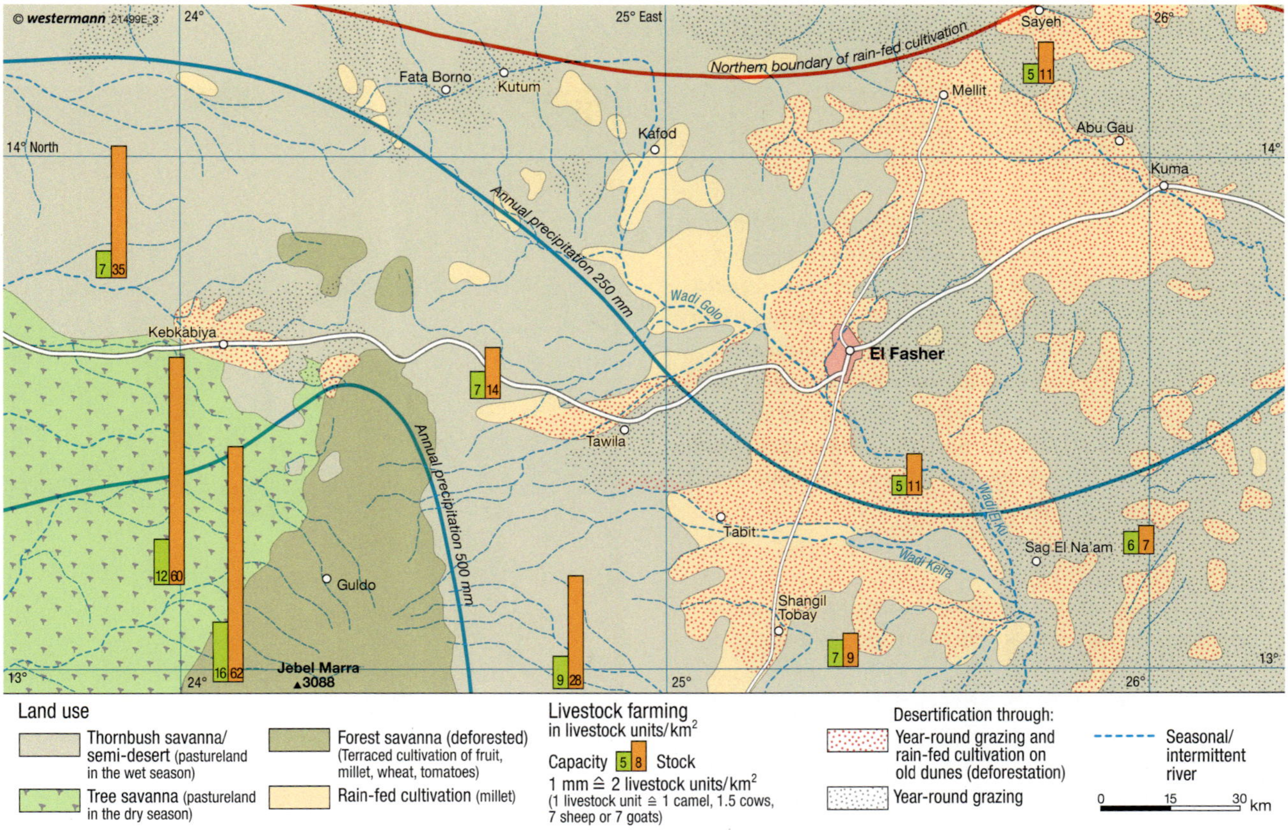

M4 Land use and livestock numbers in El Fasher (Sudan)

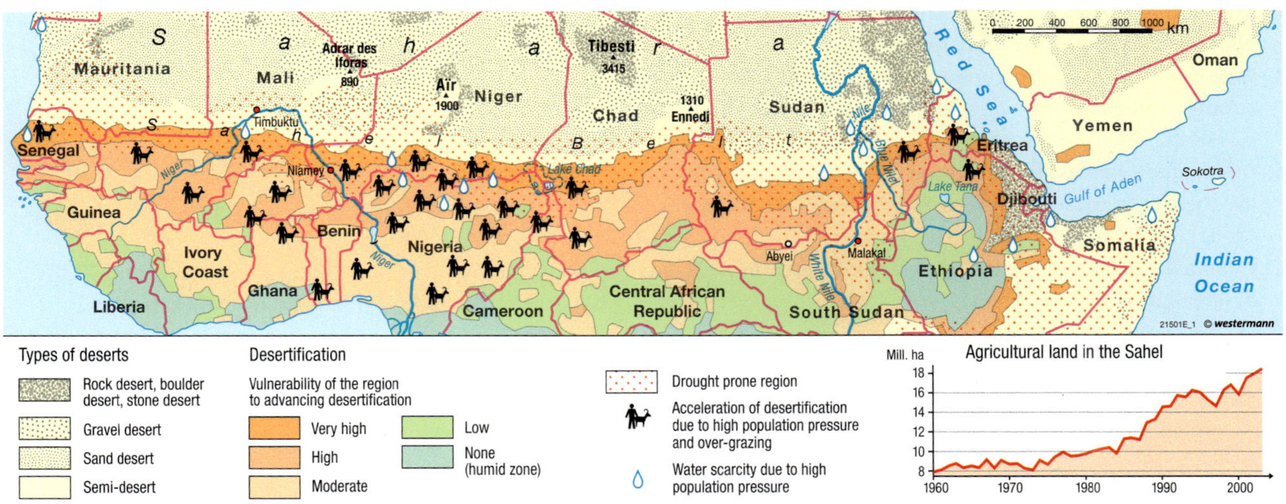

M5 Overexploitation in the Sahel Belt

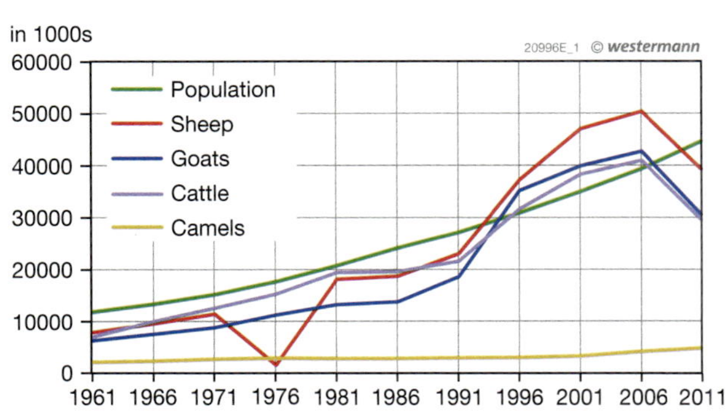

Source: FAO (Food and Agriculture Organization of the UN)

M6 Population trends and livestock numbers in Sudan

Water use for the production of 1 kg:

- Wheat: 715–750 l
- Maize: 540–630 l
- Soybeans: 1,650–2,200 l
- Rice: 1,550 l
- Beef: 50,000–100,000 l
- Cleaned wool: 170,000 l
- Milk (1 l): 1,000 l

M7 Water use for the production of foodstuffs

Causes of desertification in the Sahel

Unfavourable natural geographic conditions in El Fasher (Sudan)

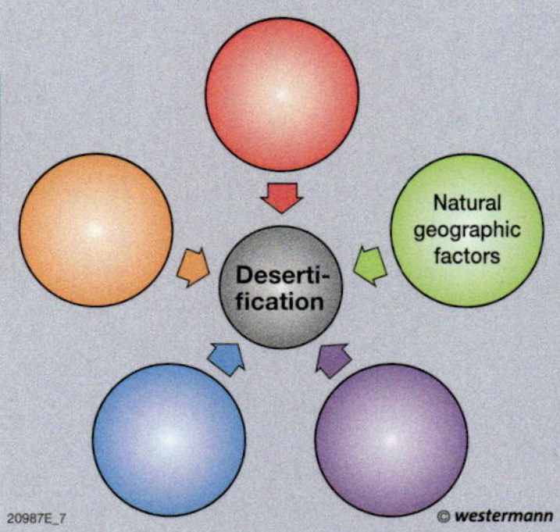

The Sahel Belt is located in a transition zone between a region with favourable agricultural conditions in regard to water availability in the south and desert in the north. This makes it a fragile landscape belt. Droughts are not predictable and can destroy entire crops. It is not clear how climate change will influence the likelihood of drought.

Ⓞ **1.** Locate El Fasher and characterise the natural geographic conditions for agricultural land use in the region (M1/M3–M7).
   A Use an atlas to help you arrange the conditions according to the geofactors (cf. p. 8).
   B Describe the location of El Fasher, and then, based on the geographic location, go on to describe the climate of El Fasher. Finally, describe the soil, terrain and vegetation. Consider the implications of the conditions for human land use.

2. Explain the impacts of drought in the Sahel and its consequences for land use (M2, → definition).

Ⓢ **3.** Research (e.g. internet) occurrences of severe and catastrophic drought. Write a brief fact sheet about these occurrences.

Ⓢ **4.** Discuss the vulnerability of El Fasher. How can people who live there take action in a future-oriented way? What can we do?

5. Expand the branch of "natural geographic factors" in your desertification mind map.

Fallow land
→ Agronomic dry boundary

→ Agronomic dry boundary
The agronomic dry boundary is the dry boundary of rain-fed cultivation. It is a dry boundary zone that runs between the dry savanna and thornbush savanna with about 8.5 arid months in the tropics and between the zones of Mediterranean vegetation and thornbush steppe with about 8 arid months in the subtropics. Annual precipitation is between 250 and 1,000 mm with a very high rate of evaporation. Drought-tolerant forms of agricultural farming are barely viable in this zone.

M1 Irrigation of planted shrubs

Period with above average precipitation	Period with below average precipitation
Increase in herd size Advance of millet cultivation beyond the boundary of rain-fed agriculture Increase in sedentism of nomadic people	Conflicts over land use caused by nomads migrating south in drought years Alternative forms of income are sought: Tree clearing for the production of wood charcoal Increased use of existing deep wells Severe drought: Millet crop failure leads to famine

M2 Influence of precipitation levels on land use

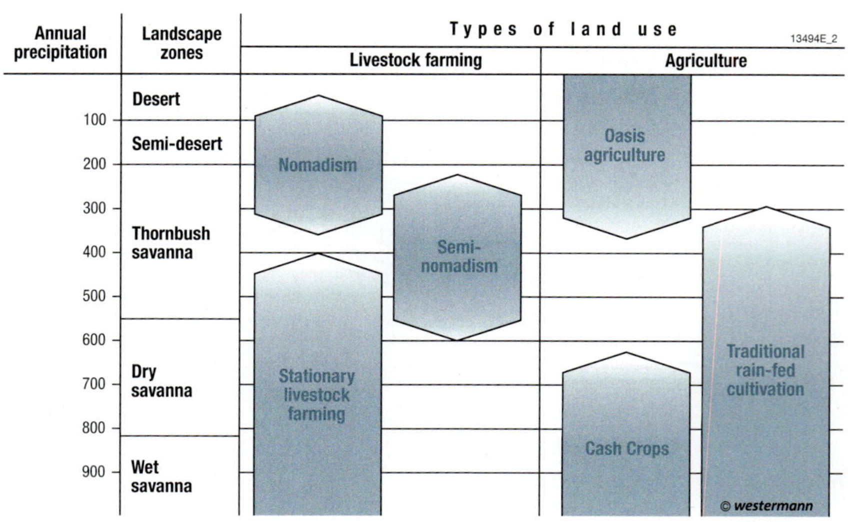

M3 Traditional agricultural land use in desert and savanna regions subject to precipitation (in mm)

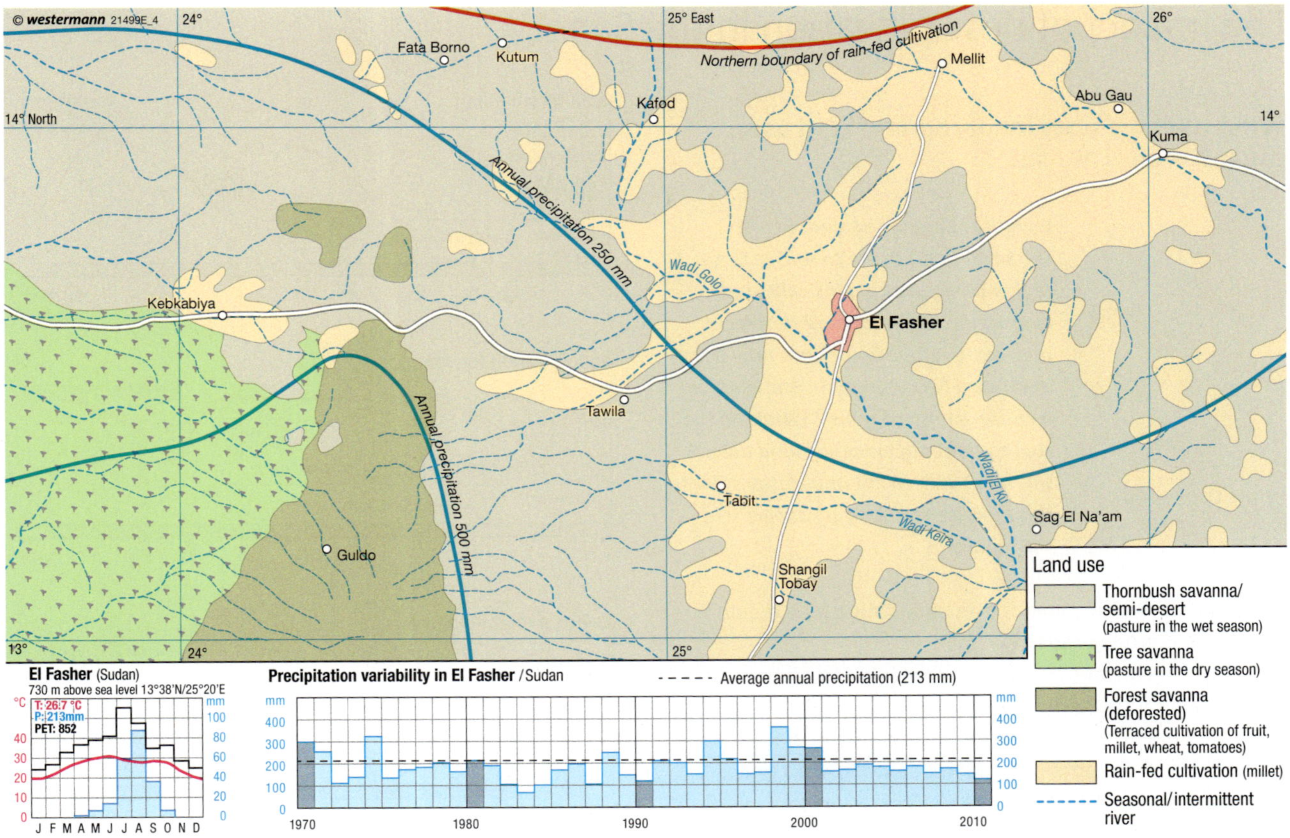

M4 Climatic threats to El Fasher

Droughts at the boundary of rain-fed agriculture are a major problem.

Even in relatively drought-safe regions such as northwest India, 950 mm of rain will fall in one rainy season and only 20 mm in another. Reserve cultivars such as [...] tubers and feeding trees and shrubs are vital.

Wild fruits can also help and migration is essential. [...]

Because the currency situation and infrastructure of most countries does not allow for the accumulation of food reserves, unfavourable precipitation trends increasingly result in severe food shortages and catastrophic famines.

Based on: B. Hornetz, R. Jätzold, 2003

M5 Droughts in rain-fed agriculture

	Tropical rainforest	Wet savanna	Dry savanna	Thornbush savanna	Semi-desert
Precipitation/ year (mm)	>1,500	1,500–1000	1,000–500	500–100	<100
Intensity of chemical weathering	extremely strong	strong	moderate	low	very low
Trace mineral content	very low	low	moderate	high	very high
Risk of salinisation	none	low	moderate	great	very great
Humus layer	moderate	moderate	low	very low	non-existent
Prevailing clay minerals	1:1 clay minerals (e.g. kaolinite)		2:1 clay minerals (e.g. montmorillonite)		

M7 Criteria for fertile tropical soils (see also Chapter 1)

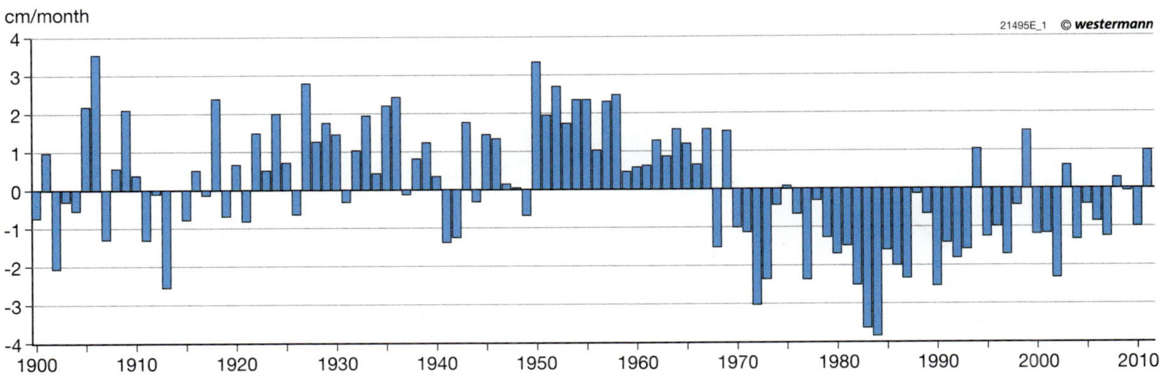

M6 Precipitation variability in the Sahel

The impact of desertification in the Sahel

What damage does inappropriate land use cause a region at risk?

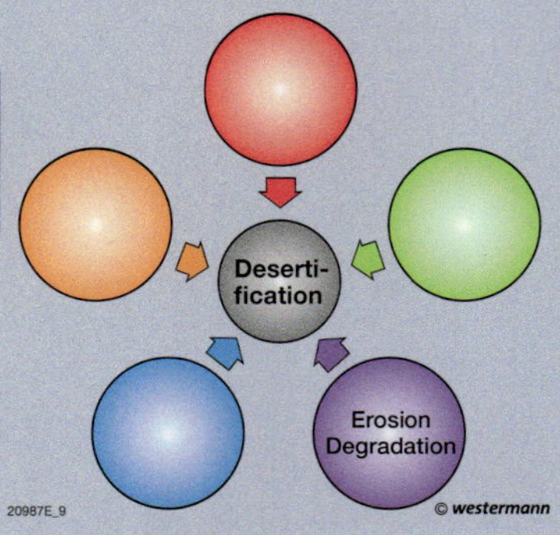

Significant population growth in the Sahel Belt has led to soils in the region being used extensively (and excessively). Areas of land that were previously refuge areas have now been developed for cultivation. But when is a region being overused? What are the consequences of excessive land use in the Sahel Belt? What implications does this have for the carrying capacity of the region?

1. Explain the implications of overexploitation for a region's potential carrying capacity.
   - **A** Describe M2 and explain the necessity of a fallow period for the level of yield.
   - **B** Describe M1 and explain the impact of population trends on the potential carrying capacity of a region.
   - **C** Analyse the statements in the introductory text on this page using M1, M2 and the → Definition.
2. Analyse the impact of excessive land use in the Sahel Belt based on the example of degradation. Back up your assertions with suitable material.
3. Search e.g. on the internet or in newspapers for significant images of soil degradation. Note the location and look up the climate zone of the location in an atlas. Formulate indications of possible vulnerability based on the images and the geographic location.
4. Using an atlas, find out which countries in the Sahel Belt (M4) are affected by erosion and soil degradation. In what way and to what extent are they affected?
5. Expand the branch of "erosion and degradation" in your desertification mind map.

→ Carrying capacity, Degradation, Erosion

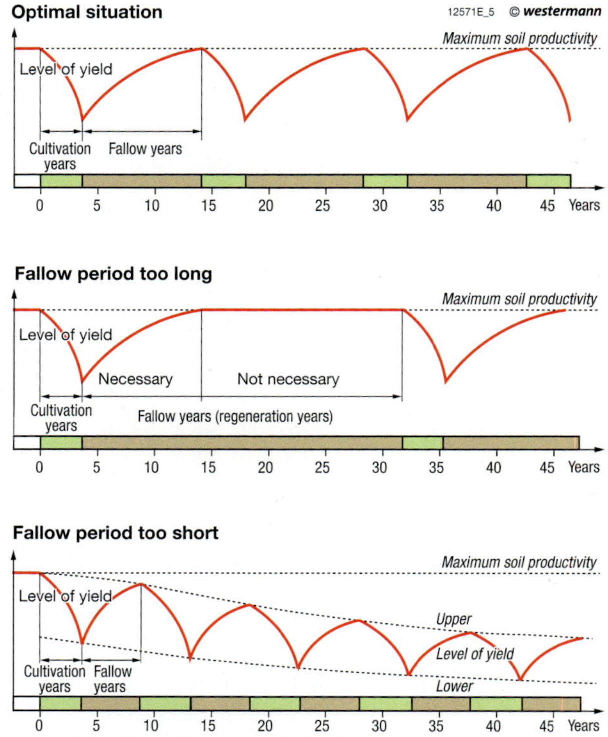

M2 Soil fertility subject to periods of fallow and cultivation

→ Carrying capacity

The term carrying capacity generally refers to the capacity of a habitat to sustain life. We differentiate between agricultural, natural and total carrying capacity.

The agricultural carrying capacity pertains to the carrying capacity of a habitat for plants and animals. This is dependent on the natural-spatial potential (natural carrying capacity). The agricultural carrying capacity in turn indicates the capability or carrying capacity of a region and its ecosystem for human use. The number of people in a region determines whether the carrying capacity is overloaded. All these factors give the overall economic (total) carrying capacity of a region. The carrying capacity assumes that the regenerative capacity of the ecosystem is ensured.

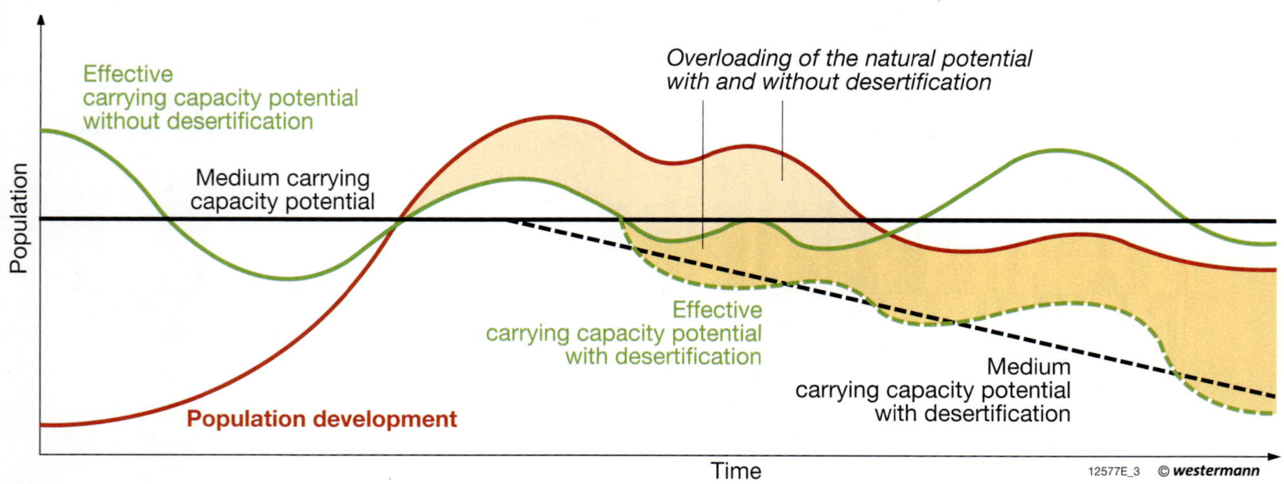

M1 Development of the carrying capacity potential

M3 Erosion in the Sahel

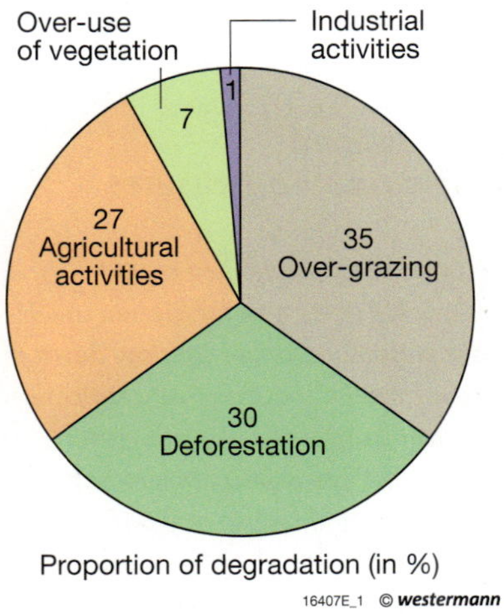

M6 Causes of soil degradation

→ (Soil) degradation → Erosion

Erosion is a collective term for removal processes. Erosion is essentially caused by the deflation, transportion and accumulation of soil particles by rain and wind. Consequently, erosion can physically contribute to the degradation of soil, i.e. reduced fertility. Soil degradation can have chemical (e.g. soil salinisation) or natural causes but can also be attributed to inappropriate human land use.

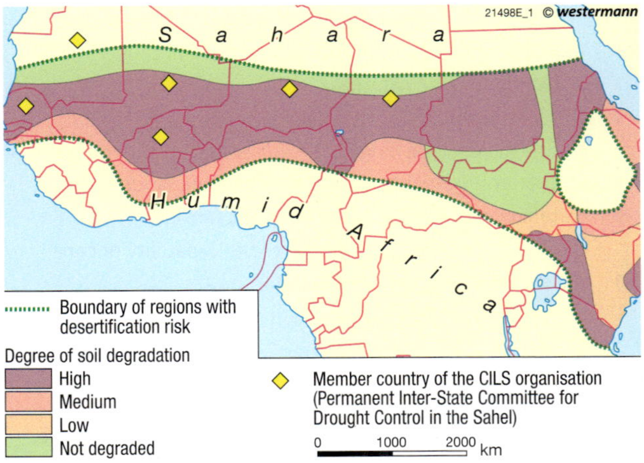

M4 Soil degradation in the Sahel

Continent	Proportion of degradation	
	absolute[1] (in %)	relative[2] (in %)
Asia	38.1	20
Africa	25.2	22
South America	12.4	14
Europe	11.1	23
North America	8.0	8
Oceania	5.2	13

1 based on areas of degraded soil worldwide

2 based on the proportion of continental land area

M7 Soil degradation worldwide

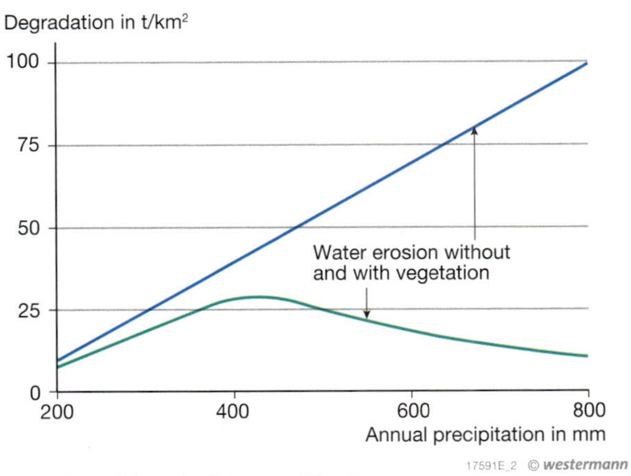

M5 Erosion with and without cultivation

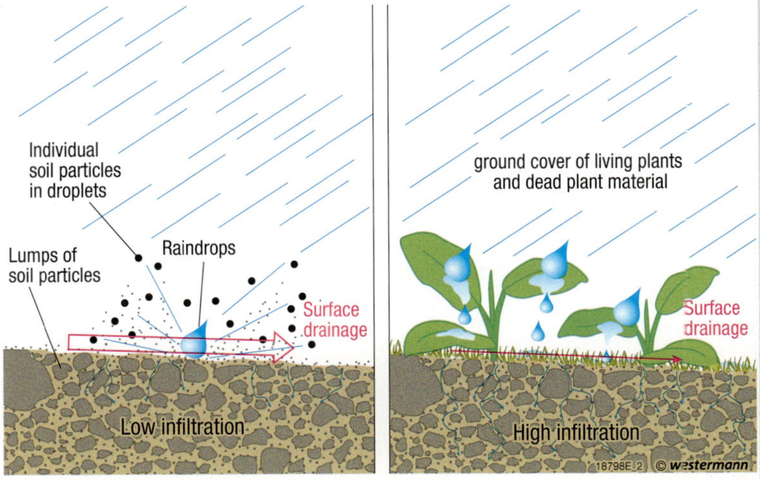

M8 Erosion with and without ground cover

The impact of desertification in the Sahel

Can desertification be stopped? Approaches for possible countermeasures

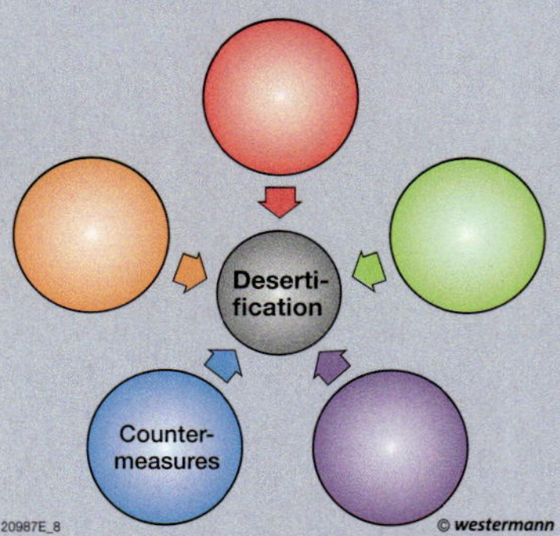

A completely desertified region is irreversibly destroyed for human use. It is therefore important to slow down or even stop the process of desertification as early as possible. For this reason, people are attempting to combat desertification collectively and transnationally. But how effective are these strategies? And are there solutions that do not require a lot of money?

1. Explain the different measures for combatting desertification visible in M2 (M1–M4).
2. Research how mulch tillage works and explain its effectiveness in combatting desertification (M3).
3. (O) The key to combatting desertification is transnational communication and global awareness.
   - **A** Present the key statements made by Serigne Mbodji on this topic (M5).
   - **B** Note further questions that you would like to ask Serigne Mbodji regarding solutions for combatting desertification (M5).
4. Explain the function of the FAO in the fight against desertification (M6).
5. Find further information about the Great Green Wall project (other than M5 and M6) and produce a fact sheet.
6. Expand the branch of "possible countermeasures" in your desertification mind map.

Erosion, Sustainability, Sustainable development

Building dry walls is one possibility for combatting soil degradation. The walls are built, as far as possible, parallel to the contours of the terrain.

The walls reduce leaching of humus-rich topsoil by slowing down runoff rainwater and preventing the formation of fast-flowing streams. Additionally, a dry wall functions like a dam. Water-borne material accumulates in front of the wall.

Dry walls are made of loose and piled up stones. The gaps between the stones are filled with mortar and cement so that the rainwater can run off through the small cracks preventing puddles from forming when there is too much rain.

Furthermore, the joins and cracks provide a habitat for small animals and plants.

Based on: C. Olehowski, M. Haspel, In Praxis Geographie 06/2009

M1 Dry walls to prevent erosion

M2 Different measures for combatting desertification

Measures against salinisation and soil protection measures:

- Drainage for seepage control
- Cultivation of salt-tolerant plants (sugar cane, cotton, millet)
- Drip irrigation
- Irrigation at night
- Sustainable methods of soil cultivation
- Mulch tillage
- Site-adapted crop rotation
- Other, e.g. contour cultivation
- Limitation of slope length
- Soil loosening and prevention of soil consolidation
- Installation of windbreak hedges and fences

M3 Measures against soil degradation

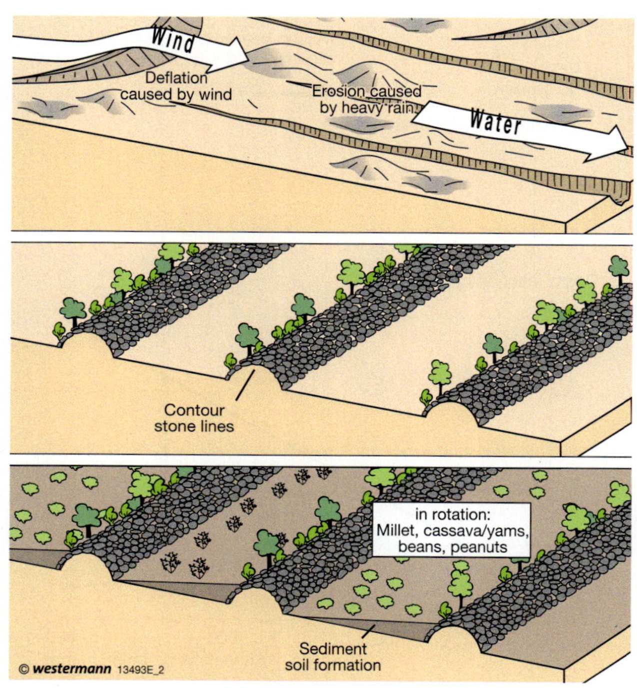

M4 Protective measures for agricultural land

When did the Great Green Wall project (GGW) begin?
One of the first meetings took place on 1 and 2 June 2005 in Ouagadougou, Burkina Faso. Today, eleven countries in the Sahara-Sahel region are mobilised in the project: Senegal, Mauritania, Mali, Burkina Faso, Nigeria, Niger, Chad, Sudan, Ethiopia, Eritrea and Djibouti.

What are the main objectives?
The GGW project is a reaction to one of the greatest concerns of our continent: degradation of the natural environment and drought. The goals of the GGW project are to combat poverty and to establish enhanced and, most importantly, sustainable development and healthy management practices in regard to natural resources and the environment.

What measures have been implemented to date?
Seedlings are grown for reforestation in seven different breeding stations. Annually about 2 million.
An area of about 5,000 ha is reforested every year. The timber industry has also been improved with the aim of ensuring the protection and sustainable development of this resource. Furthermore, so-called multi-purpose gardens have been established. Sustainable methods of cultivation and correct irrigation are implemented here.

How successful has this been?
Politicians from the participating countries have assumed a good leading role. In addition, employees of the multi-purpose gardens are paid about US $100 a month so that students can finance their university degrees and even support their parents. Other remarkable phenomena are a decrease in migration and an increased interest in literacy.

How do people in the Sahel react to these changes?
Harvest and sales of produce from the multi-purpose gardens particularly benefit the women who work in these gardens. Over and above this, the quality and variety of the local people's diet has increased due to the sale and consumption of the crops (lettuce, potatoes, tomatoes). In particular, farmers are interested in improving pastureland for their livestock.

M5 Interview with Serigne Mbodji (Employee of the National Agency for the Great Green Wall in Senegal)

An African partnership to tackle desertification and land degradation
Great Green Wall for the Sahara and the Sahel Initiative

Desertification and land degradation challenge the lives of people in the Sahel and the Sahara, home to the world's poorest populations. In 2007, African Heads of State and Government endorsed the Great Green Wall for the Sahara and the Sahel Initiative to tackle the detrimental social, economic and environmental impacts of land degradation and desertification in the region.
The initiative aims to support the efforts of local communities in the sustainable management and use of forests, rangelands and other natural resources in drylands. In doing so, the initiative seeks to improve the food security and livelihood of the people, while contributing to climate change mitigation and adaptation.
A harmonized strategy for the Great Green Wall was adopted in September 2012 by the African Ministerial Conference on Environment (AMCEN).
According to AMCEN, it is a flagship programme that contributes to the goal of the UN's Conference on Sustainable Development, or Rio+20, of "a land degradation neutral world".
From the initial idea of a line of trees from east to west through the African desert, the vision for a Great Green Wall has evolved into that of a mosaic of interventions addressing the challenges facing the people in the Sahel and Sahara.
The overall goal of the Great Green Wall initiative is to strengthen the resilience of the region's people and natural systems with sound ecosystems management, sustainable development of land resources, the protection of rural heritage and the improvement of the living conditions of the local population.
Based on: www.fao.org, 16.12.2013

M6 Report on the homepage of Food and Agriculture Organisation of the United Nations (FAO)

METHOD Concept Map

Individual aspects of desertification often complement and reinforce one another. Many results of desertification become new causes. For example, increased agricultural land use due to a growing population is one cause of desertification. When a certain population density is reached, people are forced to migrate leading to desertification at another location. A concept map takes a mind map one step further through the addition of connecting lines and explanations. A map of terms is created. It serves to interconnect your knowledge. A concept map is not a presentation tool but it is suitable for repetition e.g. before exams.

Four steps for producing a concept map

STEP 1

→ You need a blank A3 sheet of paper, small cards with terms written on them (e.g. specialised terms used in this chapter), glue, a pen and ruler.

STEP 2

→ Arrange the terms on the A3 sheet of paper.
→ Place terms that are related closer together.
→ Place any terms that you are unsure of to the side.

STEP 3

→ After gluing the terms onto the A3 sheet of paper, draw connecting lines between them.
→ Consider which direction the arrows should point. Not all terms have to be connected.
→ Groups of terms can also be arranged according to colour (see M1).

STEP 4

→ Now turn your attention to the unknown terms. Research their meaning.

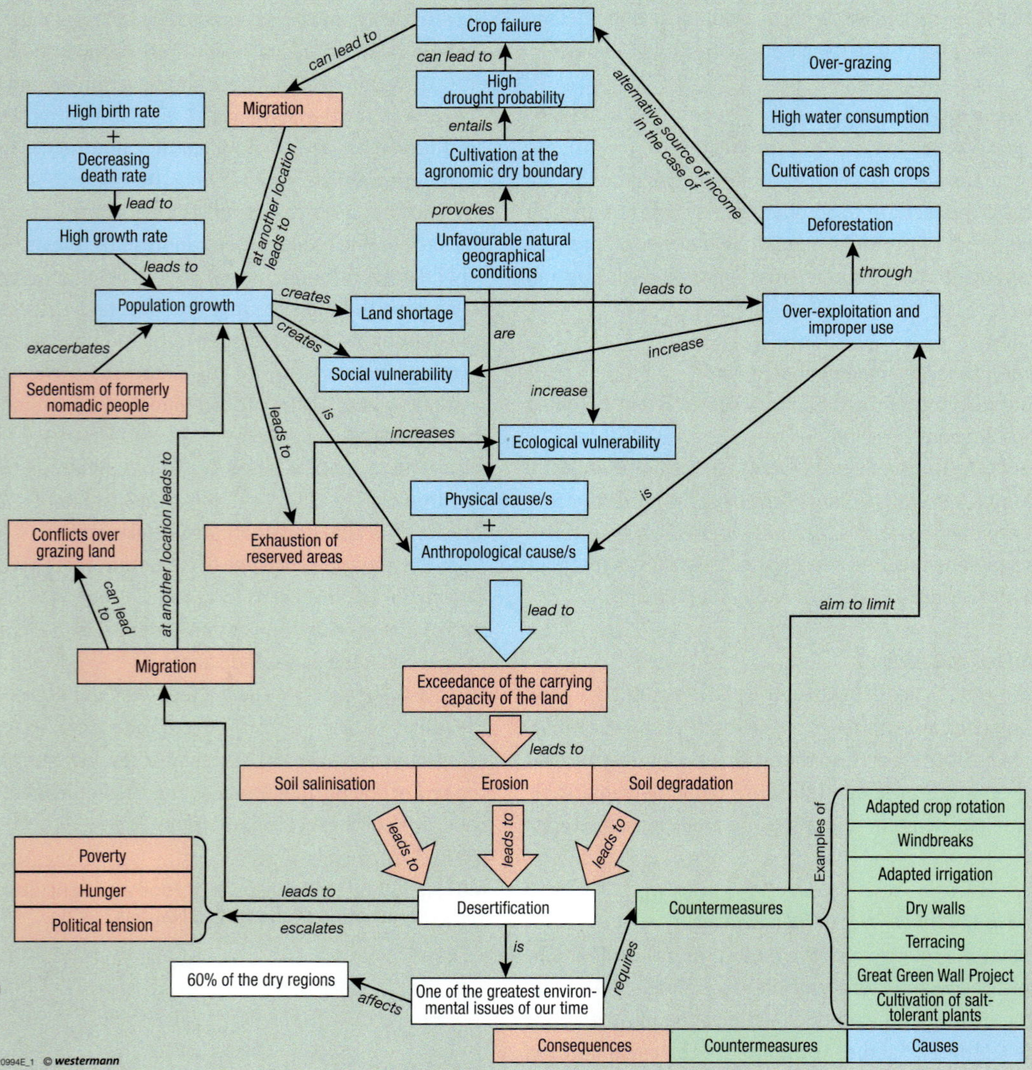

M1 Concept map on the issues of desertification

1. Turn back to p. 55 and compare: Check whether you have included all aspects of desertification (M1) in your mind map. Add missing aspects if necessary and note any questions if something is not clear. Give reasons why this set of issues has been presented in the form of an hourglass.
2. Convert your mind map into a concept map by adding arrows and explanations.

PROJECT Internet project desertification

The topic of desertification is very present in the media. This is even more so in Germany because the Secretariat of the United Nations Convention to Combat Desertification is based in Bonn. This internet project allows for a more detailed insight into the situation in regions at risk of desertification. The project consists of a film analysis and project work, in which you can set your own key topics.

Work assignment for the film analysis

Go to the video "Losing ground ..."
at ***www.desertifikation.de/video-en.html***

During the film

1. Write a protocol (M5).

After watching the film:

2. Discuss the content of the film with your fellow students and complete your table.
3. Check whether your mind map is complete and expand it using examples from the film.

Work assignment for the internet project

Research

Look at the websites listed in M2 to get an overview of topics and regional case studies.

Selecting a key topic/project phase

1. Choose a key topic that you would like to research. This can either be a geographic aspect (e.g. A country under scrutiny: Desertification in Mali) or a thematic aspect (e.g. Combatting desertification – Case studies).
2. Decide on a suitable form for presenting your topic (e.g. poster, digital presentation, chart etc.).

Presentation

Present your project to the group.

GIZ stands for Deutsche Gesellschaft für Internationale Zusammenarbeit and is a provider of international cooperation services for sustainable development. It acts on behalf of the German Federal Ministry for Economic Cooperation and Development (BMZ).

www.giz.de/fachexpertise → Environment and climate → Managing natural resources → Combatting desertification

- GIZ microsite for desertification

www.desertifikation.de

- Video: Watch the film "Losing ground ..." on this website
- Downloads & links: Useful facts

www.bmz.de → Environment → Soil

M2 Collection of links for the project phase

M3 Logo of the United Nations Decade for Deserts and the Fight Against Desertification

M4 Top: "The desert. Coming soon?" Poster for the International Year of Deserts and Desertification. Bottom: Screenshot from the film "An Boden verlieren" ("Losing ground ...").

"Losing ground ..." – a film protocol			
Causes	Impact	Countermeasures	Region

M5 Making a film protocol

The following topics are suggestions for formulating your own project topic:

- "What role does the BMZ play in combatting desertification?"
- "Combatting desertification in Central Asia – Regional case studies"
- "Combatting desertification at all levels – The example of Brazil"
- "Committed to the fight against desertification – The German CCD project"
- "Organisations fighting against desertification – What organisations are there, what do they do?"
- "How do we become aware of desertification? An overview of German publications and reports"

M6 Suggestions for key topics

Floods – natural phenomenon or natural disaster?

Floods in Passau 2013

"Not again!" sighed Ludwig Lober. He lives in the old town of Passau and in June 2013 his house was inundated by a "100-year flood" for the second time. Only ten years had passed since the last "100-year flood". This time the water level reached a record high of 12.89 m, surpassing Passau's previously highest recorded flood of 1501. Damages are estimated at several hundreds of millions of euros.

1. Note questions that come to mind when reading the text about the "100-year flood" in Passau. Work through this chapter to find answers to your questions.
2. Explain why floods are not uncommon in Passau (M3).
3. ⓞ In Passau a natural phenomenon turned into a natural disaster.
   - **A** Explain the relationship between the two (M2/ M7/ M9).
   - **B** Produce a causal diagram to illustrate these relationships (M2/ M7/ M9).
4. ⓞ Analyse the newspaper article in the context of how the media can turn a flood into a flood catastrophe (M7).
   - **A** Comment on the manner of expression in particular.
   - **B** Analyse the article in regard to information content.
5. Ⓢ Using a map, explain which regions of Germany are prone to flooding (atlas).

→ Floods, Natural disaster, Natural event

→ Floods

Floods are natural phenomena. Water runoff in rivers is increased e.g. by heavy rainfall in the source region causing water levels to rise.

→ Natural event, natural disaster

A natural event turns into a natural disaster when it causes vast material and economic damage and social life is disrupted. This can occur during a flood when residential areas close to the banks are flooded.

The German spring of 2013 was particularly wet and grey. An almost continuous succession of low-pressure areas passed over Germany. Their course was extraordinary, moving away in an arc across the Mediterranean. It was unusual weather resulting in particularly high rainfall.

In Passau the water quickly rose to record levels, large parts of the central city were flooded, the water reaching as high as the first floor in some houses. Electricity and water supplies were turned off and the economic life of the city came to a standstill.

Over 800 buildings were affected. The furnishings of restaurants, hotels and shops were destroyed and shop merchandise was rendered unusable. The structure of buildings was damaged.

The infrastructure was also affected; streets and train lines were flooded.

M2 Floods in Passau 2013

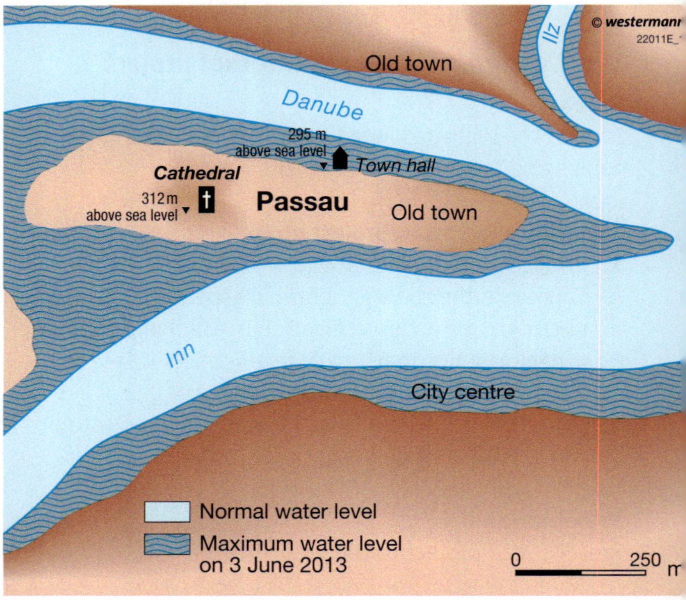

M3 Location of Passau and flooded areas of the town 2013

M1 Passau before the floods

M4 Floods in Passau 2013

M5 Flooding in the old town of Passau

M8 Disaster relief with rubber dinghy in the pedestrian zone

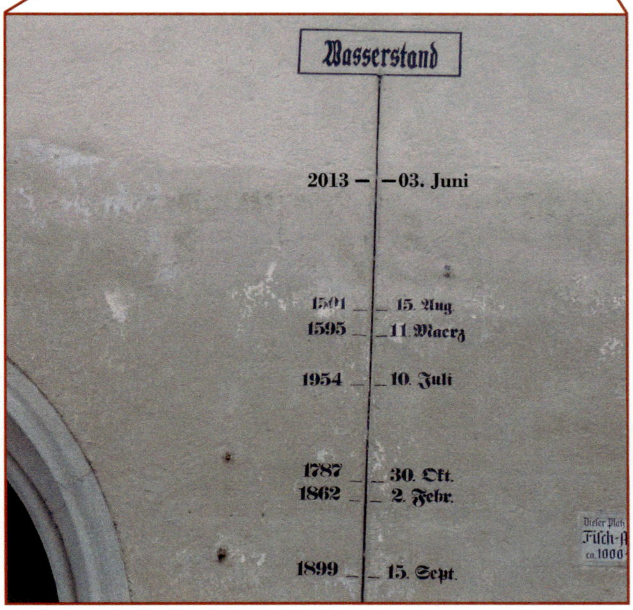

M6 The still damp wall of the town hall and the marker showing the water level in June 2013.

Estimated costs for repairing the infrastructure (Information from the Mayor of Passau)	7.6 mill. euros
Total estimated costs of flood damage in Passau (Estimation from a city planning officer)	over 100 mill. euros
Immediate flood relief for the Passau district	3 mill. euros
Total estimated costs in Bavaria (Estimation from the reinsurer Munich Re)	1.3 bill. euros
Estimated economic losses Germany-wide (Estimation from the reinsurer Munich Re)	12 bill. euros

* Reinsurance companies insure the insurance company. They have to bear part of the compensation costs that insurance companies pay out to flood victims.

(Sources: www.br.de; www.pnp.de; www.freiewelt.net; www.munichre.com)

M9 Estimated cost of damage caused by the 2013 floods

Floods in Passau: *"Oh God, I hope it holds!"*

Sheer horror in Passau.

Regardless of how versed the residents of Passau were previously in dealing with floods – this disaster has surpassed everything that went before it.

The 100-year flood of 2002? The floods of 1954? Forgotten. In the three-river town the water has reached levels not seen since 1501. And it is rising.

"My house was built in 1870 and it was never flooded", says journalist Hubert Denk. "But now the cellar is underwater and the river is outside my window. If the levels continue to rise, the water will push up through my floorboards."

He raises his voice as he speaks. He had a catastrophic night. "I only had three hours sleep." There is no drinking water. Stadtwerke Passau has turned off the water supply: floodwater could pollute the wells. A crisis team is working flat out to find a solution.

Günther Loibl is leaning wearily and exhausted on the wall of his house. For hours now, the 45 year-old has been at risk of losing the battle for his restaurant, his life's work. "The toilets and cold store are flooded", says the proprietor. "Only a few more centimetres until it reaches the taproom."

The fire brigade is pumping and pumping – in total there are 600 emergency personnel and 120 soldiers from the armed forces in action in Passau. "What they are doing here is brutal", Loibl explains. "They are trying to save every centimetre."

In many cases though, the helpers are powerless. The fire brigade was forced to give up on several shops in the central city when their shop windows shattered under the water pressure.

Source: A. Zoch, abendzeitung-muenchen.de, 03.06.2013

M7 Newspaper article

The drainage system of a river in the water cycle

Natural and disrupted runoff

Floods are natural events that can quickly develop into natural disasters. Regions are affected worldwide where there is an over-abundance of water as opposed to a shortage. Although extreme floods are uncommon, they are becoming more and more severe.
So-called "100-year floods" are occurring more frequently. The risk is increased by settlement in regions close to riverbanks and ever-growing populations. The question here is: To what extent are humans responsible for the increased risk of flooding and what measures can be implemented to reduce this risk? To get a better understanding, it is necessary to examine how humans intervene in the drainage system of a river.
What human interventions alter natural water runoff?

1. Formulate a hypothesis for these questions (M1/ M2).
2. Describe the natural runoff of water (M1).
3. ⓞ The flood risk of a river is subject to human intervention.
   A Give reasons for this statement (M4/ M6).
   B Present the risk of flooding for each of the human interventions described in M4 and M6 in a table.
4. Analyse the cartoon and give your opinion (M3).
5. ⓢ Investigate why several valleys in the Bavarian Alps are particularly prone to flooding (atlas).

Accumulation, Erosion, Evaporation, Lower course, Meander, Middle course, Sedimentation, Surface sealing, Upper course

The drainage system of a river is divided into the upper course, middle course and lower course. The gradient of a river decreases from the source region to the mouth and the volume of water increases as tributaries flow into it. If the source region is forested, about half of the precipitation that falls remains in the upper course of the river. It is stored in the soil and is returned to the atmosphere in the form of evaporation.
In the middle and lower courses of the river the gradient decreases. Consequently, the flow velocity also decreases. This results in the formation of river loops, so-called meanders. The riverbed often shifts due to the accumulation of sand and debris. Large volumes of water in a river, e.g. following torrential rain or melting snow in the upper course, can result in flooding.

M2 The natural drainage system of a river

M3 Cartoon

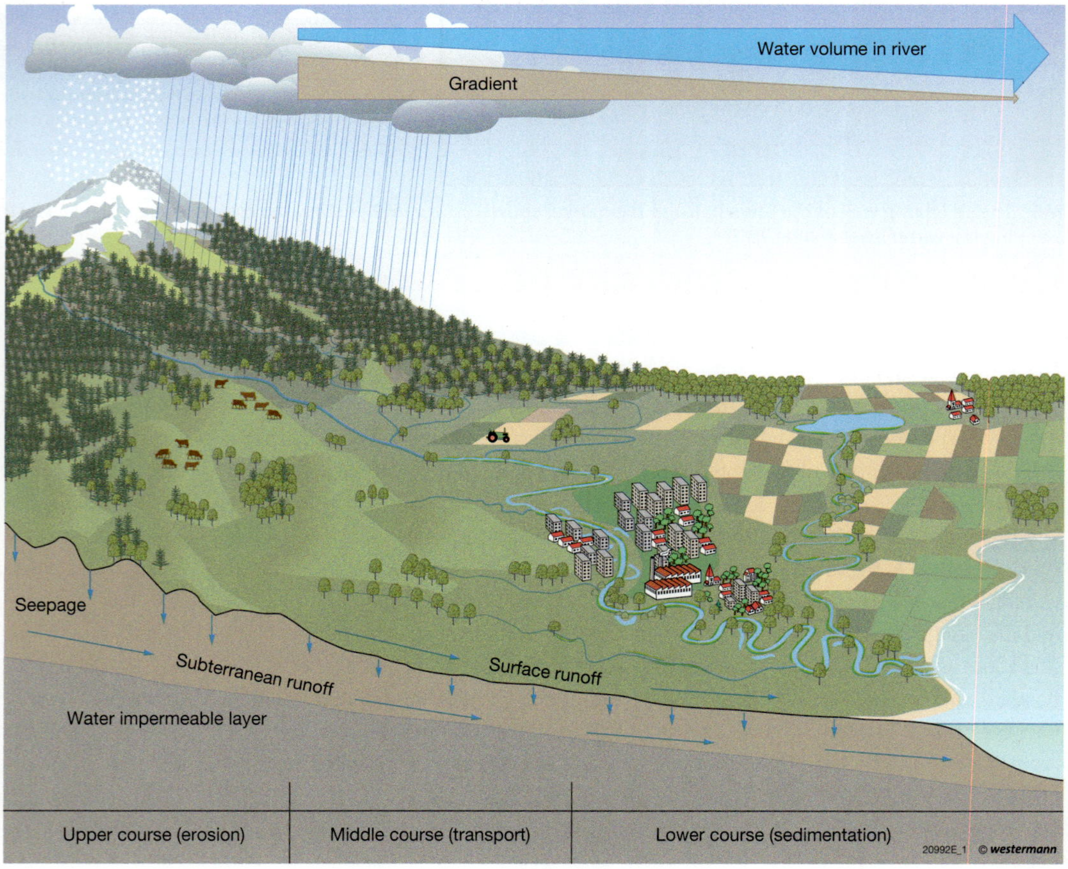

M1 Drainage sytem of a river

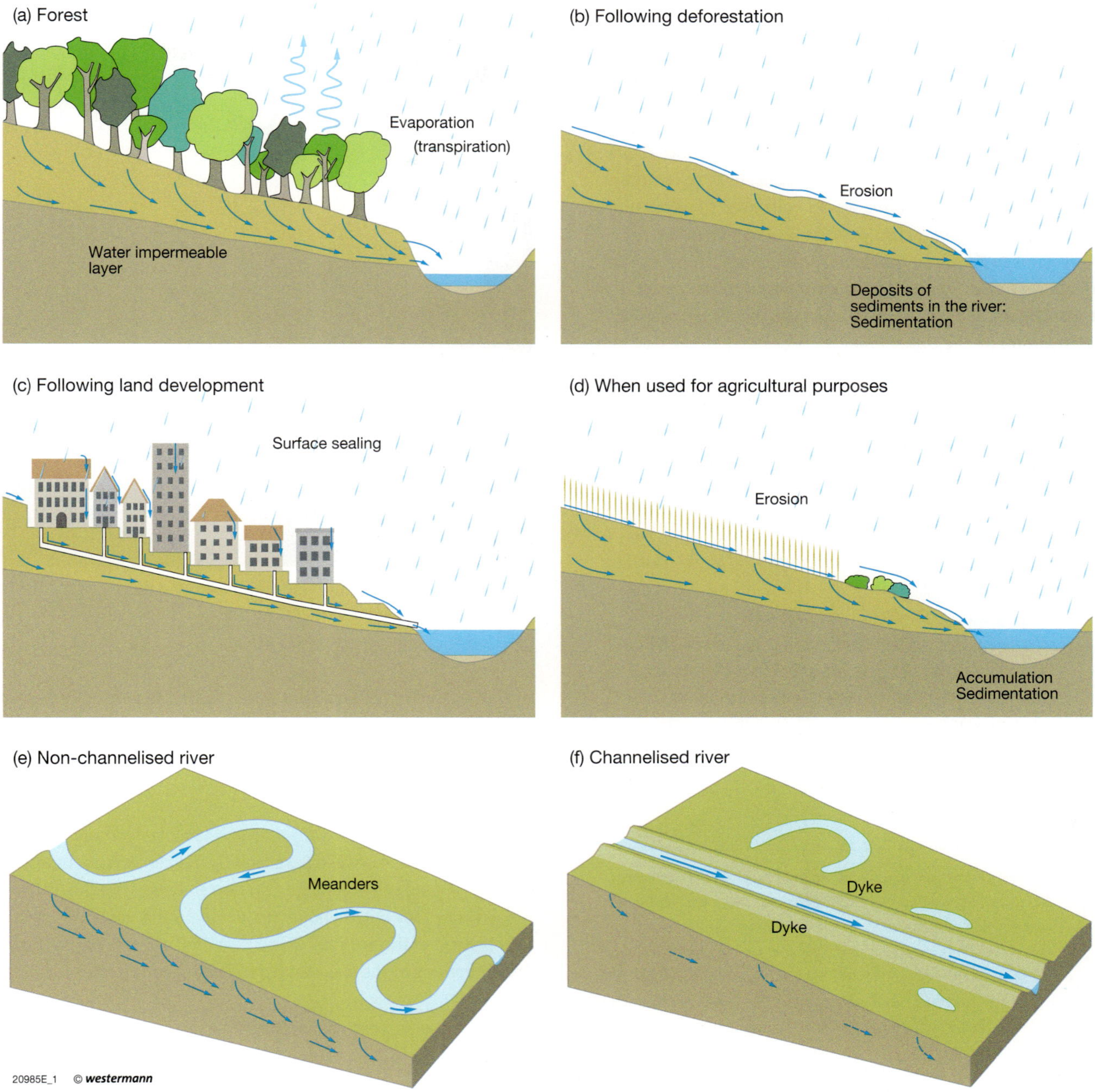

M4 Human intervention in the drainage system of a river

1991	Bangladesh (Ganges)	138,000 fatalities 10 million homeless
1998	China (Yangtze)	3,000 fatalities, 14 million homeless
2005	India (Mandakini)	1,000 fatalities 280,000 homeless
2010	Pakistan (Indus)	2,000 fatalities 20 million homeless

M5 Extremely severe flood disasters

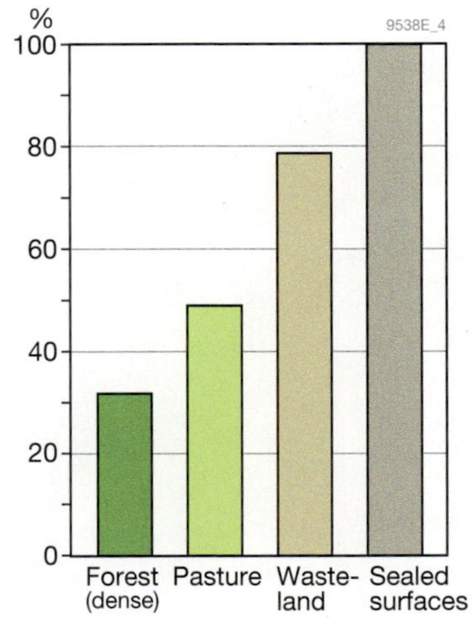

M6 Rainwater runoff in rivers, streams and sewage systems with different surfaces

The drainage system of a river in the water cycle

River realignment on the Upper Rhine

During the "100-year floods" on the Rhine in 1926, 1993 and 1995 water levels rose by more than ten metres. The city of Cologne has since spent over 260 million euros for flood control barriers alone. In 2013, another centimetre and the old town of Cologne would have been flooded again. Melting snow in the source regions of the Rhine and its tributaries in spring regularly leads to floods on the Rhine. Additionally, the torrential rain that has occurred more frequently in recent decades has caused river levels to rise. But the risk of flooding in Cologne has also been increased by human intervention on the Upper Rhine.

1. Describe and give reasons for changes in the drainage system of the Upper Rhine (M1 – M4/ M7).
2. Explain why these interventions have increased the risk of flooding in Cologne (M5/ M6/ M8).
3. Present the causes and effects of human intervention on the Upper Rhine in a causal diagram.
4. ⓞ "The 1995 floods in Cologne were caused by river realignment on the Upper Rhine."
   A Give your opinion on this statement.
   B Write a letter to the editor about this statement from the perspective of someone affected by the floods in Cologne or someone from the Upper Rhine Valley.
5. ⓢ Produce a mind map on the topic of surface sealing.

River realignment, Surface sealing

M3 Alluvial forests on the Upper Rhine around 1800 (Painting by Peter Birmann)

"Au and Daxland are in constant danger of flooding from the nearby Rhine, and Daxland most of all. In 1651 and 1652, the flooding was so severe that in the space of one and a half years over 20 houses and their foundations were undermined and destroyed by water. More than 100 fields were rendered unusable. Even the church in the middle of the village was completely surrounded by water. The coffins and their dead were washed out of the graves and floated down the Rhine."

M1 Document from the year 1747 about two townships near Karlsruhe

In 1812, the hydraulic engineer Johann Gottfreid Tulla presented Charles, Grand Duke of Baden, with an elaborate plan to eliminate flooding on the Rhine. He suggested cutting through over 30 meanders and forcing the river through a channel of reinforced riverbanks with a maximum width of 250 m.
Construction work lasted from 1817 to 1874. On completion, the course of the river between Basel and Mannheim was about 82 km shorter. Of course, the gradient was also steeper. After the realignment, water took only 24 hours to flow through this section of the river instead of the previous 64.
The shortening of the river was also advantageous for the shipping industry.

M4 River realignment on the Upper Rhine

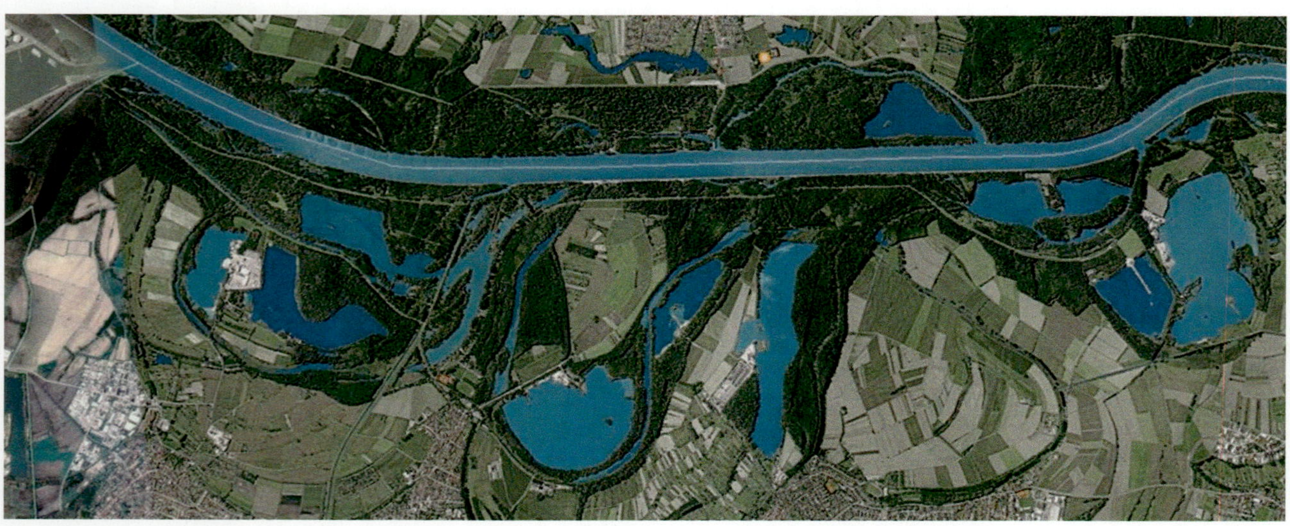

M2 Present-day situation at the Rhine near Karlsruhe

M5 Flooding in Cologne

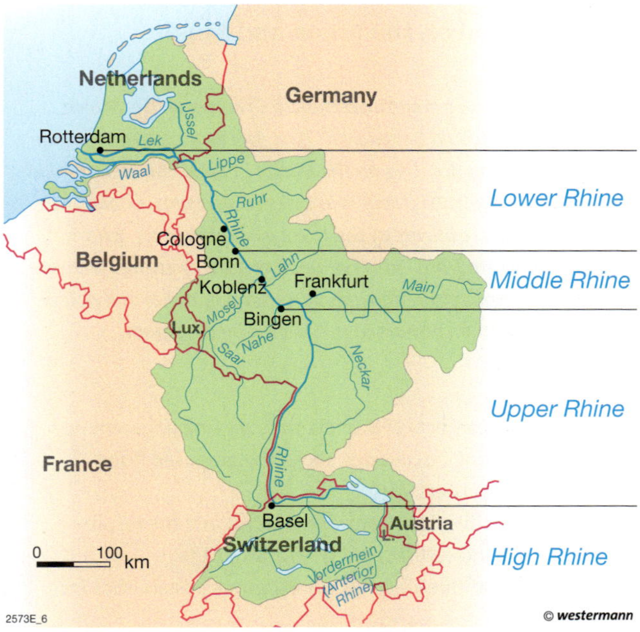

M8 Catchment area of the Rhine and its tributaries

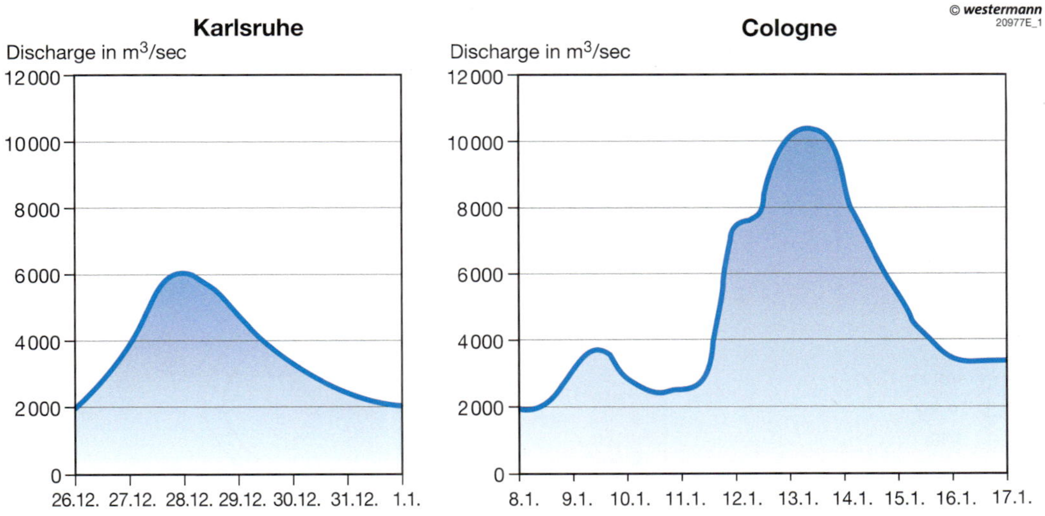

M6 Flooding of the Rhine in Karlsruhe and Cologne in 1994/1995

Flood warnings for the Rhine and its tributaries can be found *on the internet:*
→ Floods, Rhine

Information about the water levels of rivers and lakes in North Rhine-Westphalia are provided by the Landesumweltamt (federal state environmental agency):
→ Landesumweltamt NRW, water levels

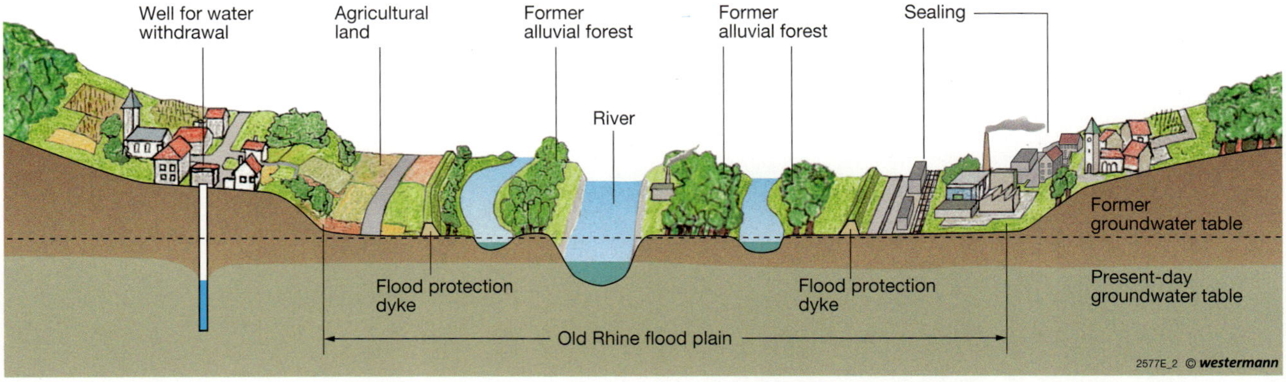

M7 Changes caused by river realignment

Flood prevention measures

Flood control on the Rhine and Mulde

Measures for flood prevention are controversial. In Grimma on the Mulde, for example, no agreement could be reached. Thus, the town was once again inundated with water during flooding on the Mulde in 2013. The cost of flood damage in Grimma after the 2002 floods was 250 million euros. Could Grimma have been saved? Many residents of the town asked the same question, for in Eilenburg (about 30 km from Grimma) flood control measures were effective (construction of flood retaining walls, relocation of dykes and polder construction). In 2013, the town remained untouched by the floods.

1. Explain the conflicts of interest that became apparent in Grimma during discussions about the proposed flood prevention barriers (M6).
2. Discuss possible flood control measures and the resulting conflicts of interest (M1/ M3/ M4/ M6/ M7).
3. (O) In the case of flooding there are various possibilities for protecting houses and residents.
   A Explain this in text form (M7/ M8).
   B Evaluate the different possibilities (M7/ M8).
4. Consider why some residents deliberately flood their cellars when the water table rises. Draft a pamphlet for residents explaining why they should flood their cellars (M7 A/ M8).
5. (S) Design a poster to present your opinion as either an opponent or supporter of polders.

Dyke, Polder, Retention area
→ Flood plain

Interstate flood prevention measures on the upper course of the Rhine are coordinated between the federal states. The most important measures are:

- Maintenance and strengthening of the main Rhine dykes.
- Construction of flood retention areas (flood retention areas are areas used for retaining floodwater, thus lowering the crests of a flood wave. Retention areas are created either through the relocation of existing dykes and renaturation or through the construction of polders).

M3 Flood control measures

Citizen's action groups are formed everywhere that polder construction is planned on the Rhine. Over the last few decades, residential areas have frequently been built too close to the Rhine and in low-lying areas in the Upper Rhine region. Planning errors of the past are catching up with us today.
Then, of course, there are still several sports grounds in potential flood areas. In light of the damage caused by floods it is incomprehensible that clubs are not more generously compensated. Forest owners and farmers should also be compensated.
Source: vorort.bund.net, 27.07.2013

M4 Conflicts of interest on the Rhine

→ Flood plain
A flood plain is a low-lying area near the banks of a river. Flood plains are submerged during floods.

Five million cubic metres of water can be stored in the Flotzgrün polder. When the high-water mark is reached the weirs are opened and the polder is flooded. The weirs remain open allowing the water to flow out of the polder as the flood wave recedes.

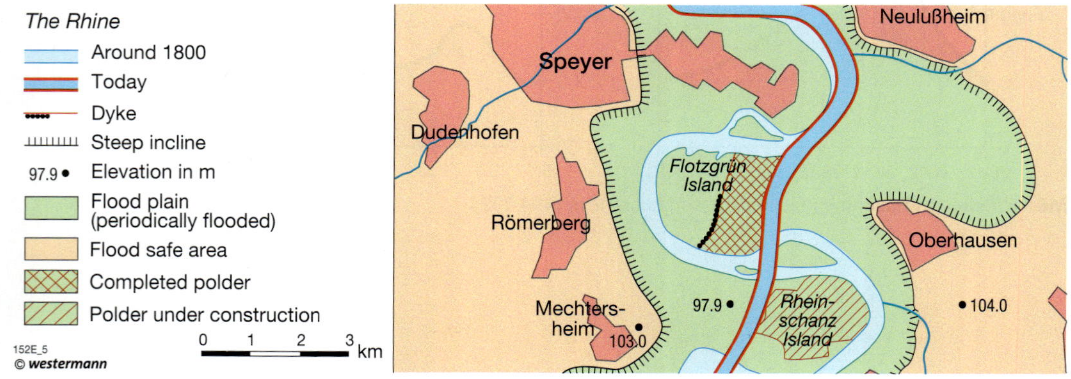

M2 The Flotzgrün polder near Speyer

M1 The Flotzgrün polder (the dyke can be seen on the left-hand side of the photograph)

M5 Construction of the flood retaining wall in Eilenburg (Mulde)

After the flood ...

... is before the flood – prevention against flooding is still insufficient

Grimma's beautiful old town is completely underwater – and it could have been prevented. The people of Eilenburg built a protective barrier; in Grimma a citizen's action group was established in opposition to the "containment" of the old town.
The river has rights too, was the opponents' argument against the planned protective barrier.
Of course, the state can't deprive its people of power in regard to important building projects, as Saxony's State Premier would like to. [...] Democracy and the constitutional state don't work without opposition and complaint. Those who want to, can even oppose a flood protection barrier. But is it desirable and right when the majority have to suffer the consequences because a defiant minority think the unobstructed view of the river is more important than the safety and well-being of many? It is high time that this too is seriously debated.
Of course, barriers are not the solution against floods, they are only damage control, an aid and at most a small part of the solution. Even if disasters cannot be prevented, it is still necessary and wise to limit the risks [...]. For this reason, environmentalists have been reminding us for decades that rivers need more space, alluvial forests shouldn't be cut down and we should not build excessively on riverbanks.
Source: B. Honnigfort, Kölner Stadt-Anzeiger, 06.06.2013

M6 Commentary on flood control measures on the Mulde 2013 (excerpt)

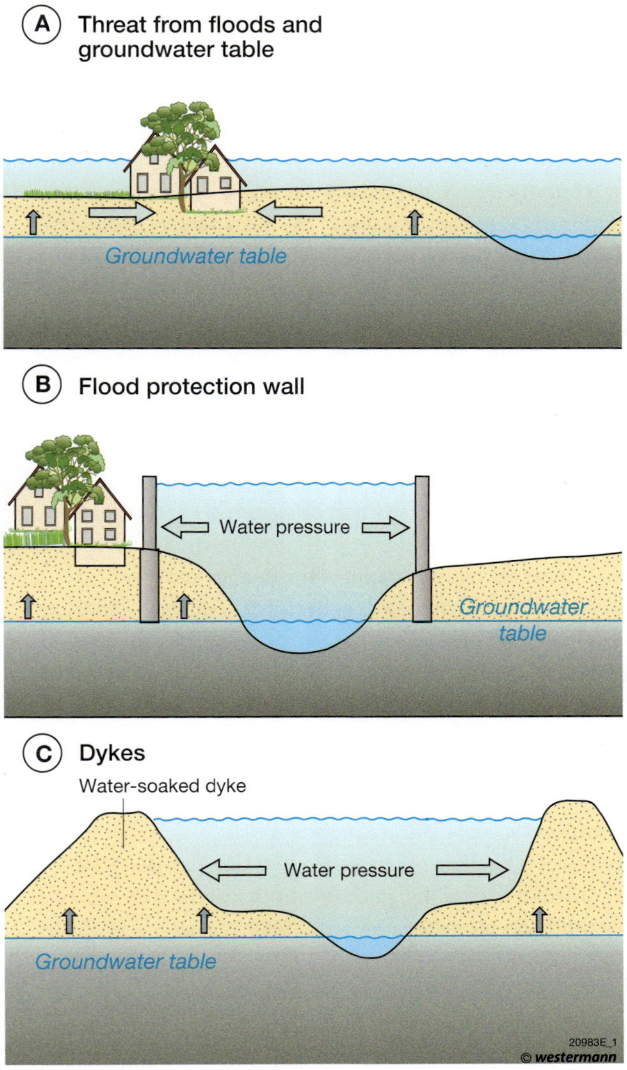

M7 Flood control measures

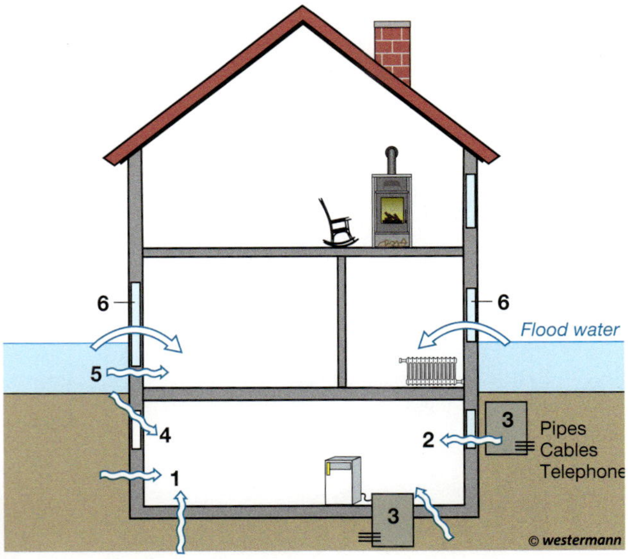

Penetration of
1 groundwater through cellar walls and floor
2 backflow water from sewerage systems
3 groundwater through house connections (pipes, cables)
4 surface water through light shafts and cellar windows
5 surface water as a result of seepage through the outside walls
6 surface water through doors and windows

20986E_1

M8 Possible points of entry for water in buildings during floods

Flood prevention measures

Bangladesh – risk of flooding and flood control

Bangladesh is a country that is particularly frequently affected by floods. With a population of 155 million, it is one of the most densely populated countries on Earth. Bangladesh is not the only country where flooding occurs frequently, but here this natural event often develops into a natural disaster.

1. Sketch a map showing the course of the Ganges and several of its tributaries as well as the countries it flows through. Enter the height above sea level at various points along the river (M1, atlas).

2. a) Analyse the river discharge diagram of the Ganges (M4).

b) Compare precipitation levels in Kathmandu with the river discharge diagram of the Ganges and draw conclusions in regard to the risk of flooding in Bangladesh (M4/ M10).

Ⓞ **3.** Explain how deforestation in Nepal affects Bangladesh (M2).

A Draft an explanatory text.

B Present the issues from two points of view: The people of Nepal, the people of Bangladesh.

4. Produce a causal diagram to present natural and human causes of flood vulnerability in Bangladesh.

5. Discuss the measures in the Flood Action Plan (M6 – M8).

Ⓢ **6.** Sketch a map of Bangladesh that shows the river systems and flood-prone regions (atlas, M9).

M3 Erosion damage in the mountains

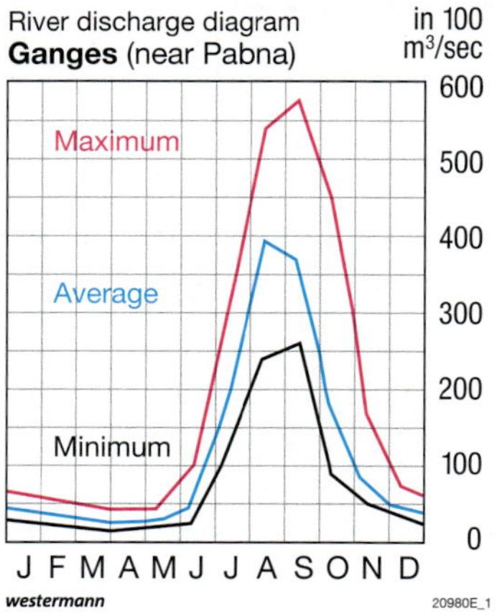

M4 Discharge diagram of the Ganges

Accumulation, Deforestation, Dyke, Erosion, Monsoon, Natural disaster, Natural event, Polder

Most inhabitants of the Himalayas are farmers. They make their living from agriculture and over 70 percent of the population heat with wood. There is no general electricity supply outside the larger towns. This was not a problem up until a few decades ago, however a sharp rise in population and mass tourism have meanwhile led to the clear felling of forests. In order to comfortably accommodate the numerous visitors, accommodation is usually built from and heated with wood. According to Greenpeace, half of the forest areas in Nepal and Tibet have already been destroyed. Deforestation results in the natural protective layer of the soil being lost. Rainwater can no longer be absorbed resulting in landslides and floods. Soil erosion occurs and desert-like tracts of land are created.

Based on: E. Mommsen, Planet Wissen, 10.06.2013.

M2 Deforestation of mountain forests in the Himalayas

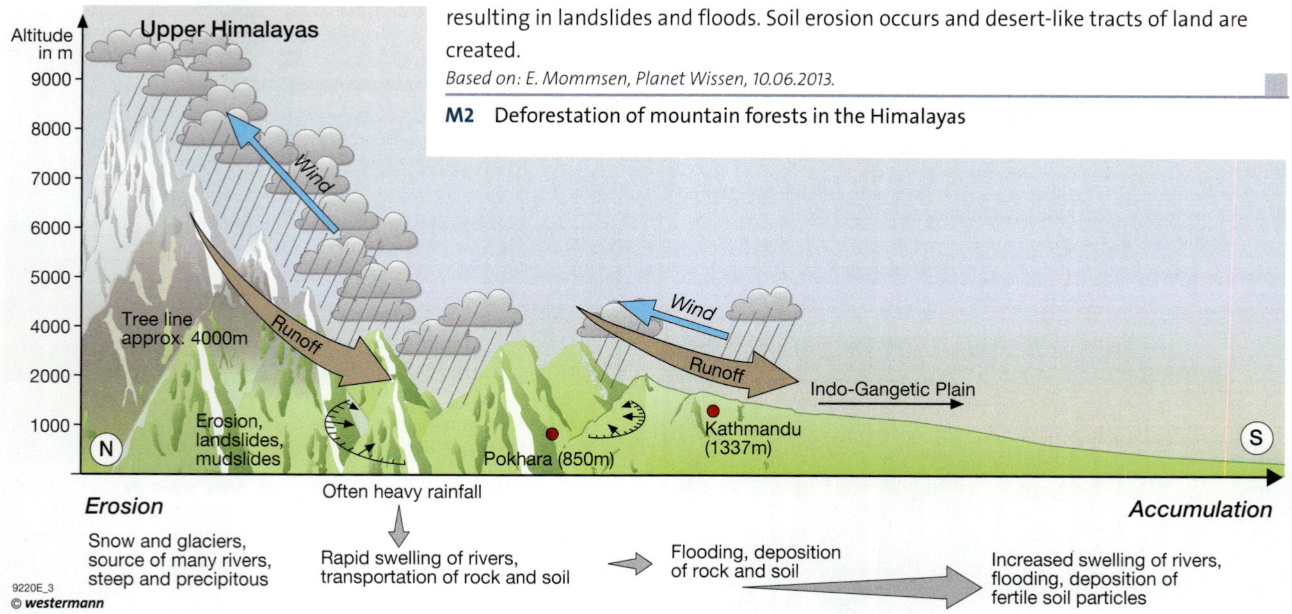

M1 Schematic north-to-south cross section of the Ganges Valley

M5 Flooding in Bangladesh

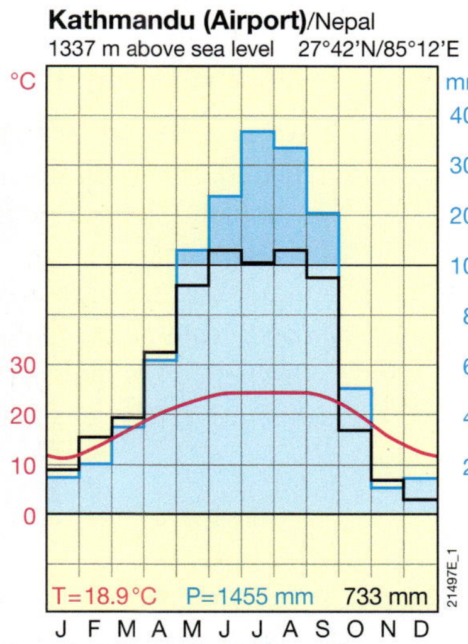

M10 The climate in Kathmandu

The FAP is fiercely debated. Dykes would disrupt natural drainage patterns, which would in turn result in increased river-bank erosion and higher water levels.
There would be an impact on both the wetlands and fishing industry in rivers and canals. Ultimately, all rivers would have to be dyked so that they can safely flow into the ocean. Supporters of the FAP point to the success in the Netherlands.
Other studies recognise the necessity to combine dyke-building with other measures. But the people of the region have developed their own adaptation methods for cultivation and lifestyle to deal with this natural threat. River dyking means loss of land. Nobody has asked them whether they really need the dykes.
Based on: I. Islam, Journal of Bangladesh Institute of Planners, 2010.

M6 Commentary on the Flood Action Plan (FAP) by Prof. Islam in a scientific journal

Flood Action Plan (FAP) for Bangladesh
(coordinated by the World Bank)

- Strengthening of coastal dykes
- Reinforcement of the Brahmaputra Dyke
- Containment of the mud wave from the Himalayas
- Construction of polders
- Flood prediction and early warning project

M8 Excerpt from the Flood Action Plan

Annual monsoon rain in summer increases the risk of floods in Bangladesh. Due to climate change this rainfall will increase even more. Extreme torrential rain will become more frequent. Additionally, the country is repeatedly hit by tropical cyclones. These too will presumably become more severe due to climate change.

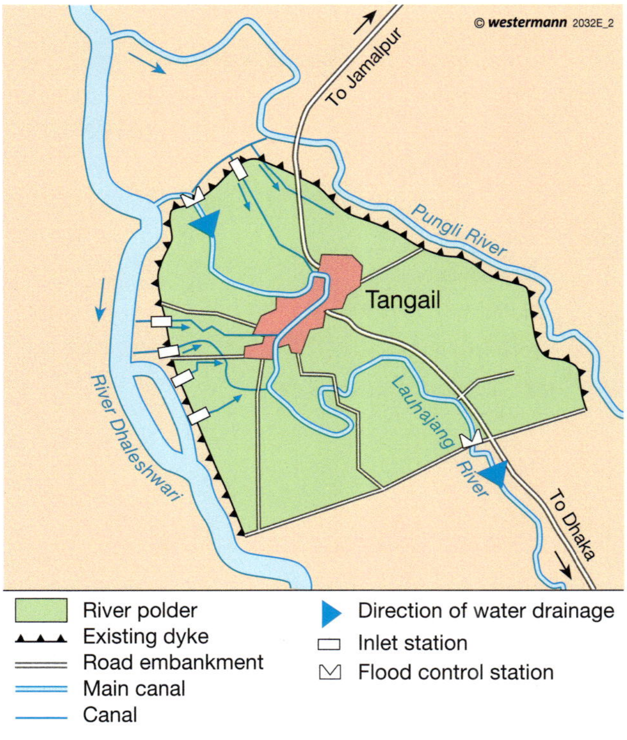

M7 River polder in the FAP project Tangail

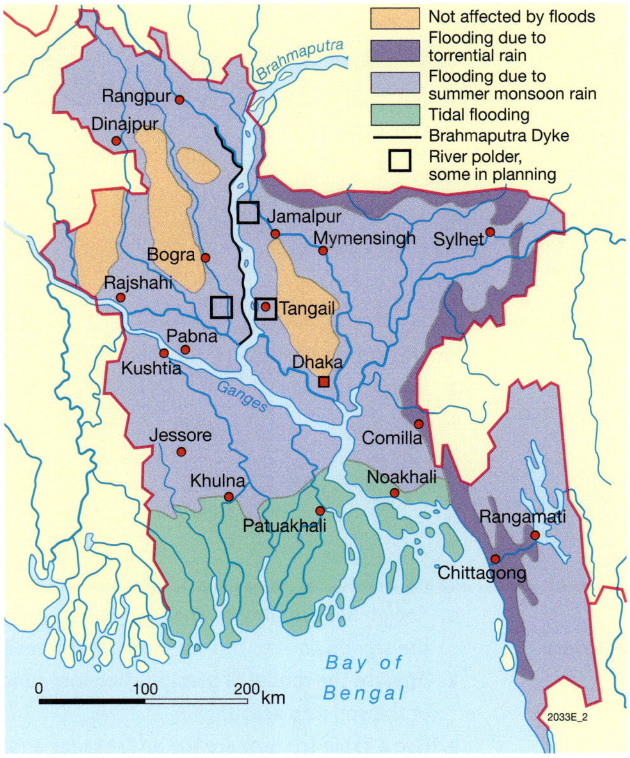

M9 Regions affected by floods

METHOD How do I interpret a model?

What is a model?

Evaporation, precipitation and runoff are all interrelated. These relationships can be illustrated schematically. The schematic water cycle in M1 is one such model.
In general, a model makes it easier to understand complicated interrelationships. It presents a simplified version of reality. Through simplification, regularities become easier to recognise.

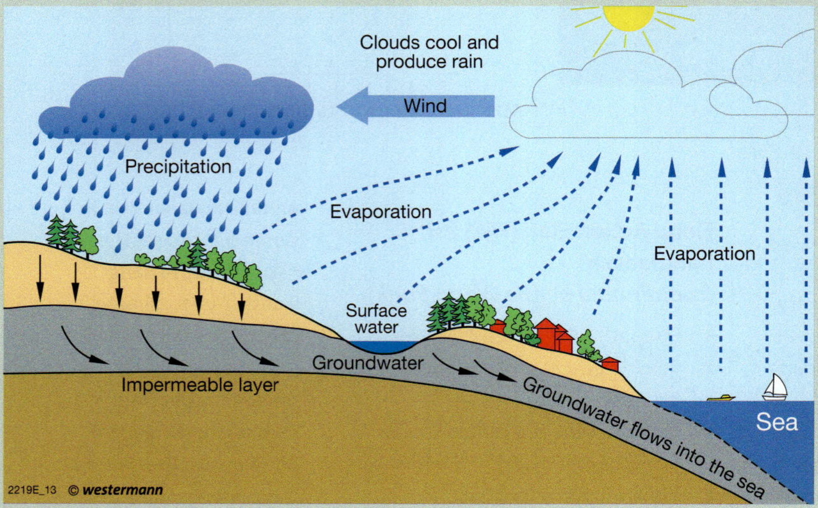

M1 Model of the water cycle

A model presents regularities that have been determined through observation and scientific research. The simpler a model is, the easier it is to understand and remember the regularities presented in it. Representations that are too general will have very little meaning.
But it can also be problematic if reality is simplified too much. If a model is simplified too much it will no longer accurately reflect the actual observations. It will also be difficult to transfer it onto similar situations. However, it should be possible to transfer a model onto similar conditions.
A further problem is that if a model is reduced too much, the thought process it evokes will be too simple, e.g. if the facts are only limited to two contrasting poles such as poor and rich. It is therefore important to analyse a model carefully. When working with models it is always crucial to establish a connection to reality and examine the transferability of the model onto specific case studies. When the findings of a model are transferred onto an individual case study, variations from the "schematic character" will become evident.
Therefore, when working with models it is essential to critically scrutinise whether the model can be transferred onto an individual case study and whether the model is even applicable.
It is always possible that a model is incorrect and that a correction or addition is necessary.

Annual evaporation of which:	**505,000 km³**
over the oceans	434,000 km³
over land	71,000 km³
Total precipitation of which:	505,000 km³
over the oceans	398,000 km³
over land	107,000 km³

www.oekosystem-erde.de

M2 Water cycle

Water reservoirs (in km³)	
Atmosphere	12,900
Oceans	1,338,000,000
Flowing water	2,100
Lakes and marshland	102,000
Soilwater	16,500
Groundwater	10,800,000
Biomass	1,100
Polar ice and Glaciers	24,000,000

M3 Global reservoirs of water (1961–1990)

Model, Water cycle

weblinks
- webgeo Water cycle
- Eduserver Water cycle

1. The global climate has changed increasingly since 1990 and global warming has been ascertained. Explain the significance of this for global water reservoirs and the water cycle (M2/ M3).
2. Analyse the model of precipitation and runoff during a tropical cyclone (M4), taking the absorption capacity of the soil into account.
3. Use a table to compare the advantages and disadvantages of models.

Six steps for analysing a model

STEP 1
The subject

→ What is the subject of the model? (What? When? Which region?)

STEP 2
The core statements

→ What relationships are presented and how?
→ What are the most important statements?

STEP 3
The explanation

→ How can the relationships and core statements illustrated in the model be explained?

STEP 4
The applicability

→ What conclusions can be drawn?
→ What issues become evident?
→ Who could the model be useful for?

STEP 5
The limitations of the model

→ What aspects of the model are too generalised so that further information is necessary when using it to examine an individual region? (Degree of generalisation, transferability)

STEP 6
The criticism

→ When is a model incorrect, not up to date, insufficient, in need of improvement?
→ Does the model fulfil its function as a memory aid?

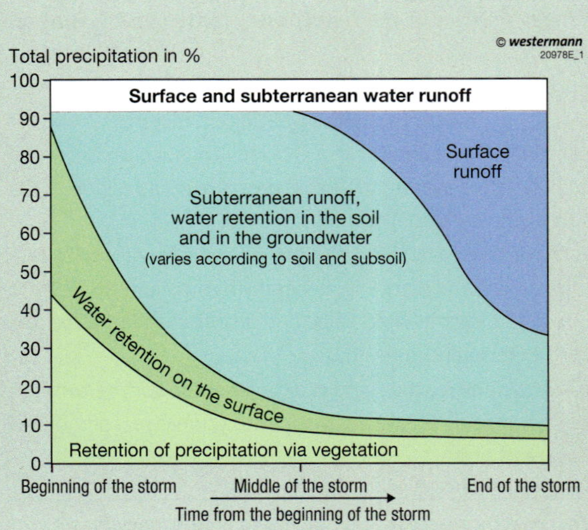

M4 Model illustrating the relationship between precipitation and water runoff during a tropical cyclone

Example analysis

STEP 1

The aim of this model is to illustrate the global water cycle, i.e. to show the distribution of water in the atmosphere and on Earth as well as the exchange of water.

STEP 2

An idealised landscape, arrows and text were chosen to illustrate the subject. The core statements are that the total amount of water on Earth is constant and that it circulates in a water cycle.

STEP 3

Water evaporates over the sea and on land. It changes from a liquid state into a gaseous state: Clouds are formed. The air masses are transported by winds. They cool as they rise resulting in condensation: Precipitation falls. Part of the rainwater seeps and runs off underground as groundwater; the remaining rainwater runs off overground, into streams and rivers to the ocean. The water cycle is a closed cycle.

STEP 4

The global exchange of water occurs in a cycle of evaporation and precipitation. Solar radiation and wind are part of the water cycle and influence it. Climate researchers can use this model e.g. when considering how the quantity of evaporated water will change as the Earth continues to heat up and how changes in the whole water cycle will affect individual areas of the water cycle, e.g. precipitation.

STEP 5

The model uses an idealised landscape to illustrate the global water cycle. Evaporation on land will change depending on the landscape type (desert/forest etc.). Vegetation, bodies of water and sealed surfaces influence the rate of evaporation. In the model however, evaporation is evenly distributed and only illustrated in an exemplary landscape type. Like evaporation, surface runoff can also differ depending on soil conditions and surface sealing. This model does not take this into account. Thus, further information is required for individual case studies.

STEP 6

The model does not show factors responsible for the formation of the water cycle: Solar radiation (evaporation) and gravity (runoff). Water storage e.g. in soil is also not taken into account. The model could also be improved in regard to surface runoff into river systems making it a closed cycle to the sea. Global water reservoirs are not shown. All arrows are identical and no differentiation is made between water quantities. Therefore, changes in evaporation and precipitation rates caused by climate change are not made visible.
However, the model does fulfil its function as a memory aid.

Tropical cyclones

Formation and distribution of tropical cyclones

MANILA, 8 November 2013 – Seven-metre-high waves, torrential rain, landslides and wind speeds of up to 379 km/h: Typhoon Haiyan (a.k.a. Yolanda) is raging on the Philippines. It is one of the strongest tropical cyclones ever recorded at landfall. The catastrophe has claimed many lives. The storm made its first landfall at the island of Samar, about 600 km southeast of Manila, on Friday morning. "Catastrophic" destruction is to be expected in the fishing town of Guiuan, the first in Haiyan's path, according to the US meteorologist Jeff Masters from the weather agency Weather Underground. The American Joint Typhoon Warning Center (JTWC) on Hawaii referred to it as a "super typhoon" because of the wind force.

A report such as this one gives rise to a number of questions:
Where does a tropical cyclone derive its energy?
How do tropical cyclones even come about? Are we at risk in Germany?
When does a natural event become a natural disaster?

ⓞ **1. A** Describe the scale of the tropical cyclone Haiyan (M1, atlas)
B Describe the scale and position of Typhoon Haiyan in the satellite image (M1, atlas).
2. Explain the formation of a tropical cyclone including the term "eye of the hurricane" (M2/ M3).
3. Describe and explain the regional distribution of tropical cyclones (M2 – M7).
4. Explain the difference between a natural event and a natural disaster focusing in particular on tropical cyclones (M5, internet).
Ⓢ **5.** Compare tropical cyclones with low-pressure areas in the temperate latitudes.

Condensation, Convection, Convergence, Coriolis force, Cyclone, Divergence, Tropical cyclone (hurricane, typhoon, willy-willy), Vulnerability

M1 Satellite images of Typhoon Haiyan (2013)

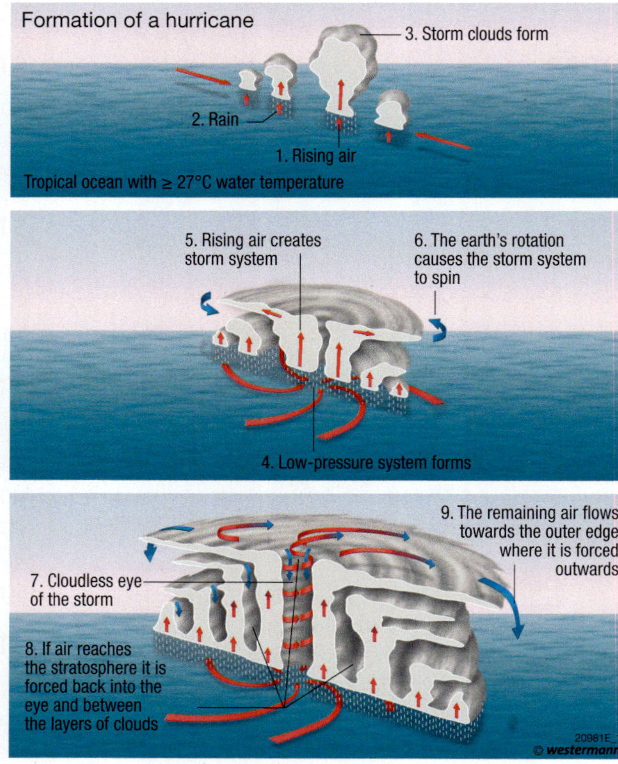

M2 Formation of a hurricane

Tropical cyclones occur in the tropical regions of all three oceans. In the Atlantic they are called hurricanes, in the Pacific typhoons and in the Indian Ocean cyclones. The following conditions are necessary for the formation of a tropical cyclone:

- Ocean surface temperatures over 26.5°C, which regularly occur in the summer months in tropical regions.
- Marked convergence at ground level and rapid outflow (divergence) of rising air at high altitudes, whereby both lifting processes and water vapour supply are controlled.
- A distance of at least 5 degrees of latitude from the Equator, otherwise the coriolis force is not sufficient to give the rising air masses the rotational impetus characteristic of tropical cyclones.

Tropical cyclones derive their vast amounts of energy from the latent energy released in the condensation processes of rising air masses.

As the pressure drops, increasing amounts of air from the surroundings spiral towards the centre creating characteristic bands of electrical storm. The diameter of a cyclone can reach up to several hundred kilometres, whereby the core pressure in the region of the "eye" is partly below 900 hPa.

The eye, characterised by cloud dissipation and calm caused by falling air movement and warming, generally has a diameter of 20 to 70 km. Despite extreme wind velocity the drift speed of a cyclone at 25 km/h is normally quite slow.

As soon as a cyclone reaches land or a cooler ocean surface the supply of warm humid air breaks away and the cyclone weakens within a short space of time.

Source: T. Fickert, F. Grüninger, Geographische Rundschau 10/2011

M3 What exactly are tropical cyclones and how do they form?

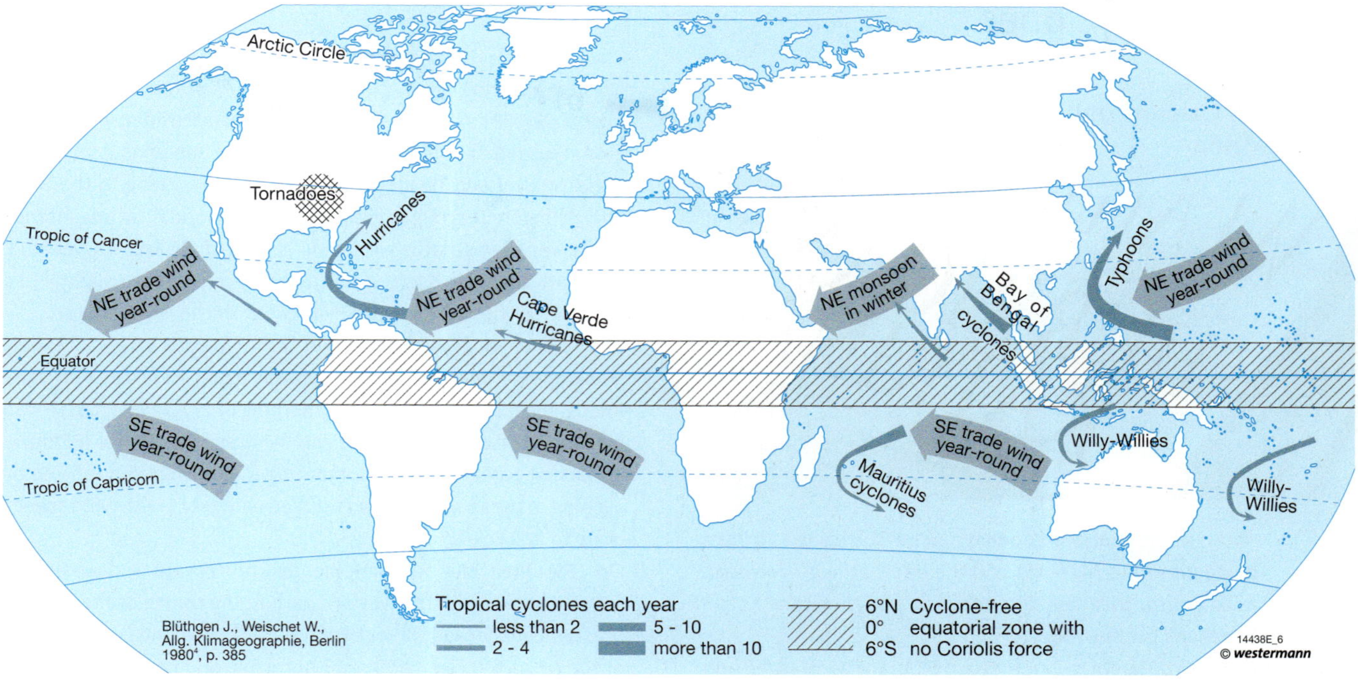

M4 Distribution of tropical cyclones

Earthquakes, volcanic eruptions, cyclones and tsunamis are generally natural phenomena that only become natural threats or catastrophes when human beings are involved. And the use of the word "natural" in conjunction with some disasters should be critically rethought. Frequently, if anything, it is short-sighted human action that is the cause of actual disasters.

For example, in the case of Hurricane Katrina in 2005, the greatest damage was caused by flooding due to errors in urban planning. Poorer countries and sections of society, in particular, are most vulnerable to natural threats.

Rising population pressure and limited financial resources lead to increased settlement in high-risk regions or mean that risk-minimising measures cannot be carried out. But increasing wealth can also contribute to the settlement of high-risk, yet "exclusive" locations, in the USA, for example, high-priced residential areas along the hurricane-prone Florida coast and the luxury villas [...] in the high fire risk area of the Hollywood Hills come to mind.

If natural phenomena occur in such densely populated regions they often result in extreme death tolls and severe material damage – a natural event becomes a natural disaster.

Source: T. Fickert, F. Grüninger, Geographische Rundschau 10/2011

M5 A question of perspective: Natural event or natural disaster

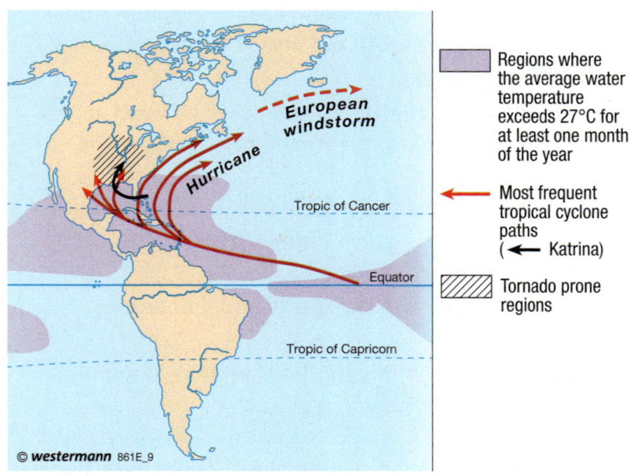

M7 Average paths of Atlantic hurricanes in the month of September

weblinks

Adresses/Organisations

- National Hurricane Center (NHC)
- NOAA Satellite and Information Service
- Eduserver Hamburg
- Webgeo

© westermann 21496E_1

Category	Wind velocity (in km/h)	Tidal wave in m when it meets land	Core pressure (in hPa)	Landfalls* USA	Average damage (in US $)	Damage potential*	Typical damage
Tropical low-pressure system	< 62	0					
Tropical cyclone	63–118	0–0.9			> 1000000	0	Almost no damage
Hurricane 1 (minimal)	119–153	1–1.5	< 960	110	33000000	1	Minimal damage e.g. to trees
Hurricane 2 (moderate)	154–177	1.5–2.5	965–975	73	330000000	10	Uprooted trees, damaged buildings, flooding of coastal roads
Hurricane 3 (extensive)	178–209	2.5–4.0	945–964	75	1412000000	50	Mobile homes destroyed, broken windows, roofing blown off
Hurricane 4 (extreme)	210–249	4.0–5.5	920–944	18	5224000000	250	Mobile homes blown away, flooding of low lying areas
Hurricane 5 (catastrophic)	> 249	> 5.5	< 920	3	15500000000	500	Catastrophic destruction, severe flooding

M6 Hurricane classification (Saffir-Simpson hurricane scale)

*Landfall: Describes when a hurricane strikes land after first moving over the ocean.

*Damage potential: Possible damage, e.g. the damage potential of hurricane 5 is 500 times higher than hurricane 1

Tropical cyclones

New Orleans under the impact of Hurricane Katrina

Ⓞ **1.** Analyse the course and behaviour of Hurricane Katrina using materials M1 – M3 and the internet (e.g. www.csc.noaa.gov/hurricanes). Draw the track of Hurricane Katrina in a blank outline map.
   A Use a different colour for each different wind speed and produce a matching key.
   B Katrina had a diameter of over 300 km in places. Draw in the approximate left and right hand boundaries of the hurricane along its path and shade the affected areas according to your key.

2. Explain the effects Hurricane Katrina had on the economic and settlement areas of the southeast of the USA (M4 – M7).

3. In light of the destruction in New Orleans, discuss possible national safeguards against future hurricanes (M5, internet).

Ⓢ **4.** Write a newspaper article about the tropical cyclone Katrina.

Ⓢ **5.** Research a recent cyclone event and present your results in digital form (internet).

Condensation, Convection, Convergence, Coriolis force, Cyclone, Divergence, Tropical cyclone (hurricane, typhoon, willy-willy), Vulnerability

M1 Damage caused by Hurricane Katrina

Hurricane Katrina began as a tropical low-pressure system on 23 August 2005, 282 km southeast of Nassau, and intensified into Tropical Storm Katrina on the following day over the Bahamas.
The storm initially headed northwest before turning to the west. Shortly before reaching the coast of South Florida, Katrina intensified to hurricane status with wind speeds of 130 km/h. Over land, the hurricane turned to the southwest and weakened slightly to a strong tropical storm during the night of the 26 August.
Katrina then moved into the warm Gulf of Mexico, where after only a few hours it reached hurricane status again. Here, the hurricane intensified rapidly so that on the morning of 27 August (CEST) it had become a Category 3 hurricane with wind speeds of up to 185 km/h. It reached Category 4 on the morning (CEST) of 28 August and by evening of the same day it had reached Category 5 with wind speeds of up to 280 km/h.
On 29 August, the hurricane headed northward and weakened slightly, primarily in the western part. In the course of the day, the hurricane made landfall south of New Orleans along the Gulf Coast of the US as a severe Category 3 hurricane and headed northward where it hit the Mississippi coast only minimally weaker.
The eye passed just to the east of New Orleans so that the "eye wall" (the ring with the greatest wind speeds) battered the city for some time. During the night of 30 August, the hurricane weakened to a tropical storm over the US State of Mississippi.
Based on: ncdc.noaa.gov

M2 The path of Hurricane Katrina

Date	Time	Coordinates Northern latitude	Coordinates West. longitude	Wind speed in km/h	Atmosph. pressure in hPa (Hectopascal)
23.08.2005	18:00	23.1	75.1	55.6	1,008
24.08.2005	06:00	23.8	76.2	55.6	1,007
24.08.2005	18:00	25.4	76.9	74.1	1,003
25.08.2005	06:00	26.1	78.4	92.6	997
25.08.2005	18:00	26.2	79.6	111.1	988
26.08.2005	06:00	25.4	81.3	120.4	987
26.08.2005	18:00	24.9	82.6	157.4	968
27.08.2005	06:00	24.4	84.0	175.9	950
27.08.2005	18:00	24.5	85.3	185.2	948
28.08.2005	06:00	25.2	86.7	231.5	930
28.08.2005	12:00	25.7	87.7	268.5	909
28.08.2005	18:00	26.3	88.6	277.8	902
29.08.2005	06:00	28.2	89.6	231.5	913
29.08.2005	18:00	31.1	89.6	148.2	948
30.08.2005	00:00	32.6	89.1	92.6	961
30.08.2005	06:00	34.1	88.6	74.1	978
30.08.2005	18:00	37.0	87.0	55.6	990
31.08.2005	00:00	38.6	85.3	55.6	994
31.08.2005	06:00	40.1	82.9	46.3	996

M3 Path of Hurricane Katrina as a data table

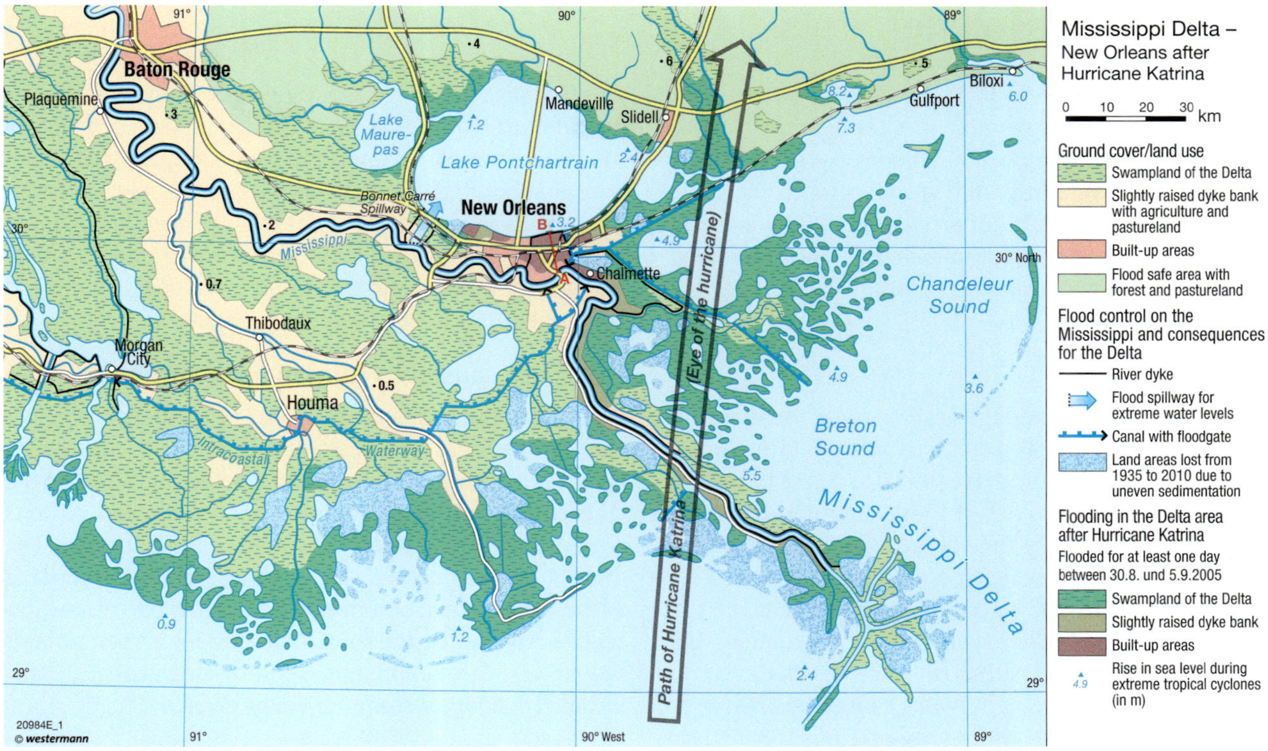

M4 Mississippi Delta – New Orleans after Hurricane Katrina

In the Gulf states (principally Louisiana and Mississippi), the loss of life is unknown but will likely reach well into the hundreds and possibly higher. It is clearly one of the most devastating natural disasters in recent US history. From Katrina's first landfall in Florida, while it was at category one strength, initial estimates suggest 11 deaths resulted.

The loss of life and property damage was worsened by breaks in the levees that separate New Orleans from surrounding lakes. At least 80 % of New Orleans was under flood water on August 31st, largely as a result of levee failures from Lake Pontchartrain. The combination of strong winds, heavy rainfall and storm surge led to breaks in the earthen levee after the storm passed, leaving some parts of New Orleans under 20 feet of water. Storm surge from Mobile Bay led to inundation of Mobile, Alabama causing imposition of a dusk-to-dawn curfew for the City. Large portions of Biloxi and Gulfport, Mississippi were underwater as a result of a 20 to 30+ foot storm surge which flooded the cities.

A major economic impact for the nation was the disruption to the oil industry from Katrina. Preliminary estimates from the Mineral Management Service suggest that oil production in the Gulf of Mexico was reduced by 1.4 million barrels per day (or 95 % of the daily Gulf of Mexico production) as a result of the hurricane. Gasoline had reached a record high price/gallon as of Monday August 30th with concerns over refinery capacity apparently driving the increase. More information is available from a Department of Energy report. Over 1.7 million people lost power as a result of the storm in the Gulf states, with power companies estimating that it will take more than several weeks to restore power to some locations. Drinking water was also unavailable in New Orleans due to a broken water main that serves the city. Power was lost to 1.3 million customers in southeastern Florida from the initial landfall on August 24th.

Estimates of damages for Hurricane Katrina are still extremely preliminary and properly assessing losses will take many months. However, the total losses as a result of Katrina are estimated to exceed $100 billion with over $34 billion in insured losses.

Both of New Orleans' airports were flooded and closed on August 30th and bridges of Interstate 10 leading east out of the city were destroyed. Most of the coastal highways along the Gulf were impassable in places and most minor roads near the shore were still underwater or covered in debris as of August 30th. Katrina also disrupted travel as it headed inland, with more than 2 inches of rain falling across a large area from the coast to parts of Ohio during the 48 hours after Katrina made landfall.

Based on: ncdc.noaa.gov

M5 Impacts of Katrina

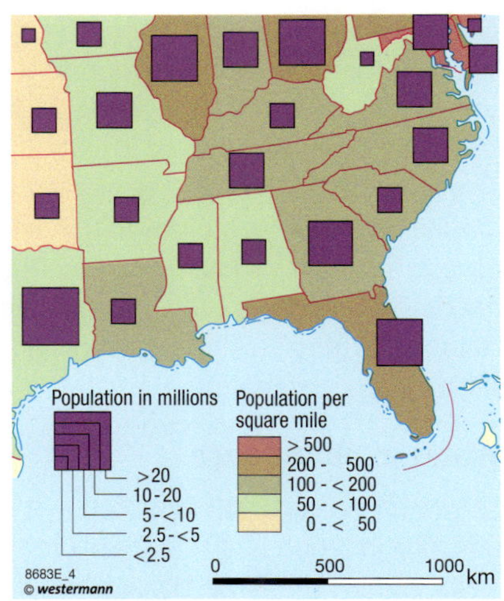

M6 Population density in the southeast of the USA

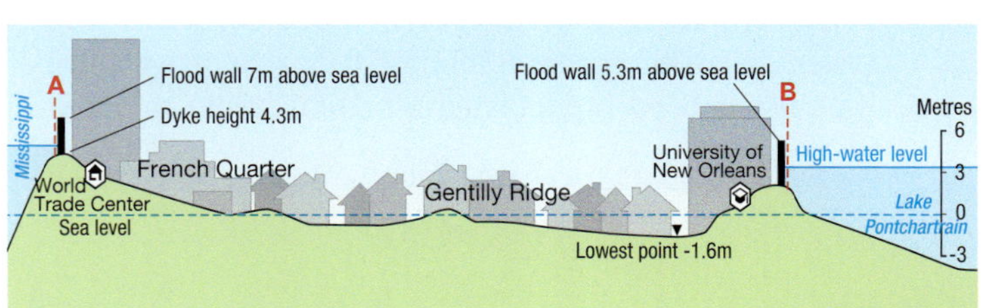

M7 Cross section showing the flood zone of New Orleans

The essentials in brief

Water, our live-giver, is in almost everything that we use in our daily lives. Virtual water describes the quantity of water that is required in the production of goods, e.g. a pair of jeans or a litre of milk. A scarcity of this vital commodity largely determines the vulnerability of a region. Not all regions in the world have a sufficient supply of water. On the other hand, an abundance of water also increases the vulnerability of a region.

Water scarcity

Desertification, i.e. anthropogenically induced desert formation, is one of the greatest environmental issues of our time. It adds to the natural factors that can lead to desert formation. Droughts often occur in desert fringe areas located in the agronomic dry boundary zone because of the high precipitation variability of these regions. A significant cause of soil degradation is excessive land use (e.g. in the Sahel Belt, at the Aral Sea). In combination with climatic fluctuations, this leads to crop failure and permanent soil infertility.

Overexploitation of land is primarily caused by population growth. More people require more food, but the areas of agricultural land can either not be extended at all or not sufficiently. The carrying capacity of the soil is therefore exceeded. Soils are left unprotected and exposed to erosion and degradation. The cultivation of cash crops intensifies the problem: if production is extended beyond self-sufficiency, in order to sell more products on the (world) market, vulnerable soils are burdened (e.g. increase of livestock numbers). If the expansion of agricultural land use is not adapted to the potential of the region, the carrying capacity of the soil is exceeded.

The fight against desertification is being conducted at various levels. Farmers can make an important contribution towards combatting desertification at a local level by implementing crop rotation and improved irrigation systems. Erosion can also be reduced through regional cooperative projects e.g. hillside terracing. At an international level, the Sahel states have joined forces and are implementing measures to combat desertification (e.g. the "Great Green Wall"). The United Nations coordinates this project providing both financial assistance and expertise.

Water in abundance

The water on Earth is integrated in a global cycle. Evaporation, precipitation and runoff ensure that the total amount of water stays the same; the individual parameters can however vary. Increased precipitation leads to increased runoff. High surface runoff in river systems can result in raised water levels and flooding. Regions in the lower course of a river are particularly at risk. A natural cause of increased runoff in river systems is melting snow in spring. If the banks of rivers/river mouths are densely populated, a natural event can quickly develop into a natural disaster. Vulnerability increases.

The amount of water vapour in the atmosphere depends on the evaporation of seawater and evaporation on land (evapotranspiration). If the temperature rises, evaporation increases. Global warming caused by climate change leads to less water in the oceans and stored in the ground. Greater amounts of water in the atmosphere lead to increased precipitation and subsequently increased runoff in river systems.

Flood risk caused by human intervention

When forests are cleared, the roots of the trees no longer stabilise the soil. Water is stored in the soil; it seeps slowly into the groundwater and runs off underground. Without the stability of the roots, erosion occurs. Soil no longer functions as a water reservoir and rapid surface runoff increases. The accumulation of eroded soil increases the risk of flooding in the lower course of a river. For the most part, rainwater seeps naturally into the soil. Surface sealing prevents water seepage. Surface runoff and runoff by means of artificially installed drainage systems (drainage pipes, discharge pipes) increases.

Water runoff in the river system is dependent on the gradient. The greater the gradient, the quicker the water runs off. The river course is linear. As the gradient flattens out towards the lower course the flow velocity of the river decreases. The river forms meanders and accumulation increases. If a river is realigned by human intervention and the water constricted by dykes, the flow velocity increases, less water can seep away and be stored, and there is more water runoff. This increases the risk of flooding in the lower course of the river.

Measures for reducing flood risk, on the one hand aim to prevent a flood turning into a disaster and, on the other hand, address the causes of flood risk. Dykes and floodwalls are built in order to prevent flooding in areas close to riverbanks. However, dykes can burst when they are soaked with water and exposed to the pressure of increased water levels for long periods of time. Also, there is a limit to how high they can be built. The construction of flood control barriers often leads to conflicts of interest between local residents.

Measures that restore a river's natural drainage system are more successful. The construction of polders, relocating dykes and creating flood plains are all part of combatting the causes. But these measures also result in conflicts of interest.

Floods caused by tropical cyclones are also a threat. Cyclones form in the tropical regions of the three oceans and can be hundreds of kilometres in diameter. When landfall occurs, extreme wind speeds coupled with heavy rainfall can lead to widespread devastation. For example, floods in the wake of Hurricane Katrina caused immense economic damage over thousands of square kilometres with devastating social consequences for the people who live there.

Skill check

All the skills that you were able to acquire in this chapter are listed here.
You can judge for yourself if you have mastered each skill: ***securely, moderately or barely.***

Explanatory skills

Can I ...		Unsure? Look it up on these pages
1.	explain the term vulnerability using drought and desertification as examples?	50/51, 60
2.	explain the relationship between ecological, social and technological vulnerability using drought and desertification as examples?	50/51, 60
3.	identify causes and consequences of desertification?	54–67
4.	identify causes and consequences of the anthropogenic threat to habitats using desertification as an example?	54–67
5.	present flooding as a natural process in the context of the water cycle?	68–73, 76/77
6.	present the increased impact of floods caused by different kinds of human intervention?	68–77
7.	describe land use conflicts in the context of flood prevention measures?	74–77
8.	describe and explain the formation, structure, paths and distribution of tropical cyclones?	80/81
9.	explain the impact Hurricane Katrina had on economic and settlement areas?	83

Methodological skills

Can I ...		
10.	analyse different thematic maps and statistical data, graphs and texts to answer questions about desertification?	54–67
11.	present the causes and consequences of desertification graphically in a concept map or mind map?	66
12.	analyse a film to find solutions for issues surrounding desertification?	67
13.	analyse different kinds of presentation and work materials such as maps, graphs and texts to answer questions about floods and tropical cyclones?	68–83
14.	substantiate statements using suitable and correct material references and quotes?	48–83
15.	use a model to describe and explain the water cycle?	78/79
16.	formulate key statements based on the model of the water cycle?	82/83
17.	present geographic information graphically, e.g. as a sketch map?	54–67

Evaluation skills

Can I ...		
18.	evaluate the suitability of the Sahel Belt as an economic and settlement area based on various geofactors?	60/61, 8/9
19.	evaluate measures for overcoming natural land use boundaries in the Sahel Belt in ecological and economic terms?	58–61
20.	evaluate possible solutions and limitations of adapting to drought and desertification?	64/65
21.	evaluate flood prevention measures from the point of view of different people affected by them?	74–77
22.	evaluate the risk potential of tropical cyclones using Hurricane Katrina as an example?	82/83
23.	discuss possible civil defence measures in light of the devastation in New Orleans ?	82/83

Action skills

Can I ...		
24.	present the results of work on tropical cyclones using appropriate terminology and in a factually relevant way?	80–83
25.	pay attention to local approaches to flood control issues and opportunities to influence decision-making processes?	64/65, 67

III Living with the Earth's Endogenous Forces

Results of the earthquake and tsunami in Japan on 11 March 2011: unimaginable devastation, the death of probably more than 20,000 people, and finally nuclear meltdown and the release of radioactivity at Fukushima.

The Earth – a dynamic planet

Continents in movement

Firm, compact and stable, that is how the Earth appears to us. But in fact it is moving and changing on a daily basis. Even in the first half of the 20th century, the notion that continents can move was regarded as a somewhat audacious theory. This theory was developed by the geologist Alfred Wegener on the basis of observations, for example that Africa and South America fit together like pieces of a jigsaw puzzle and that similar fauna and flora are to be found on both sides of the Atlantic. He assumed that the continents originated from the break-up of a supercontinent. How, if not through movement, could one explain the discovery of coal in Antarctica, for example? This coal must have been formed in what was originally a tropical region.

Wegener's theory of continental drift was an important starting-point for the development of a comprehensive model which today provides a coherent explanation for the structure and dynamics of the Earth and the Earth's surface. The fact that the Earth's surface is constantly changing over vast periods of time, and that we are living today in only a phase of transition, remains one of the most astonishing discoveries of science.

We mainly become aware of the fact that the surface of the Earth is in movement, and that forces from the interior of the Earth – endogenous forces – can have an impact on our lives, when earthquakes occur or volcanoes erupt and result in disasters.

The purpose of this chapter is to explain the dynamics and structure of the Earth: how do earthquakes arise, what are the causes of volcanic eruptions, and what are their consequences? What connections exist between the location and the frequency of such events?

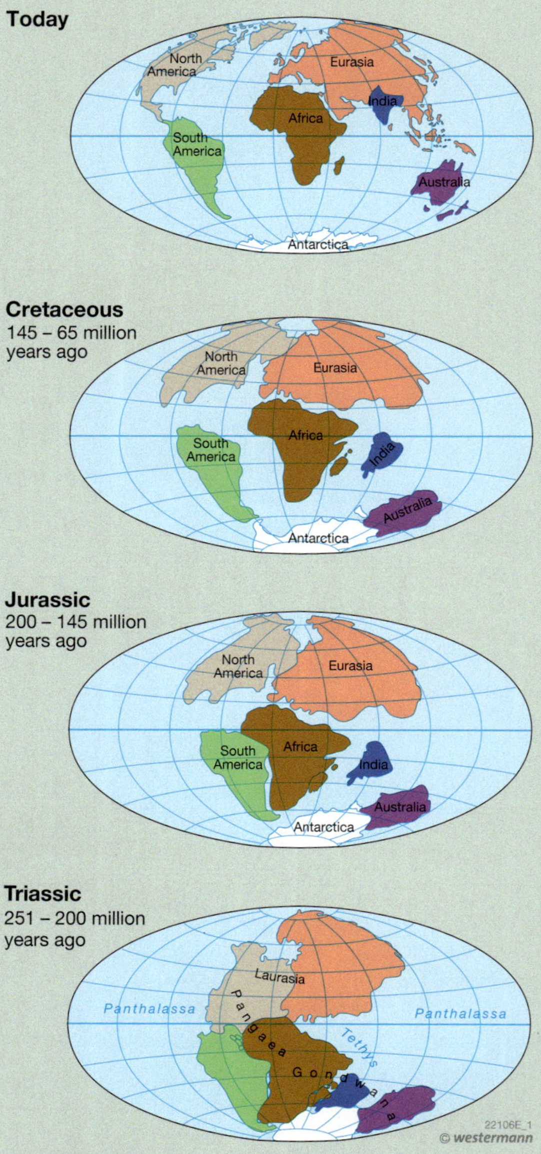

M2 Continental drift

M1 A rare natural spectacle. In 2013 a new island appeared off the coast of Pakistan. It was presumably the result of a volcanic eruption which was caused by a severe earthquake and through which mud was catapulted to the surface by methane gas. The effect of ocean waves will cause the island to disappear again

M3 "Remembering" – an installation by the Chinese artist Ai Weiwei in 2009, Haus der Kunst, Munich, in memory of the victims of the earthquake in China in May 2008 (9,000 school satchels, 100 m long, ten metres high)

Catastrophic chain reaction

Disruption in one sector of the infrastructure spreads inexorably across other, indirectly connected, systems like a cascade and thus reveals the vulnerability and interdependence of complex urban systems. The tsunami initially crippled the emergency electricity supply in the Fukushima Daiichi nuclear power station, leading already on 11 March to a meltdown in at least three reactors that were no longer being cooled.
Much too late, up to 200,000 people were evacuated from a radius of 20 km around the wrecked power station. Even today it is not clear whether and when these people will be able to return to their home communities. The resulting shutdown and technical inspection of all other nuclear power stations in Japan led to energy bottlenecks throughout the country which still endanger Japan's fragile economy today.
Energy shortages as well as the closure of important infrastructure and production facilities due to the disaster led not only to food shortages in the disaster regions, and even in Tokyo, but also caused global disruption in the supply chains of key industries.
Based on: C. Dimmer. Japan nach dem 11. März 2011. GR, 3/2013

M4 Effects of the tsunami in Japan

"What caused these schoolchildren to die?"

With this work of art, the Chinese artist seeks to commemorate the earthquake in Sichuan Province in 2009 which cost the lives of 90,000 people including thousands of schoolchildren. Using satchels, Ai Weiwei formed the sentence: "For seven years she lived happily in this world" in Chinese characters. These were the words with which a mother remembered her daughter.
It was the most severe earthquake in China for decades. It was noticeable that so many schools collapsed while the neighbouring buildings withstood the shocks. Immediately after the quake, foreign experts also discovered glaring deficiencies in the construction of the so-called "tofu schools". They had given way and collapsed like soft bean curd.
Artists who wished to expose incompetence, bungling, fraud, theft of materials and corruption in the building industry were arrested. Others, against the resistance of the authorities, sought to explain why so many children had died in the schools. With a group of volunteers, Ai Weiwei documented more than 5,000 names. In China, the dissident artist is subjected to threatening acts of repression.
Based on: Wahrheit wird zu Subversion, Süddeutsche Zeitung, 17.05.2010

M5 The Sichuan earthquake as a political issue

→ Vulnerability
Susceptibility. This refers to the impact of a natural event on the social groups affected by it, how these groups cope with the impact of natural events and how they can take precautions.

Ⓞ **1. A** The Earth – a dynamic planet. Explain this statement with reference to the various materials.
B Explain with the help of M2: Why is the discovery of hard coal in Antarctica or Spitsbergen a reason to regard continents as movable entities?
2. a) Explain why Ai Weiwei's public questioning of "what caused the schoolchildren to die?" represented a provocation for the Chinese authorities (M5).
b) Using this example, show the difference between a natural event and a natural disaster.
3. "The more technical possibilities it has available to it, the more successfully can mankind harness nature. But in doing so, it makes itself increasingly vulnerable." Discuss this thesis using the example of Japan (M4, photo p. 86/87).
Ⓢ **4.** Based on these examples, formulate theses to explain what the vulnerability of a region or country to natural disasters depends on.

Earthquakes – moving facts

The earthquake in Haiti – a devastating catastrophe

"PORT-AU-PRINCE, Haiti, 12 January 2010. At first the people were only startled by a slight judder, then the earth began to tremble, but the tremors quickly grew stronger, walls began to shake, trees swayed dangerously back and forth. The vertical and horizontal movements caused upper storeys of buildings to slide sideways, supporting columns snapped like matchsticks. People ran out into the streets screaming in panic. Within just 53 seconds – that was how long it lasted – the severe earthquake, which began at 16:53 hrs local time, had flattened large areas of the Haitian capital." This is how a journalist reported the event (H.-U. Dillmann, Aus Politik und Zeitgeschichte, 05.07.2010). The epicentre of the quake lay in the immediate vicinity of the capital city, Port-au-Prince, where over 40 percent of the population was concentrated and two-thirds of Haiti's gross domestic product was produced.
What are the causes and effects of such a natural event?

1. Explain how earthquakes arise (M1/ M5 – M8).
2. Locate Haiti (atlas) and give reasons for the earthquake risk (M1).
3. ⓞ **A** Describe the effects of the earthquake (M2 – M4).
   B Give reasons why "disaster" is an appropriate description for the earthquake.
4. Research the current situation in Haiti (internet).
5. Ⓢ Name islands in Central America that are threatened by earthquakes in a similar way to Haiti (atlas).

Earthquake, Epicentre, Hypocentre, Magnitude

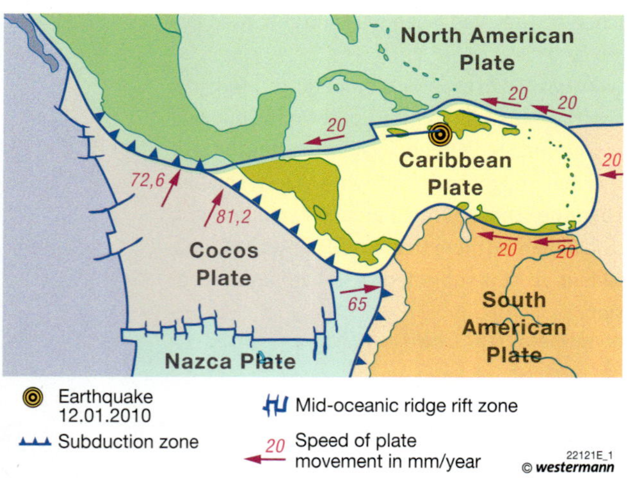

M1 The geological situation in the Caribbean. The Septentrional fault, part of the boundary between the Caribbean und North American plates, runs along Haiti's northern coast. Another, the Enriquillo-Plantain fault, runs along the south of the island. At these two fault lines the tectonic plates are moving in opposite directions.
Evidently the plates had become hooked into each other at the southern fault, at least since the last major earthquake in 1860. Since then, the plates should have shifted relative to each other by several metres. Now the tension was released at a stroke. The result was an earthquake of magnitude 7.0 with its epicentre in the region of the capital city. At a depth of some 13 km, the hypocentre lay relatively close to the surface of the Earth.

M2 After the earthquake: helpers in the destroyed Port-au-Prince

Balance sheet of the earthquake

According to the International Committee of the Red Cross, up to three million people have been directly or indirectly affected by the earthquake. This represents one-third of Haiti's population. Some 1.3 million people were made homeless because their houses – numbering 97,294 according to the Haitian government – were either completely destroyed or so badly damaged that they had to be demolished.
Since the middle of January, 800,000 people have been living in one of over 800 temporary camps that have been erected on almost every piece of open space in Port-au-Prince and its environs. Within Haiti there are more than half a million so-called Internally Displaced People (IDP) – earthquake victims who have left their former residence and are living in their home communities in the northern, eastern and western provinces, mainly with relatives.
Some 4,000 school buildings are no longer usable. The Ministry of Education has counted 1,300 dead teachers. There are no official figures for the number of dead schoolchildren and students, but they must be counted in their thousands.
School classes had to be interrupted until early April, and the hastily-installed classrooms are still insufficient for the approx. 350,000 schoolchildren. Many private and public universities have also been unable to fully resume their activities due to a shortage of teaching accommodation.
Between 250,000 and 300,000 people died; the UN speaks officially of 225,000 deaths. But in the first few days the dead were buried in mass graves without their data being recorded: anyone finding a body would lay it by the side of the road, from where it would be collected and removed, mainly at night. For days, more and more new, swollen, lifeless corpses were piled up on the pavement outside the Hospital Chirurgical de la Trinité in the city's Bel-Air district while, a few metres further on, badly injured people were receiving treatment on the street.

Source: H.-U. Dillmann: Als die Möbel "zu tanzen begannen" – Szenen aus Haiti. APuZ, 28/29, 2010

M3 Effects

M4 Temporary camp for earthquake victims

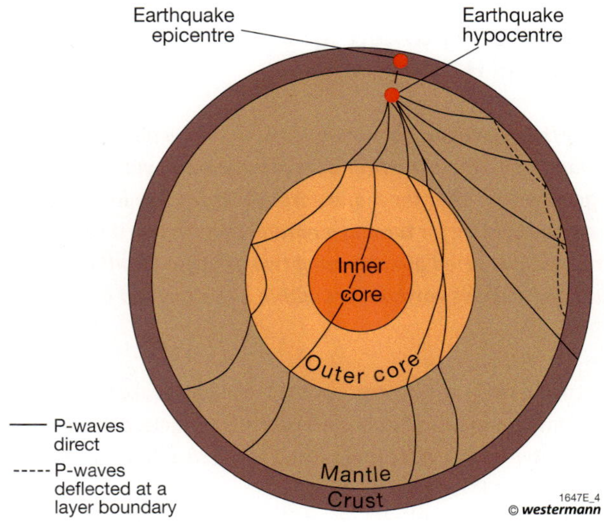

M6 Path of seismic waves after an earthquake

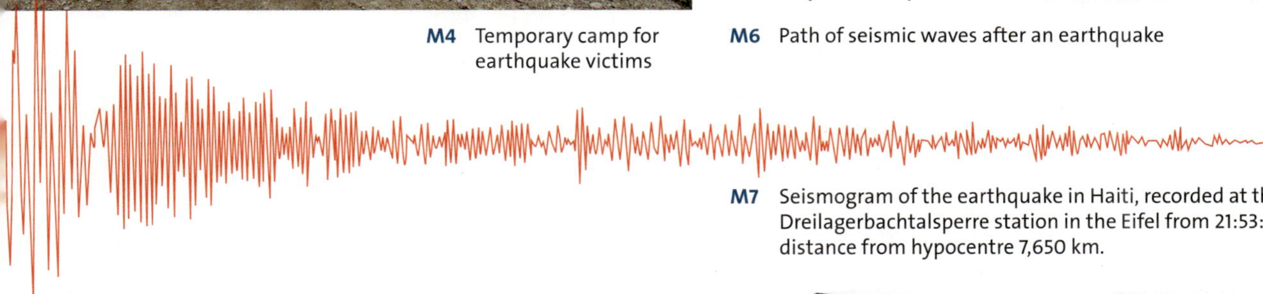

M7 Seismogram of the earthquake in Haiti, recorded at the Dreilagerbachtalsperre station in the Eifel from 21:53:09 UTC, distance from hypocentre 7,650 km.

How earthquakes come about

Earthquakes are tremors in the Earth's crust. They mainly originate at breaks in the crust when tension is released suddenly and spasmodically.

The tremor starts from a deep hypocentre (seismic focus), and the energy that is released travels through the Earth in the form of waves. Vertically above it on the Earth's surface is the epicentre of the quake, the point at which the waves first reach the surface.

A distinction is made between pressure waves (P-waves) and shear waves (S-waves). With a pressure wave, the medium in which it is transmitted oscillates back and forth in the direction of propagation, like a punch ball. With a shear wave, it oscillates up and down perpendicular to the direction of propagation, like a rope when it is shaken in a wave motion.

When the Earth is caused to vibrate at any particular point, it reacts partly like a punch ball and partly like a rope. But pressure and shear waves propagate at differing speeds. Pressure waves are the faster movers and shear waves arrive later. But fluids can only transmit pressure waves, not shear waves.

In addition, tremors also set in motion surface waves which spread along the Earth's surface like the waves caused by throwing a stone into a pond. The waves do not follow a straight path; they are interrupted or reflected at places where there are changes in the elasticity behaviour or thickness in the Earth's interior.

The strength of earthquakes is usually expressed in terms of magnitude on the Richter scale, developed by the seismologist Charles Richter. It is measured by means of a seismograph which registers ground movements on the Earth's surface. The magnitude is expressed on a logarithmic scale, meaning that an increase of one unit, e.g. from 5 to 6, represents a tenfold increase in the ground movement.

Based on: V. Kaminske, C. Keipert: Bau und Dynamik der Erde, 2006

M5 The fundamentals of earthquakes

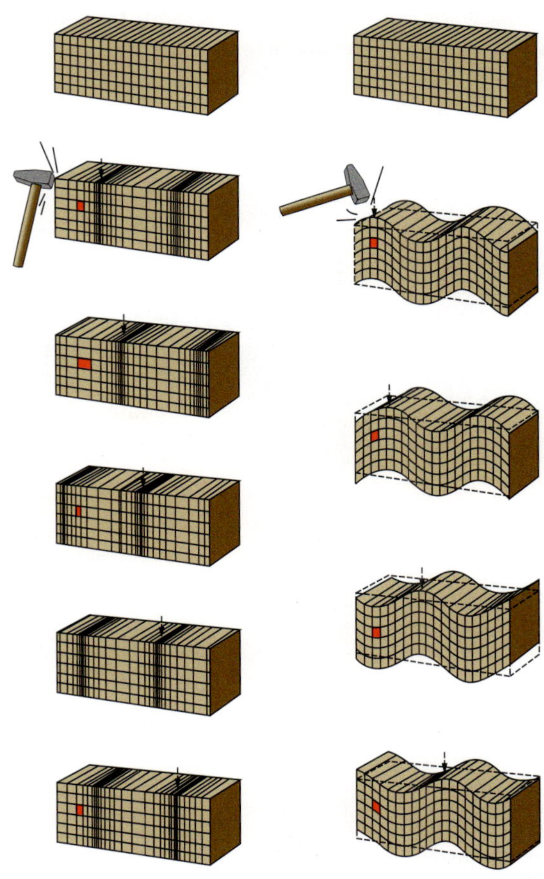

M8 Earthquake waves. A pressure wave occurs when a block of fixed particles is struck with a hammer in a longitudinal direction. The wave is propagated by each particle oscillating back and forth.
Shear waves occur when the hammer strikes the block in the transverse direction. This causes each particle to oscillate up and down, and transmit this movement to its neighbour.

Earthquakes – moving facts

Susceptibility to earthquakes – Chile and Haiti

"Earthquakes by themselves do not kill anybody. And they also do not cause any damage when there is nothing there to be destroyed," concluded an American geophysicist after the major earthquake in Chile in 2010. But what factors are responsible for turning a natural event into a disaster? This question can be answered by comparing two countries, Haiti and Chile, which both suffered massive earthquakes within a short period in 2010.

1. Locate Chile and give reasons for its earthquake risk (M1).
2. Compare the earthquake events, their effects, and the respective political, economic and social situations in the two countries (M2/ M4).
3. Ⓞ **A** Chile and Haiti are susceptible or vulnerable to natural disasters in different ways. Explain.
   B Summarise: What turned the earthquake in Haiti into such a great disaster?
4. Explain the structure of the Earth and characterise its individual shells with reference to their important features (M6).
5. Ⓢ After the melting of large ice masses, as for example in Scandinavia at the end of the Ice Age, compensatory movements take place. Explain why historic seaports there are now situated many metres above sea level.

Continental crust, Core, Crust, Lithosphere, Mantle, Oceanic crust

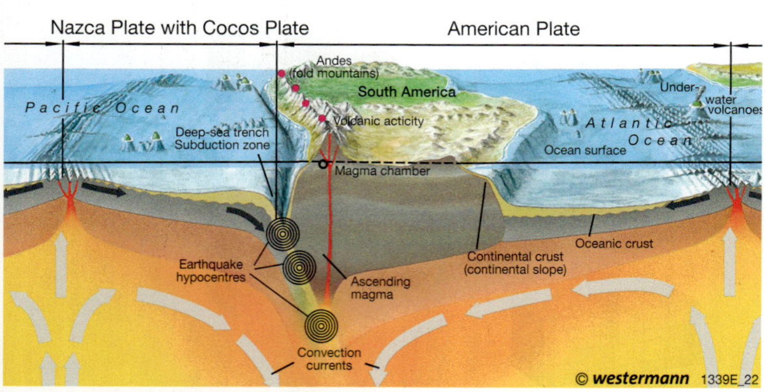

M1 Geological situation of Chile

	Chile	Haiti
Strength of earthquake on the moment magnitude scale	8.8	7.0
Number of deaths	700	220,000
Estimated material damage	30 bill. $	8 bill. $
Population	17 mill.	10 mill.
Gross National Product (adjusted for purchasing power)	243 bill. $	11.5 bill. $
GNP per capita (adjusted for purchasing power)	14,311 $	1,151 $
Public health spending (per capita)	863 $	58 $
Corruption Index: Position in country ranking (Position 1 = least corrupt)	21	146

Source: Bündnis Entwicklung Hilft, 2012

M2 Comparison of earthquake regions

M3 Temporary tent camp for victims of the earthquake disaster in Haiti

The earthquake in Chile was significantly stronger than that in Haiti, yet it was in the small Caribbean country that considerably more people lost their lives. How come?

The answer is very simple: Chile is far better prepared for such disasters. Latin America's most prosperous country has strict building regulations, which are also generally adhered to. And there was also a disaster plan in existence, which was put into operation immediately after the severe earthquake. In Haiti, no such plan exists. It is not the strength of an earthquake that causes buildings to collapse, but above all their building structure. In the past years, and particularly in the public housing sector, many houses have been built in accordance with strict earthquake safety provisions. Thus the poorer districts in particular were spared from massive destruction – in stark contrast to Haiti.

In Chile, not only the architects and building owners but also the public authorities including the rescue services are prepared for earthquakes. In May 1960, the Andean country suffered what was then the strongest earthquake ever recorded anywhere in the world, and the February 2010 quake was its third with a magnitude above 8.8. In Haiti, on the other hand, the last comparably severe quake was 250 years ago. And finally, Chile is rich enough to enable the population affected by the earthquake to provide for themselves. Haiti, by contrast, will be dependent on international aid supplies for months, if not years. Its notoriously poor infrastructure and governance are responsible for this.

Based on: F. Bajak, Associated Press, 28.2.2010

M4 Two earthquakes with differing consequences

Dangers

[...] The first danger from such an earthquake arises from the effects of the ground movements.

Along with the direct damage to buildings from the tremors themselves, the ground on which the buildings stand may also be damaged. [...].

As a consequence of the ground tremors, landslides and mudslides can also be triggered.[...]

The second danger comes from the permanent displacement of ground after very strong quakes, from which buildings located in the immediate vicinity of faults may sustain severe damage.

The third danger consists of flooding. Earthquakes can destroy dams or damage dykes [...].

The fourth source of danger is fire. Earthquakes can destroy supply and disposal lines. [...]

Fires become an even more serious problem when no water is available for fire-fighting on account of burst water pipes. [...]. The real dangers for people come from being struck by parts of buildings, drowning in floods from broken dams, being buried under landslides or trapped by the ensuing fires.

Source: www.seismo.uni-koeln.de

M5 Dangers from earthquakes

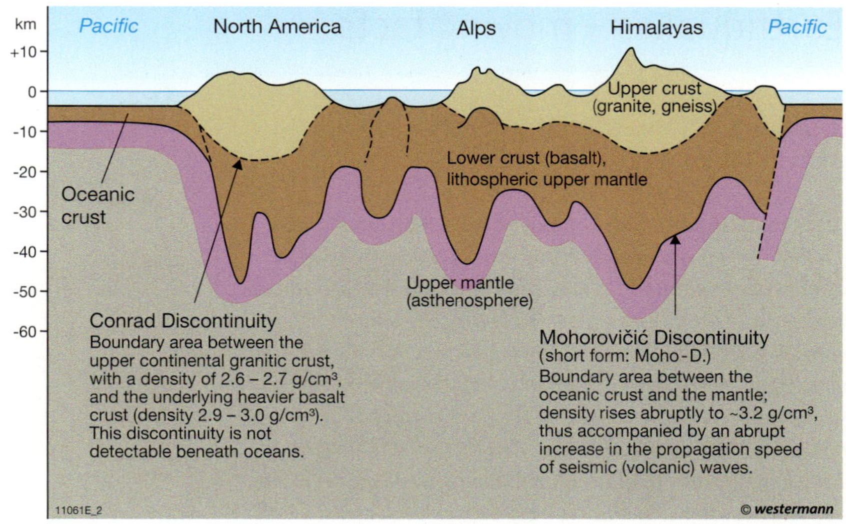

M8 Profile along the 45th parallel of the Northern Hemisphere. The Earth's crust and the uppermost part of the mantle together form the lithosphere, the solid rock shell of the Earth. This floats like a thin skin on the viscous asthenosphere and is pushed, distorted, broken and shifted by forces from the Earth's interior. Portions of the lithosphere (plates) are immersed into the asthenosphere like icebergs floating in the water. The continental crust immerses more deeply into the denser mantle.

The speed of the waves depends on the ruling density of the rock or the physical state of the material.

Higher-density rock transmits seismic waves faster than rock with a lower density. Fluids transmit the waves slower than solid bodies or not at all.

When the speed and direction of the seismic waves abruptly change, it can be concluded by scientists that boundary layers, so-called discontinuities in the Earth's structure, are present.

Based on: V. Kaminske, 2005

M7 Seismic waves

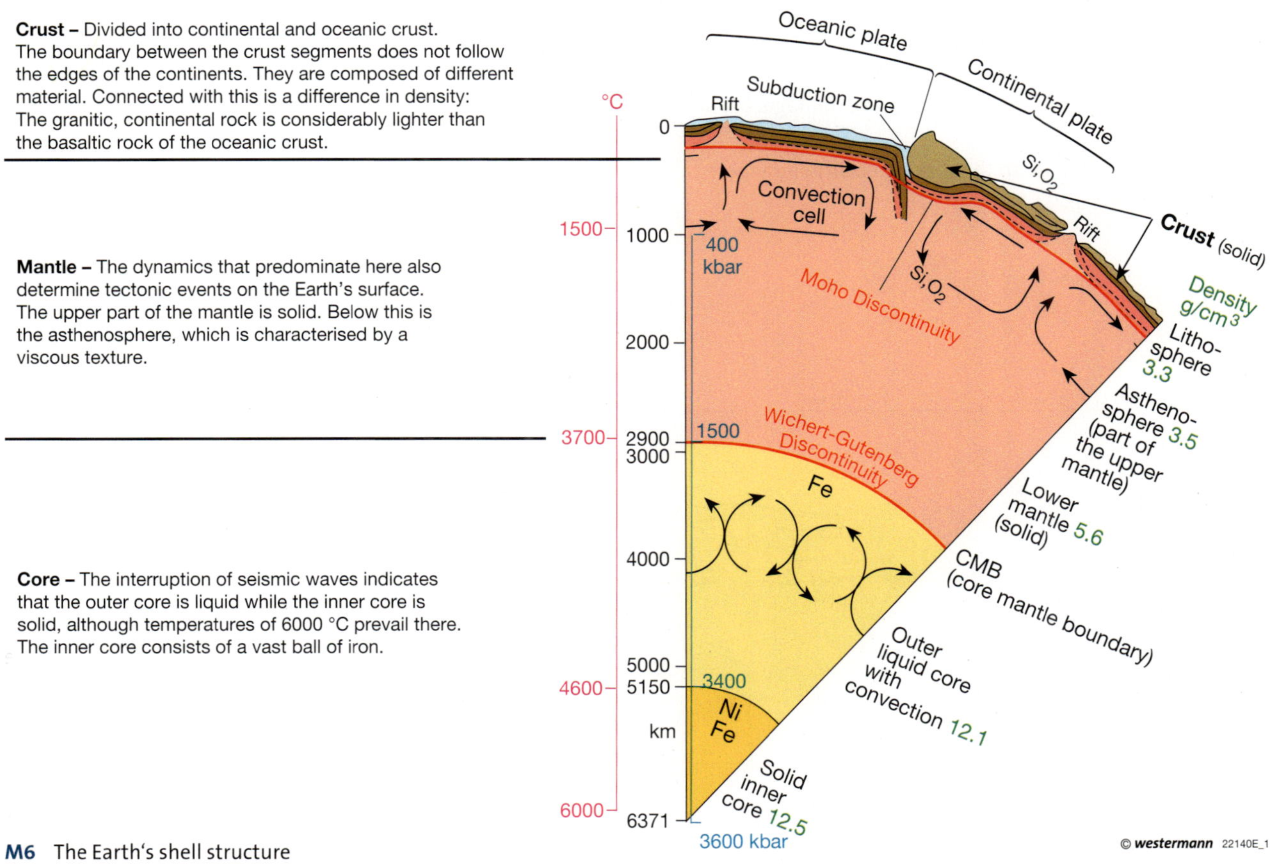

M6 The Earth's shell structure

Earthquakes – moving facts

Tsunamis – a threat to the coasts also in Japan

On 11 March 2011, a severe earthquake of magnitude 9.0 occurred off the east coast of Japan. The epicentre was 129 km east of the port city of Sendai. After the main quake at 14:46 hrs local time, more than 30 aftershocks with magnitudes of at least 6.0 were recorded in the following 48 hours. A tsunami warning was issued for all the coastal districts of the Pacific region. Tsunamis (jap. "harbour wave") occur due to movements on the sea floor which cause water to be displaced. They are triggered e.g. by seaquakes, plate tectonic shifts in the sea floor or major rock and ice falls. Shortly after the first quake, a tidal wave up to ten metres high devastated 500 km of the Pacific coastline. In narrow, deep bays the wave ascended as far as 40 m up the hillsides. In the flat plain near Sendai, the water masses penetrated five kilometres into the interior of the country. In the Fukushima nuclear power station, the tsunami resulted in explosions in the reactors which were no longer controllable, and ultimately led to a meltdown. A large number of inhabitants had to be evacuated from the surrounding areas (see also p.89, M4).
There are certainly very few other countries where such a potential for natural risks exists as in Japan. But there is also no other country in the world that is so well prepared against events such as earthquakes. So how does Japan protect itself? How do tsunamis arise? And what were the consequences of the 2011 earthquake?

1. Explain how tsunamis arise (M1).
2. Give reasons for the earthquake risk in Japan (M2, atlas).
Ⓞ 3. **A** Give a tabular overview of the consequences of the Tohoku earthquake (M3 – M6, p. 86).
B Using the data from M3, compare the consequences of natural disasters in Japan, Haiti and Chile (M2, p.94).
4. The balance sheet of the earthquake would have been even worse if decisive strategies and forms of behaviour had not contributed to minimising the earthquake damage. Explain (M6).
Ⓢ 5. Locate regions of the Earth that are especially endangered by tsunamis (atlas).

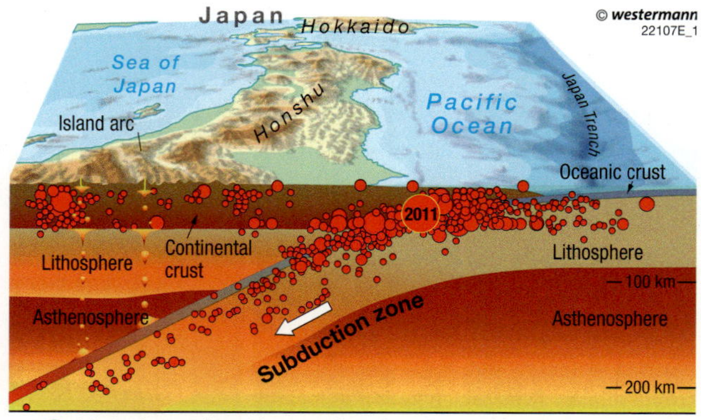

Historic earthquakes at the plate boundary

M2 Japan, geological situation

Consequences of the earthquake

The earthquake mainly affected the Pacific coastal strip in the Tohoku region: c. 16,000 dead and nearly 3,000 missing, some 130,000 buildings completely destroyed, infinite personal suffering, incalculable damage to property.
The media's focus on the massive wall of water sweeping inland left the impression that large parts of Japan were destroyed. But "only" some 400 km² along the east coast was directly devastated by the tsunami, i.e. 0.1% of the country's land area; the destruction was disastrous locally but was not spread over a wide area. That the primary impact did not result in any particularly large-scale damage in the 130 km distant city of Sendai, with over a million inhabitants and the centre of the Tohoku region, borders on a miracle. Japan's core economic regions were spared.
The damages are put at around 150 to 250 billion US dollars out of a GDP of 5,459 billion US dollars. But this calculation is only valid if the damages caused by the nuclear accident at Fukushima are left out of the account. Japan can cope with the economic consequences of the earthquake and tsunami, but the consequences of the reactor disaster remain uncertain.
Based on: W. Flüchter: Das Erdbeben in Japan 2011, GR 12/2011

M3 Balance sheet of the Tohoku earthquake

Tsunami

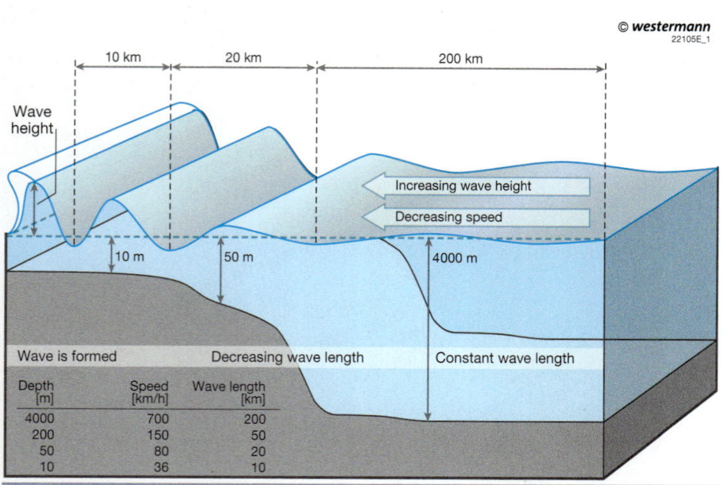

A tsunami spreads out in all directions from its point of origin at a speed that depends on the depth of the water. The wave is often very long, but on the open ocean it may not be very high.
In contrast to a "normal" wave which is triggered by the wind and thus only affects the topmost layer of water, a tsunami affects the entire column of water from the sea floor to the surface.
Due to the great length of the wave, it is not noticeable on the open sea. Only when it reaches the coast is the front of the wave slowed down while a gigantic mass of water is pushing from behind at high speed. As a result, the wave piles up and can sometimes reach a height of many tens of metres when it runs out onto the coast.
Based on: R. Glawion u. a.: Physische Geographie, 2012

M1 Formation of a severe tsunami

M4 Water masses at Miyako in Japan from a tsunami caused by the earthquake on 11.3.2011

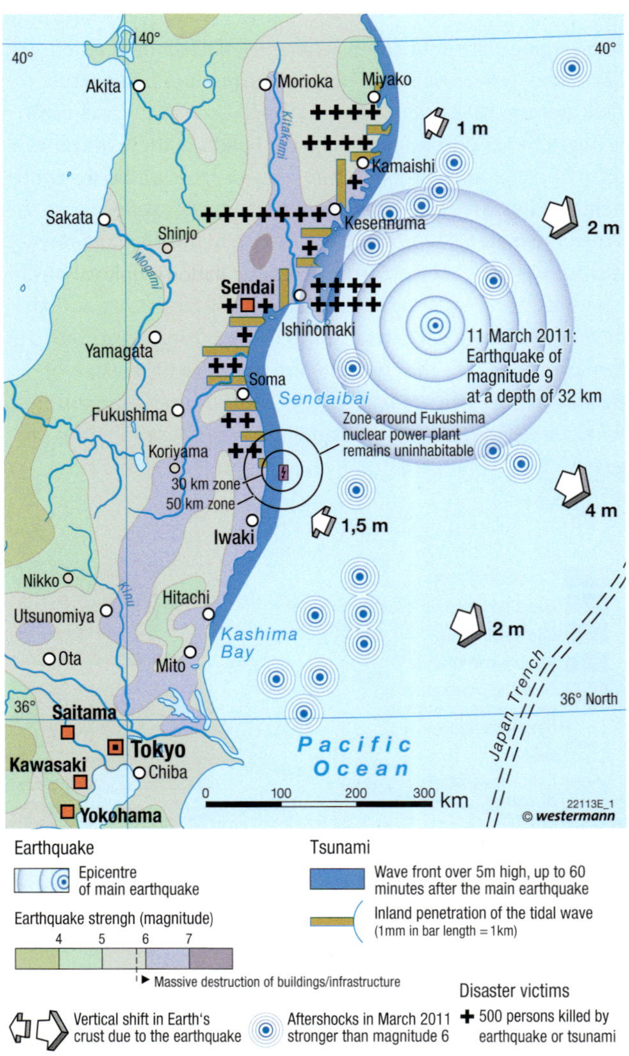

M5 Tohoku earthquake off Japan on 11.3.2011

Seismic early warning

- **Earthquake:** It is currently impossible to reliably predict the timing and location of earthquakes. But in the event of an earthquake, an efficient early-warning system comes into play. This is based on the differing speeds of the P-waves and S-waves. The first signals of the P-wave mean an early-warning potential of 40 seconds in the case of the metropolis of Tokyo, 370 km from the hypocentre, or 14 seconds in the case of the 130 km distant Sendai: valuable time. Express trains could be halted automatically, flood gates opened, bridges closed off. In spite of the permanent danger of earthquakes, the Shinkansen high-speed train has not suffered one single fatal accident.
- **Tsunami:** The fact that tsunami waves move much more slowly that seismic waves makes it possible to obtain adequate early warning with the aid of a close-knit network of seismic sensors, gauges and buoys. According to reports, the tsunami took 10 to 30 minutes to reach the nearest affected coastal region – valuable time for saving human life. Tens of thousands took advantage of this opportunity.
- **Earthquake-resistant and earthquake-flexible building technology:** In terms of building technology, Japan with its internationally first-class vibration monitoring and insulation technology represents the model for cities with a similar degree of vulnerability, e.g. San Francisco and Istanbul. But even this exemplary country still needs to do a great deal to make its great mass of private wooden housing both more earthquake-resistant and more fire-resistant. In Japan, the secondary effects due to fires are at least as dangerous as the destructive power of the earthquake.
- **Conduct and reaction of the population in an emergency:** In Japan through the ages, people have been accustomed to natural disasters. Respect for the powers of nature is deeply rooted in the consciousness of the population. Disaster prevention exercises are held at least once a year, ritualised on 1 September in memory of the Great Kanto earthquake of 1923.

Based on: W. Flüchter: Das Erdbeben in Japan 2011. GR 12/2011

M6 Disaster precautions

Plate tectonics – a model

It's all on the move!

In the last 50 years, scientific understanding of the Earth has changed completely. It was originally assumed that the Earth was largely rigid and fixed, but today we know that the Earth's crust and the interior of the Earth are in constant motion. The new insights have been brought together into the theory of plate tectonics. This hypothesis enables science to give plausible answers to important questions. It can explain the causes of earthquakes, of volcanic eruptions, or of the formation of mountains.
At its heart is the realisation that the lithosphere is divided into different plates, six large and many smaller ones, which can move in relation to one another. The plates are set in motion by convection currents circulating within the asthenosphere. Some plates divide and form new plates, while others slowly disappear.
But this takes place over enormously long periods of time. Plate tectonic processes mainly affect mankind in the form of events at the plate boundaries. It is at these zones of weakness in the lithosphere that earthquake hypocentres and volcanoes are concentrated.
The observation that, in the course of the Earth's history, continents must have drifted across its surface, gave the primary impetus to revising our theories of the Earth's structure. Today it is believed that the continental plates come together again and again to form a "supercontinent" which subsequently breaks up, drifts apart, and is then pushed together again.

1. Summarise in an overview the occurrences and processes at the various plate boundaries.
2. Classify the tectonic situation (M1) in Haiti (p. 90 M1), Chile (p. 92 M1) and Japan (p. 94 M2). Using these examples, explain the processes at the plate boundaries (M2).
3. Ⓞ With the aid of the plate tectonics model, describe and give reasons for:
   A the global distribution of earthquake regions (M3 – M5, atlas).
   B the distribution of younger fold mountains (M3 – M5, atlas).
4. Ⓢ Put together in a table the major plates, and the continents and oceans belonging to them.

Convection, Convergence, Divergence, Fold mountains, Island chain, → Rift zones, Subduction zones

- ***Collision of plates (convergence)*** – When an oceanic and a continental plate meet, the thicker oceanic plate is pushed down (subduction). When two continental plates collide, they thrust over and under each other. The plates are compressed, deformed or folded. The result is a thickening of the lithosphere. This is how the Himalayas came into being.
- ***Separation of two plates (divergence)*** – Here the lithosphere moves apart along a line of separation. The plate boundary is designated as constructive because new lithospheric material is formed there.
- ***Two plates slide past each other – (transform movement)*** Plates can shift past each other horizontally along the plate boundaries or fracture zones within the plates. [...]

Source: V. Kaminske &. C. Keipert: Bau und Dynamik der Erde, 2006

M1 How plates can move at the plate boundaries

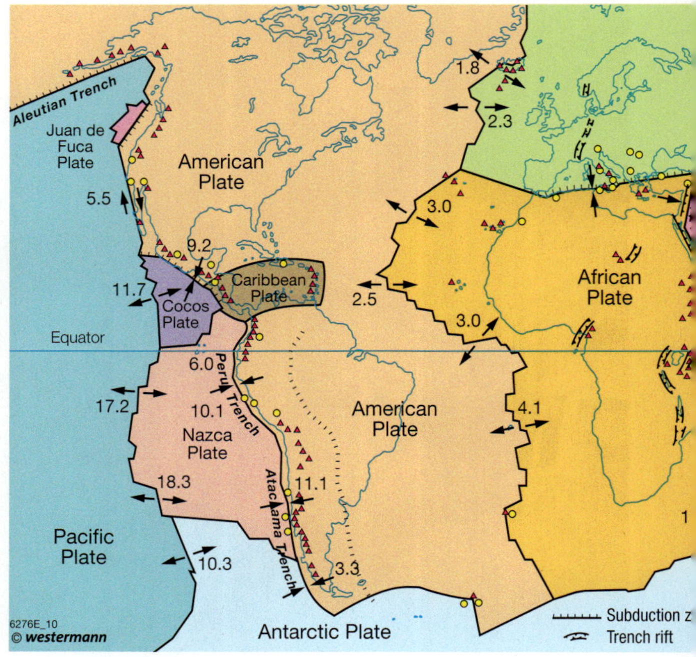

M2 Movement of lithosphere plates

→ Rift zones

At diverging plate boundaries, where the plates drift apart on both sides, new crust is constantly being produced. As a result, fault systems called rift zones are formed, such as the mid-oceanic ridges. Series of volcanoes along the length of these rift zones form fractures and barriers. Rift zones also occur within the continental plates, as for example the East African Rift Valley.
When the material brought to the surface at the rift cools down, it preserves traces of the Earth's magnetic field. Inversions, which have occurred frequently in the course of the Earth's history, can be recognised as a pattern of stripes in the oceanic crust. The pattern corresponds to the chronological sequence of reversals of polarity in the Earth's magnetic field. It can be seen from this that, for example, the Atlantic is expanding by several centimetres per year.

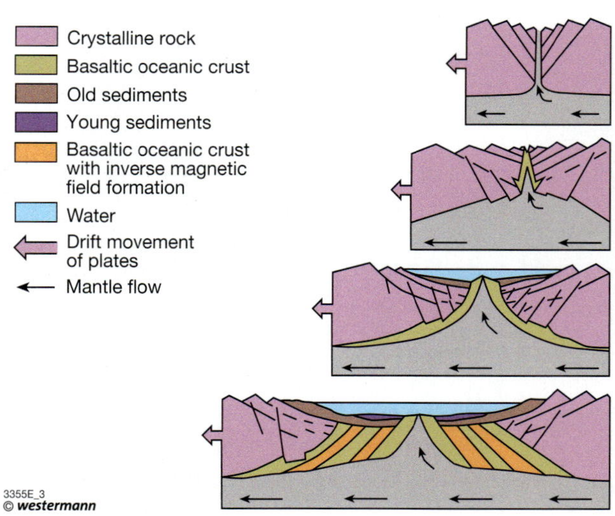

M3 Separation of two plates

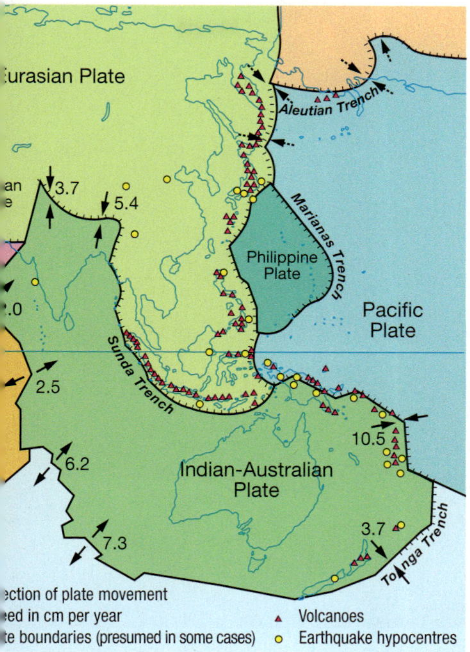

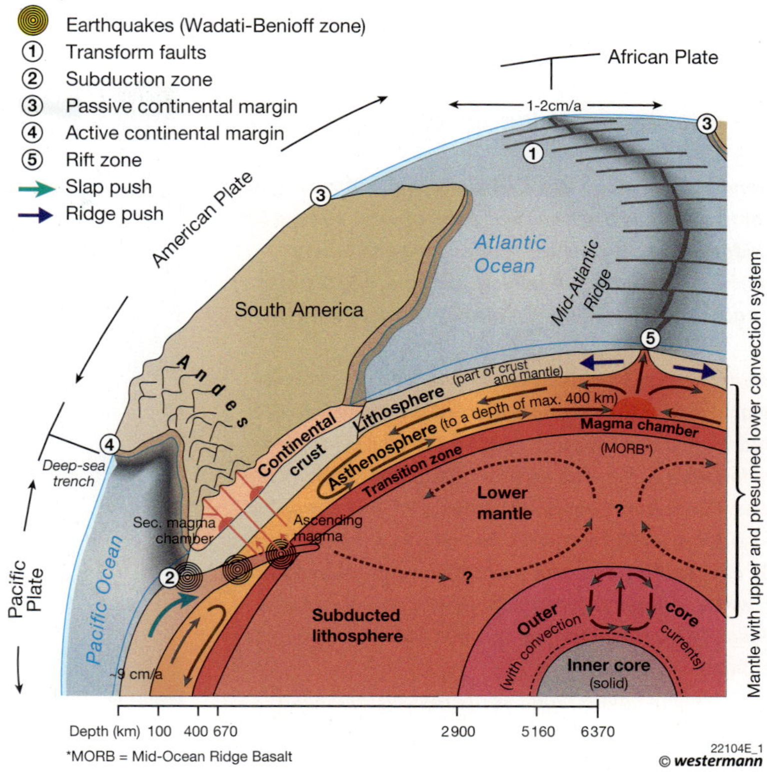

M5 Model of plate tectonic processes

→ Subduction zones

The zones where oceanic crust sinks down are situated especially in regions at the edges of large oceanic basins. These are referred to as subduction zones, which can be seen as representing the counterpart to the rift zones and which, along with the deep-sea trenches, are the places with the greatest ocean depth. When two oceanic plates collide, the older of the plates sinks down underneath the younger one on account of its greater density. Island chains are thus formed (Marianas Type). When an oceanic and a continental plate are pushed against each other, the oceanic plate sinks down because of its greater density. The pressure causes fold mountains to form on the continental plate (Chilean Type). Thus, oceanic crust that was formed in the rift zones is destroyed again in the subduction zones and the sunken material melts down into the Earth's mantle.

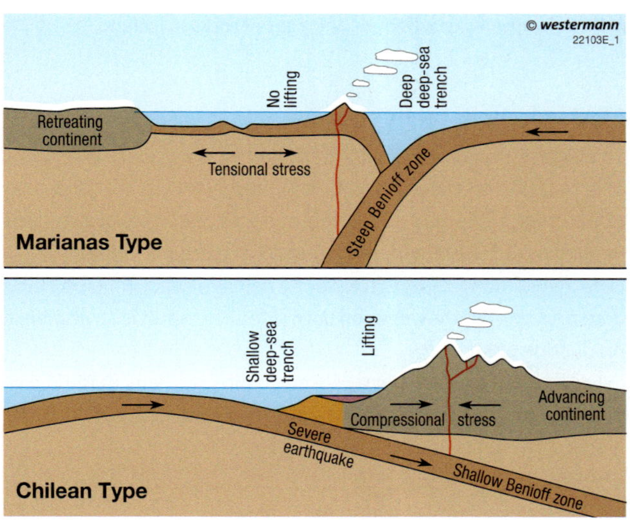

M4 Collision of plates

What drives the plates?

Even today, science has no clear answer to this question. Different forces are assumed to be at play.
Major drivers for the plate movements seem to be convection currents in the Earth's interior. In the upper mantle, hot magma rises under the diverging plate boundaries and cooled material sinks down in the subduction zones. The heat energy for this comes from the core or the lower mantle.
In the interior of the Earth there seem to be two separate convection circulation systems. The heat in the mantle is responsible for breaking the lithosphere into plates and for the plate movements. Further driving forces, the importance of which is a matter of controversy among scientists, are:

- *"Slap-push":* The heavier oceanic lithosphere plates sink down in the subduction zones. The sinking plate edge pulls the rest of the plate down after it.
- *"Trench suction":* When one plate sinks down it causes suction on the other plates at a subduction zone.
- *"Ridge push":* The plates slide sideways away from the bulged mid-oceanic ridges by the force of their own weight, and are pushed aside additionally by the rising magma.

Based on: K. Stüwe: Einführung in die Geodynamik der Lithosphäre, 2000

M6 Plate movements: an explanatory approach

Volcanic activity – blessing and danger

Iceland – one step from Europe to America

In some places in Iceland one can really stand with one foot in America and the other in Europe, at least from a plate tectonic point of view. Iceland lies on the Mid-Atlantic Ridge at the rift zone between Europe and America. The oldest parts of the island are only four million years old, and it is still constantly growing. Against this background, volcanic activity on this island is ever-present.
However the eruption of the Eyjafjallajökull volcano under a glacier in Iceland in March 2010 was unusual: the eruption affected the entire Northern Hemisphere. In Iceland itself, although some people had to be evacuated and some roads were inundated, the local effects were not particularly dramatic. In past centuries, this volcano was rarely active, and now it erupted ash particles up to a height of eight kilometres. But why was the whole Northern Hemisphere affected? Why was an island formed at precisely this place in the Atlantic, when material is constantly rising along the whole of the Mid-Atlantic Ridge? How is magma formed, and how does it find its way to the Earth's surface?
Why do the volcanoes in Iceland have a high risk of erupting, and what were the consequences of the Eyjafjallajökull eruption?

1. Explain how far "the pathway must be provided by tectonics" (M1) in Iceland, and give reasons for the volcanic activity on the island (M1/ M3/ M4).
Ⓞ 2. A Present the consequences of the outbreak of Eyjafjallajökull clearly in a schematic diagram.
   B Present the consequences of the outbreak of Eyjafjallajökull in the form of a newspaper report.
3. Describe how far one can speak of vulnerability in connection with this eruption.
Ⓢ 4. Work with the atlas: Produce an overview of regions that are especially severely affected by volcanic activity. Explain the term "Pacific Ring of Fire".

Eruption, Lava, Magma

Need one only drill deep enough? ...

What is fluid on the surface is not necessarily fluid inside the Earth. Pressure is a decisive factor. The greater the pressure bearing upon a rock, the greater the temperature that is required to liquefy the rock. [...]
Fundamentally, all volcanoes have their original source in the mantle. From anomalous hot zones at different depths, masses of molten rock are formed which are stored in great reservoirs in the mantle [...]. Every volcanically active region has its own peculiarities. It is not the case that, anywhere in the world, one would only need to drill deep enough to find magma in the borehole. [...]
With a temperature of over 1,200 degrees Celsius, magma is very hot [...]. But the energy contained in molten rock rising through a vent is not sufficient to burn its way through many kilometres of rock like a cutting torch.
For this to happen, in most cases the pathway must be provided by tectonics. A definition: Magma that has reached the surface is called lava.

Source: U. Schreiber: Vulkane, 2011

M1 The origin of magma

M2 Lava rising from different vents of the Eyjafjallajökull volcano underneath the glacier melted the ice of the Eyjafjallajökull Glacier. The water coming into contact with molten rock in the vents created great explosive power. This produced the eruption column which rose up to 8,000 m.

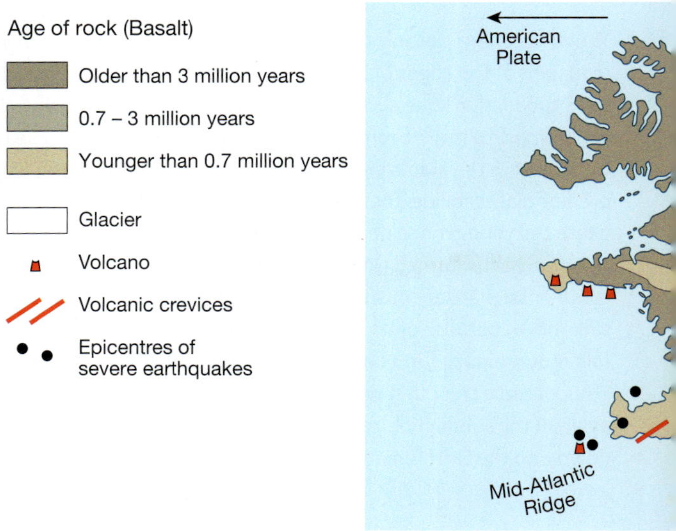

M3 Volcanically active zones and ages of rocks in Iceland

Most volcanoes cannot be seen

The largest area of the globe is taken up by the oceans. If we were able to push the water aside it would reveal an equally large area of the crust, composed exclusively of basalt, fed from lava streams flowing out of the long fault systems of the oceanic ridges.
Here, new oceanic crust is constantly being formed, which is being drawn away from the ridges on both sides. The result is an immense production of lava [...].
In Iceland, this process can be observed directly. Here, magma produced by an oceanic ridge joins with that produced by a hot spot [particularly active region, see p.100].

Source: U. Schreiber: Vulkane, 2011

M4 Volcanic activity at mid-oceanic ridges

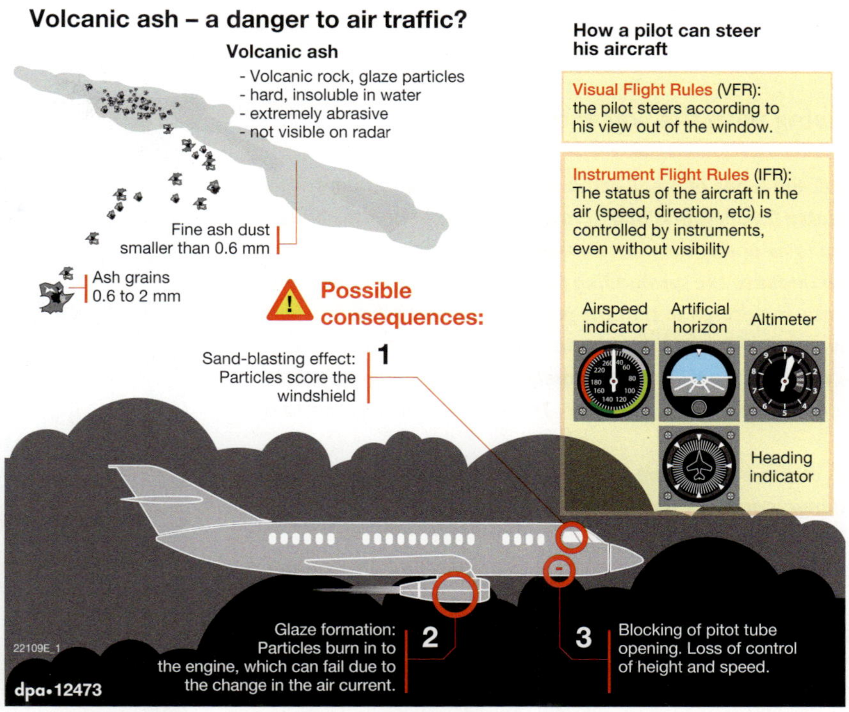

M6 Volcanic ash – a source of danger for air traffic?

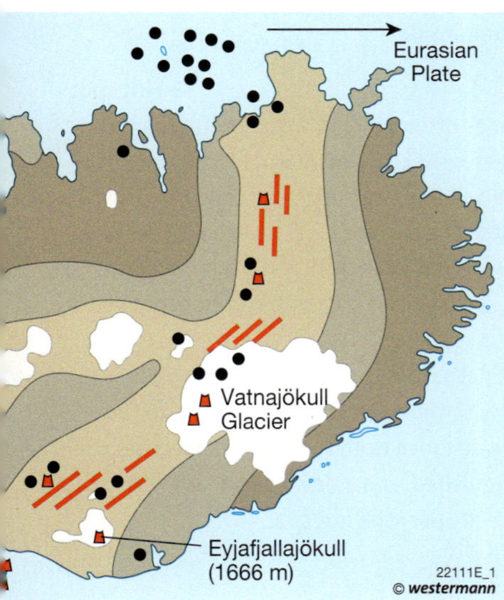

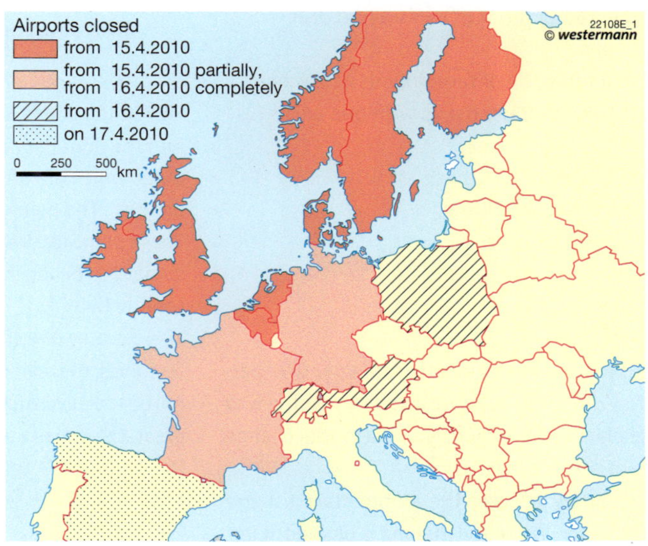

M7 Closure of European airports in April 2010

Damage amounting to billions

The unprecedented ban on flying cost billions. More than 100,000 flights in Europe's airspace were cancelled.

The airline companies alone estimated losses in sales amounting to 1.3 billion euros. Production downtime was reported in manufacturing industry because urgently-needed air freight deliveries did not arrive on time.

At German airports alone, nearly three million passengers could not be handled on account of the days-long closure of airspace. An estimated 50,000 tonnes of air freight could not be moved.

Based on: otr/Reuters/dpa/apn at www.spiegel.de, no date

M5 Economic consequences

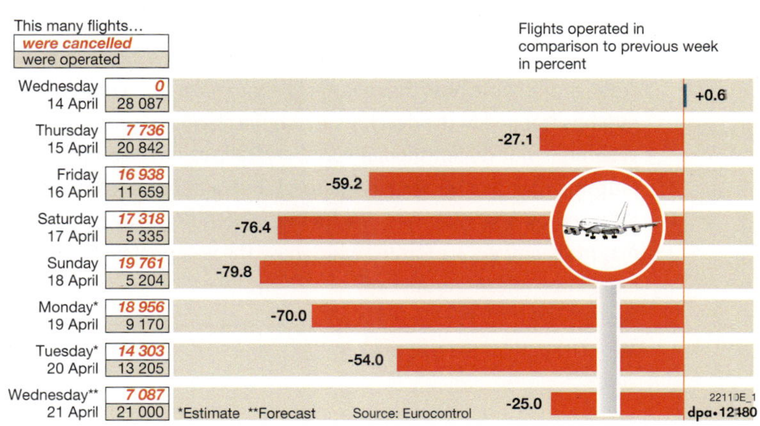

M8 Development of air traffic in April 2010

Volcanic activity – blessing and danger

Living with the volcano

The Merapi volcano is one of many active volcanoes in the comparatively densely populated Indonesian island of Java. Due to its unpredictability, it is considered to be one of the most dangerous volcanoes in the world, representing a threat to the surrounding population. Therefore the mountain is constantly monitored so that the population can reach safety with an advance warning of up to eight minutes. Because, despite the danger posed by the volcano, its environs are densely populated and people have even settled on its flanks again and again. The large city of Yogyakarta is nearby.
How does this contradiction come about? And what other types of volcano – and what dangers arising from volcanic eruptions – exist?

1. Give reasons for volcanic activity in Java and the formation of Merapi (M3/ M5, atlas).
2. Contrast with each other the favourable location and the threats from nature for the settlements in the vicinity of Merapi.
3. Explain the process by which magma is formed and how it reaches the Earth's surface (M1, p. 98 M1).
4. (O) **A** Explain the age and process of formation of the Hawaii Islands (M6).
   B Explain how far the existence of hot spots (see → Definition) can be accepted as proof of continental drift (M6).
5. (S) Summarise clearly in a table the different kinds of volcanic activity and contrast their characteristics with one another.

Island arc, Lahar, Magma chamber, Plume, Pyroclastic flow, Volcano, → Hot spot

M2 Rice cultivation on Merapi. The magma in the approx. 3,000 m high volcanic cone is viscous, so that it has developed a "dome" on the crater. When this collapses, the material slides down the mountainside.

On the way to the surface

In certain circumstances the magma can climb from a depth of perhaps 70 to 50 km in a few days. But there are also stops on the way to the surface, at those places where there is a substantial change in the density relations of the environs.
This is the case at the boundary between the continental crust and the mantle at a depth of 30 km. When the magma is blocked, it immediately leads to a change in its composition. Its temperature falls, special minerals crystallise, the relative proportion of easily volatile materials increases, and the succeeding molten rock mixes with the old. This can happen more than once on the way through the crust. Tectonic forces distend a limited section of the lower, middle or upper crust; at the same time the magma fills the space thus created and remains trapped in the developing magma chamber.
When it is not replenished, it cools, crystallises out and becomes plutonic rock (e.g. granite or gabbro). But when it is constantly followed by new magmas in a further distending crust, gigantic magma chambers are formed which have the potential for the most dangerous volcanic eruptions.
Source: U. Schreiber: Vulkane, 2011

M1 The path of the magma

→ Hot spot

Within the plates, volcanoes often occur whose conduit reaches into the boundary zone between core and mantle. Such conduits, which are often many kilometres thick, are called "mantle plumes". The point of exit at the surface [...] is called a hot spot. These point-like heat sources remain practically stationary for some 100 million years [...]. Hawaii is an example of hot spot volcanic activity. [...]
There is only one conduit path from the mantle. Due to plate tectonics, the solid plate with the oceanic crust wanders to the northwest. Eventually the distance from the conduit becomes too great, the activity is extinguished, and a new volcano is formed to the southeast – again directly over the hot spot.
Source: U. Schreiber: Vulkane, 2011

The subducted plates are drawn down into the depths at different angles. There, pressure and temperature increase significantly with every kilometre. The minerals in the sinking packet of rock change, water is released and rises into that part of the mantle which belongs to the neighbouring plate that is pushing it down. This is precisely the position above which the volcanoes of the active continental margin and island arcs are found. The mantle below the volcanoes is just as hot as below the rest of the continent. [...] But decisively, in the subduction zones, water is added. This significantly lowers the melting point of the mantle rock and creates the conditions for the formation of volcanoes.
From this process, magmas are formed which feed a quite different type of volcano. [...] The water and other highly mobile components ultimately lead to explosive volcanic activity, which can result in the most dangerous eruptions that we know.
Source: U. Schreiber: Vulkane, 2011

M3 Volcanic activity at subduction zones

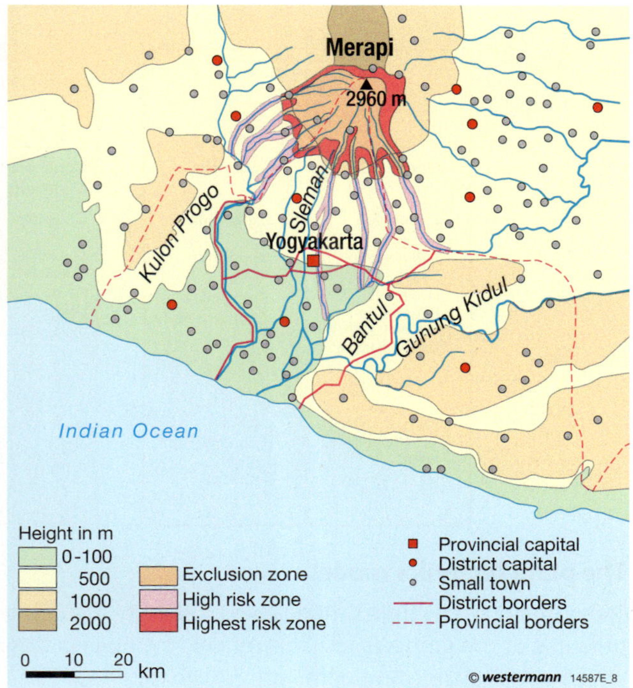

M6 Risk zones on Merapi – Great danger comes from hot avalanches, so-called pyroclastic flows, which have again and again caused devastating destruction. With temperatures of up to 800°C, the avalanches, consisting of a mixture of ash, lapilli and gas, race down the mountainside into the valleys at more than 100 km/h, often reaching far into the surrounding land. Mud and debris flows, called "lahars", are also a threat. These are formed after tropical rainfall when ash mixes with rainwater. Often, very fine ash particles in the atmosphere act as condensation nuclei, so that volcanic eruptions may be accompanied by heavy torrential rain.

Stratovolcano – conical, phases of ejection of ash and cinders alternate with phases of ouflow of viscous, acidic lava. Also explosive eruptions.

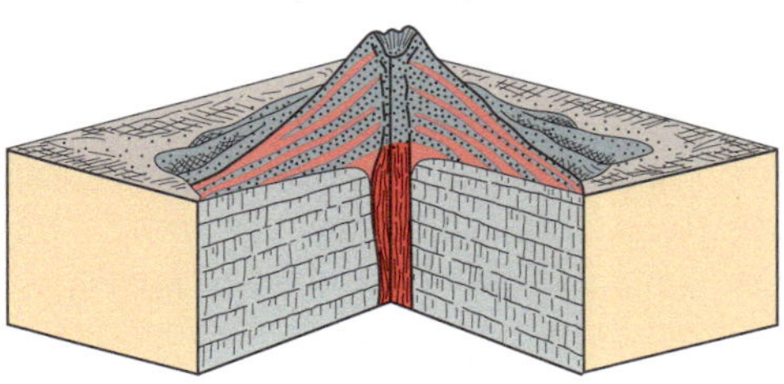

Shield volcano – created by rather steady outflows of very fluid, basic lava from a central vent.

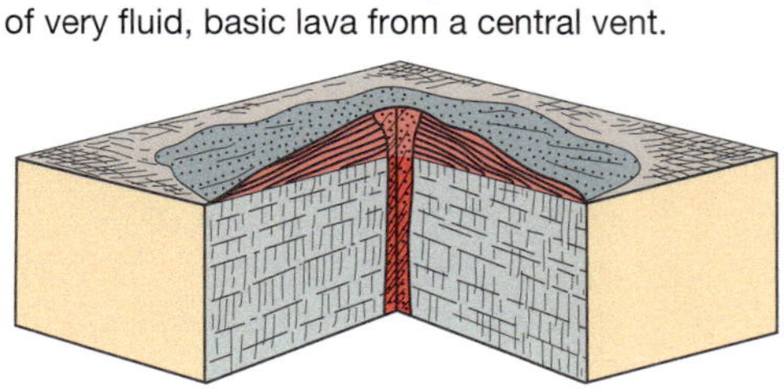

M4 Types of volcano

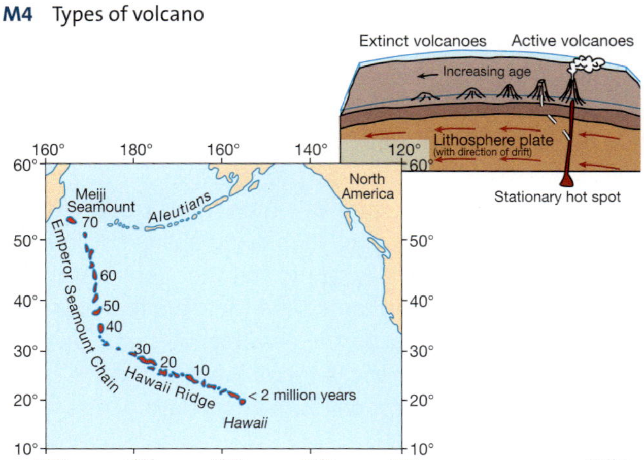

M5 Wandering volcanic islands

Is volcanic activity a blessing for agriculture?

Most volcanic rocks have a composition that provides many inorganic nutrients for plants. The hard rock, the lava, weathers slowly. Depending on the climatic conditions, it may take decades or centuries before a soil can develop with nutrients that are sufficiently available to plants.

The process is faster with ash and lapilli deposited on the surface, which have better water retention, can be more easily penetrated by root matter, and are more quickly etched by organic acids and rainwater. In warm moist climates, this quickly develops into a very fertile soil which, even after decades, continues to deliver nutrients for agricultural use from the slowly dissolving deposit of minerals. Well-known examples are soils in Java and Bali, which were produced by ash from active volcanoes.

It is undoubtedly their fertility that has made it possible for paddy field rice cultivation to develop here. With up to three harvests per year, this has provided the population with its basis of existence for over 2,000 years.

Source: U. Schreiber: Vulkane, 2011

M7 Fertile soil

The essentials in brief

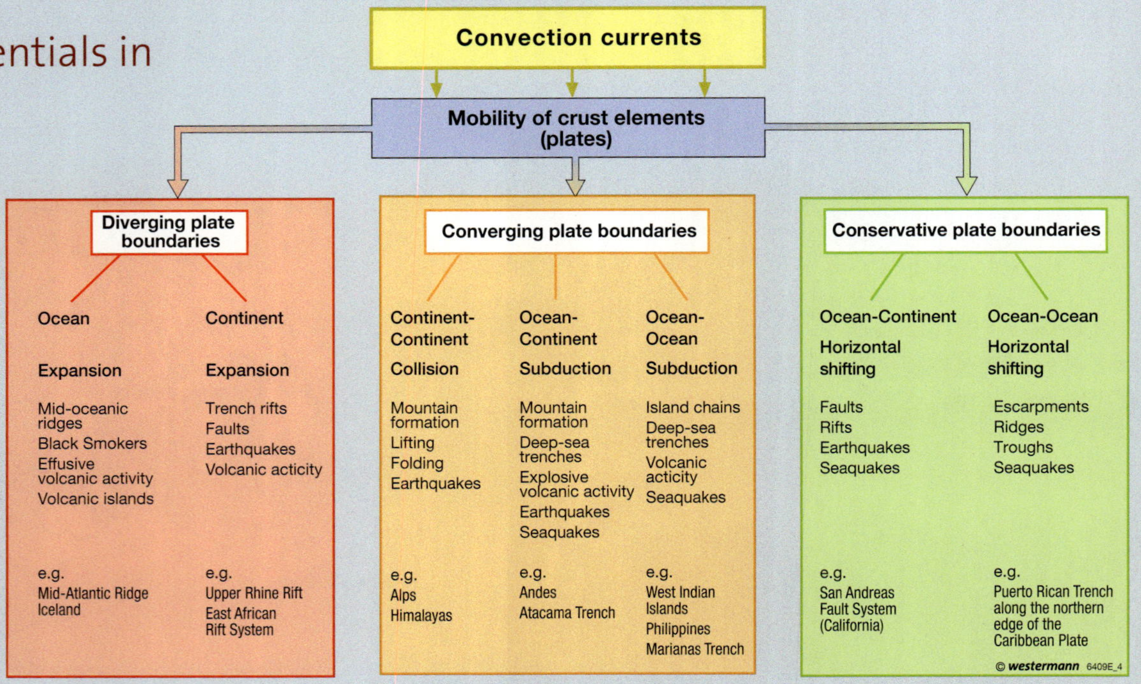

M1 Overview of movements of the Earth's crust

The plate tectonics model

Today, the plate tectonics model can be seen as a guiding principle of the geosciences. This model of mobile plates, which was a further development from the theory of continental drift, allows a multiplicity of geological phenomena such as earthquakes, volcanic activity, mountain formation or the creation of island arcs to be explained and put in context.
According to current research, the Earth's crust comprises some 20 large, mobile lithosphere plates which carry the continents and which move in relation to one another. The motor for plate movements has been identified as convection currents in the Earth's interior, aided by pressure and drag movements. Several times in the course of the Earth's history, large continents have been formed, the most recent being the supercontinent of Pangaea which broke apart 150 million years ago. This produced today's pattern of continents, which however is constantly changing.
The occurrence of earthquakes and volcanoes is concentrated in the zones of weakness in the lithosphere. These are plate boundaries, i.e. areas where two plates collide, move apart or slide past each other, and fault zones within the plates. At the mid-oceanic ridges, new crust material is constantly being created which is later destroyed and melted down at the subduction zones. This occurs where an oceanic plate is pushed down beneath the lighter continental plate.

Volcanoes and earthquakes

Stratovolcanoes are typical of volcanic activity at the subduction zones. At a certain depth, part of the water-bearing crust is melted and forced towards the Earth's surface under high pressure.
At the mid-oceanic ridges, magma can rise relatively unobstructed and the lava spills out in long fault systems.
Volcanic activity at hot spots within the plates occurs at magma exit points, e.g. at volcanic islands. Because the magma's conduit path to the surface remains practically stationary for very long periods of time, the lithosphere plates can wander across it. By the time of the next eruption, its contact to the previous exit vent may have been lost and a further volcanic island is formed. A chain of islands is thus created.
Although volcanoes represent a source of great danger, for example from pyroclastic flows, people have settled on the slopes of many active volcanoes up to great heights. The background to this is the extraordinary fertility of volcanic soils.
Earthquakes occur when pent-up tensions in the lithosphere are suddenly released. This mainly takes place at plate boundaries. The seismic waves spread out in all directions from a hypocentre. Therefore earthquakes can be measured even at a great distance. The yardstick for the intensity of the quake is its magnitude, calculated as the maximum acceleration of ground movements. Earthquakes cannot be predicted in terms of either their timing or their precise location. They can trigger tsunamis, which can cause devastating destruction at the coasts.

Natural events and disasters

The effects of such natural events are very different in different parts of the globe. While in developing countries the most striking aspect is often the number of dead, in industrialised countries high levels of economic damage are often also recorded. These countries, however, are in a position to afford such high costs thanks to their economic power and affluence. In developing countries the cost of disasters, measured as a proportion of GNP for example, is significantly greater than the proportion in the industrialised countries. And in a long-term perspective, more than 80 percent of deaths due to natural disasters occur in the developing countries.
A reason for the comparatively high levels of damage in developing countries is their high vulnerability. Poverty and a lack of alternatives lead to the fact that people settle in risk zones, while at the same time there are few possibilities for them to adapt to the possible risks or cope with disasters. The greater the vulnerability, the greater is the risk that a natural event will turn into a natural disaster.

Skill Check

All the skills that you were able to acquire in this chapter are listed here.
You can judge for yourself if you have mastered each skill: ***securely, moderately or barely.***

Explanatory skills

Can I...		Unsure? Look it up on these pages
1.	explain the structure of the Earth and the driving forces of plate movements?	92/93, 96/97
2.	explain the plate tectonics model in terms of its essential features and processes at the plate boundaries?	96/97
3.	explain with reference to the model the origins of earthquakes, tsunamis and volcanic activity as consequences of tectonic processes?	98–101
4.	show relationships between their frequency and geographical distribution?	90–101
5.	explain, using examples from individual regions, the dangers, effects and consequences of earthquakes, tsunamis and volcanoes?	96–101
6.	derive consequences for settlement and provision for regions that are subject to natural risks?	88–101
7.	give reasons for differing vulnerability with regard to different regions and countries, using examples?	88–101

Methodological skills

Can I...		
8.	identify problem-related geographical facts and develop corresponding questions, by for example setting out theories to explain the vulnerability of a country to natural events?	88/89, 90–101
9.	analyse different kinds of presentation and working materials such as e.g. maps, pictures, statistical data, graphs and text, and use these to deal with questions about the origins of natural risks and the consequences of natural events?	88–101
10.	explain plate tectonics as a model for explaining seismic and volcanic activity as well as the formation, movement and destruction of the Earth's plates?	96/97
11.	using correct terminology, present plate tectonics and natural events and their effects orally and in writing in a way that is problem-oriented, factually and logically structured, and relevant to the assignment, operators and materials?	88–101

Evaluation skills

Can I...		
12.	evaluate the potential dangers from earthquakes, tsunamis and volcanic activity?	88–101
13.	using examples, evaluate different societies or countries in terms of their level of vulnerability and degree of susceptibility?	88–95, 98–101

Action skills

Can I...		
14.	present the results of work on earthquakes, volcanic activity and plate tectonics using appropriate terminology and in a factually relevant way?	88–101
15.	develop strategies and forms of behaviour to reduce the level of damage, e.g. from earthquakes?	90–95

Examination training – presenting the results of your work

The danger of earthquakes – waiting for "the Big One" in California

1. Locate San Francisco and the San Andreas Fault.
2. Explain the cause of the earthquake risk in California.
3. Assess the possibilities and limits of disaster provision in California in comparison to Japan.

These additional materials are required to solve the assignments:

M1 Diercke Weltatlas, 2015, map: California – earthquakes, p. 207

The next "Big One" will come

At dawn on 18 April 1906, shortly after 5 a.m., California was torn apart for half of its length. The Earth's crust broke along a seam running from north to south for a length of 1,280 kilometres. According to conservative estimates, 3,000 to 4,000 people died, hundreds of thousands were made homeless. [...]

Today, if the "Big One" were to strike in the afternoon, according to official estimates there could be 18,000 dead, 250,000 injured and more than 700,000 homeless in Los Angeles. San Francisco expects a catastrophic disaster of similar proportions. The insurance company Munich Re calculates damages of up to 200 billion euros for both cities. The withdrawal of American capital could lead to a crisis in global financial markets. The chances that San Francisco will be struck by a severe earthquake in the next 30 years are twice as high as the chance that the city will be spared, the US Geological Survey concludes. [...]

Experts consider that evacuation is not a realistic option, so other kinds of measures are taken. In the last 15 years, according to the California Seismic Safety Commission (CSSC), amounts equivalent to some 20 billion euros have been spent on building renovations. But as an earthquake precaution, this is far from adequate.

According to the CSSC report, countless buildings are not earthquake-proof: high-rise buildings, four-storied urban motorways, buildings on slopes or on reclaimed building land. It is unclear whether they will stand up to a quake. [...] The problems that could arise in an emergency were revealed by the earthquake at Northridge in 1994. This only occurred at the edge of Los Angeles, but put 23 hospitals out of action.

Many Californians seem to be scarcely concerned by the inevitable disaster. Appeals by the authorities to keep stocks of medicines, clothing, batteries and preserved foods for the emergency go largely unheeded. Thomas Jordan, Director of the California Earthquake Center, chooses [...] clear words: "San Francisco," he says, "is on the fast track to destruction."

Based on: A. Bojanowski, www.spiegel.de, 13.04.2006

M2 The effects of earthquakes in California

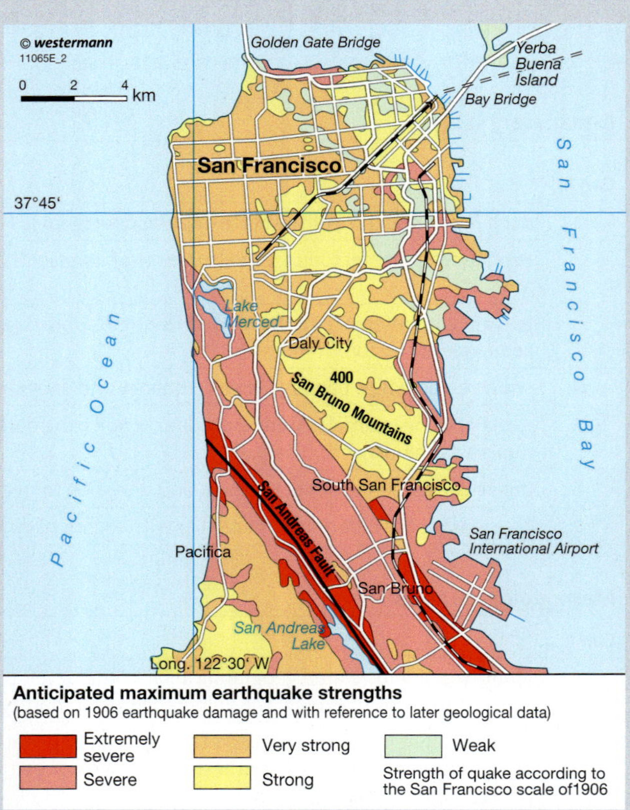

M3 San Francisco: the earthquake danger

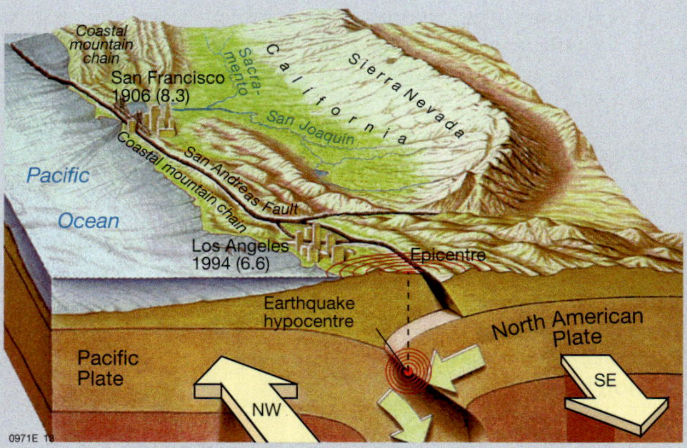

M4 Cross-section through California

Facilities, infrastructure	Facilities with a high probability of being damaged by earthquakes in the next 30 years: Number/ Percent of all facilities
Hospitals	76/ 76
Police, firefighters, local authorities	2970/ 75
Elementary schools	987/ 72
Middle schools	164/ 69
High schools	233/ 70
Colleges, universities	62/ 81
Bridges	2721/ 63
Railways (km)	150 km/ 89
Roads (km)	30,305 km/ 57

M5 Effects of earthquakes in the San Francisco region

M6 Satellite image with a digital elevation model of the San Andreas Fault

How can I express myself factually using the correct terminology?

It goes without saying that the correct terminology should be used. But a geography examination also requires factual analysis and argumentation, and factually-justified opinions.
Colloquial formulations and emotionally-coloured statements are to be avoided.
It is also important to keep the aspects of content and form separate.

Examples:
Wrong: It can be seen in M1 that the population has increased since 1999.
Or: M1 shows that the population has increased since 1999.
In these examples, content and form are mixed.

Correct: The population has increased since 1999 (M1).

From: Kreuzberger, N. & C. Kreuzberger: Zentralabitur Erdkunde – NRW, Brunswick 2012

M9 Terminology

The quality of presentation also counts!

In Abitur assignments, the quality of presentation counts for 20 percent of the overall grade. So it is worth paying attention to this.
Incidentally, requirements such as precise formulation, or the delivery of a text that is readable and understandable, cannot of course be separated from the quality of the content.
Even when it is the content of an examination that is being assessed, text passages that remain unclear cannot be given a grade – even when the "right answer" was intended.

M7 Presentation

References to source materials and quotations

References to materials and citations should be given in brackets at the end of a sentence or paragraph (M2). Quotations from materials as evidence for statements are superfluous. Real verbatim quotations should be used very sparingly, i.e. only for important, central points or as a thesis for discussion. The material reference in brackets should be placed at the end of direct quotations or of indirect quotations that are closely based on the wording of the source. Passages quoted verbatim should be identified by using "quotation marks". Reproductions of the general meaning of a source text, for example, may often cover very long passages. In this case it is sufficient to provide the reference to the source material at the end of the relevant paragraph or section.

M10 Use of evidence

Important aspects of the requirements for quality of presentation in the Abitur are:

Requirements	Tips
Text should be structured in a coherent, stringent and intellectually clear way (here, stringent means a strict, logical train of thought)	Avoid parentheses! Structure your work before making the fair copy! Don't forget a clear thread, short introductions and (interim) summaries for the assignment sections
Descriptive, interpretative and evaluative statements should be logically related to one another	When evaluating materials, pay attention to the conclusions that can be drawn from them
Texts should be formulated in a precise and conceptually differentiated way, using correct terminology	Use specialist terminology and an appropriate form of factual expression (M9)
Statements should be substantiated by appropriate and correct evidence	Do not forget the references to materials! (M10)
Writing should be linguistically correct (grammar, orthography, punctuation) and in a sure style	Pay attention to correct language! Spelling and punctuation are important

Based on: Kreuzberger, N. & C. Kreuzberger: Zentralabitur Erdkunde – NRW, 2012

M8 Requirements for quality of presentation in examinations

IV Production and Use of Fossil Fuels

Between economy and ecology

Neft Dachalari – an oil town on the Caspian Sea in Azerbaijan

Fossil fuels – finite raw materials in heavy demand

Limitless hunger for energy

More and more people in the world consume more and more products and use more and more services. A flight to the Balearic Islands for the weekend, a new flat-screen TV for the living room or a new smartphone – many people take all these things for granted. Billions of smartphones are charged every evening, countless aircraft are refuelled several times a day. And that is only the energy that is needed for using the devices. On top of that is also the so-called "grey energy" – the energy consumed in the production, storage, delivery and recycling of a product. The energy that is required to manufacture a car corresponds roughly to the electricity consumption of a four-person household for ten years. In recent decades, economic development and increasing affluence have resulted in a very major increase in demand for energy in many countries around the globe. But energy demand differs widely by region depending on the level of development. It is the countries with the largest population that will be decisive for future global energy demand.

weblinks

- International Energy Agency
- Federal Institute for Geosciences and Natural Resources

Fossil fuels

Fuel raw materials can be divided into fossil and renewable fuels. The term "fossil" refers to the conditions and the original materials out of which the fuel was formed. In some cases the raw materials were formed over a period of millions of years and were originally living entities (plants or animals). In the course of time, deposits of the most important fossil fuels (hard and brown coal, natural gas and oil) have formed due to changes in temperature and pressure. For this reason one also speaks of raw or primary energy, since this is stored in natural fuel materials.

Deposits, Fossil and renewable fuels, Grey energy, Primary energy, Reserve, Resource

In contrast to the so-called renewable energies, the deposits of fossil fuels will eventually be exhausted. Even though not all available resources of fossil fuel deposits are currently known, it is certain that at some point in the future the reserves will be completely used up. Fundamentally, the availability of fossil fuels plays an important role in the economic development of a region. The growing need for fuels and the search for additional raw materials are often a trigger for political disputes and even for armed conflicts.

Energy extraction – not without risks

Already today, the extraction and use of energy raw materials has far-reaching consequences for mankind and the environment. We are frequently confronted by headlines about incidents in nuclear power stations, or tanker or mining accidents. Moreover, their extraction often represents a major intervention in the natural balance and the lives of the inhabitants. But the availability of energy is essential for the economic development of countries and regions. Our dependence always becomes clear to people when the flow of energy is interrupted and the heating goes off, or when there is no electric power available.
It can be assumed that the demand for energy will continue to increase in the coming years. Knowledge of the most important fossil fuels (their origins, deposits, etc.) is necessary in order to assess the regional impact of the use of fossil fuels from an economic, ecological and social point of view. Therefore this chapter focuses on the following key questions:
What deposits exist in the world?
How are energy raw materials extracted and used?
What are the economic, ecological, social and political consequences of the extraction and use of fossil fuels?

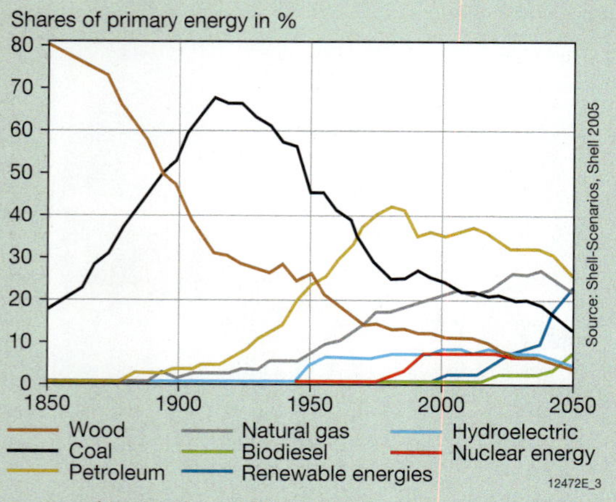

M1 Trend in the share of individual energy sources in global energy consumption

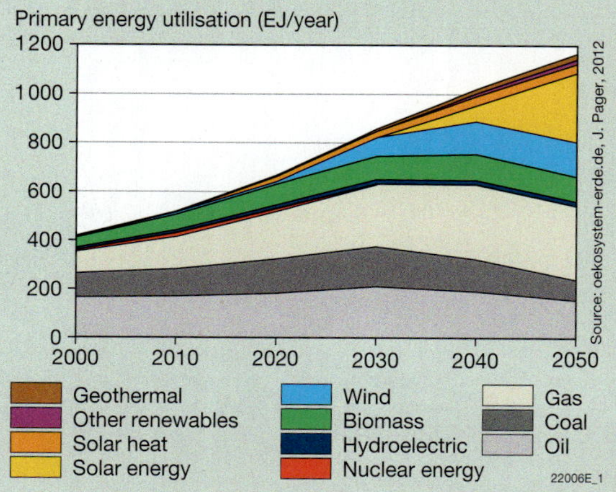

Source: Päger, Jürgen 2012

M2 Scenario for the global energy mix

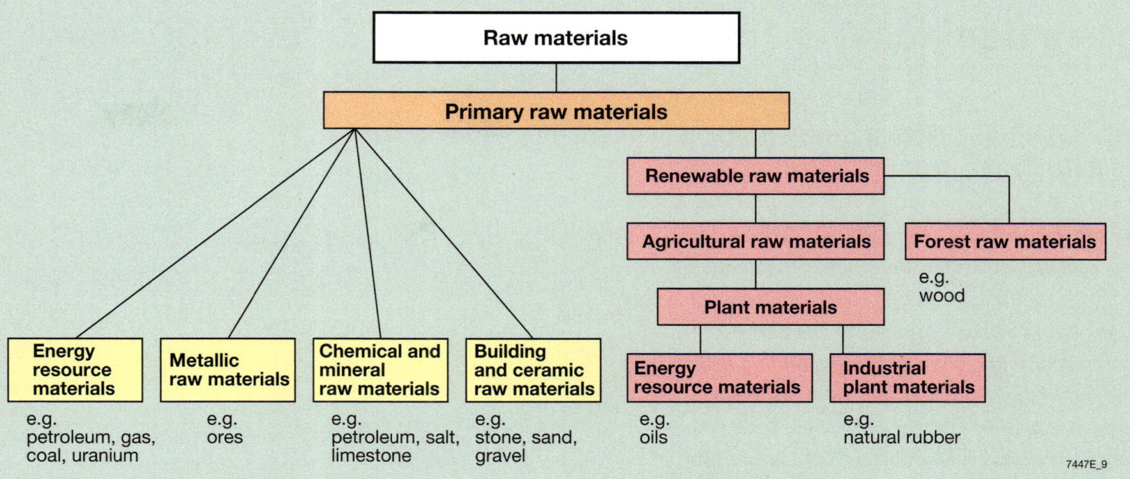

M3 Classification of primary raw materials

Energy consumption in million t oil equivalent	2002	2012
North America	2,741.4	2,725.4
USA	2,295.5	2,208.8
Central and South America	474.9	665.3
Europe (incl. Russia)	2,852.0	2,928.5
EU	1,743.1	1,673.4
Russia	628.2	694.2
Germany	334.0	311.7
Middle East	464.3	761.9
Africa	291.9	403.3
Asia and Oceania	2,773.7	4,992.2
PR China	1,073.8	2,735.2
India	310.8	563.5
Japan	513.3	478.2
World	**9,597.8**	**12,476.6**

Source: BP 2013

M4 World energy consumption by continent and region

M6 Power lines near Bochum

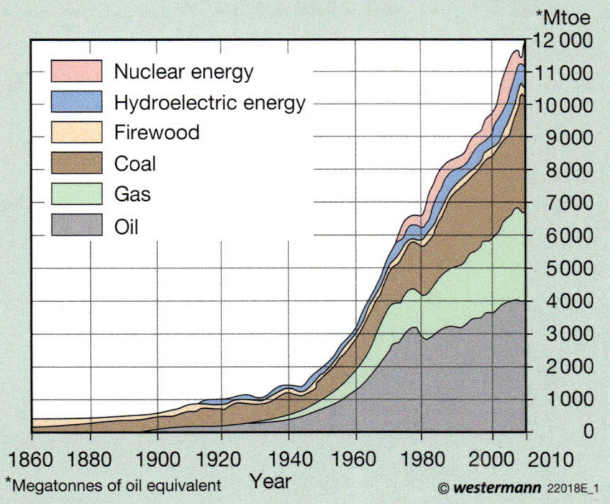

Source: International Energy Agency, 2010

M5 Development of worldwide consumption of energy resource materials

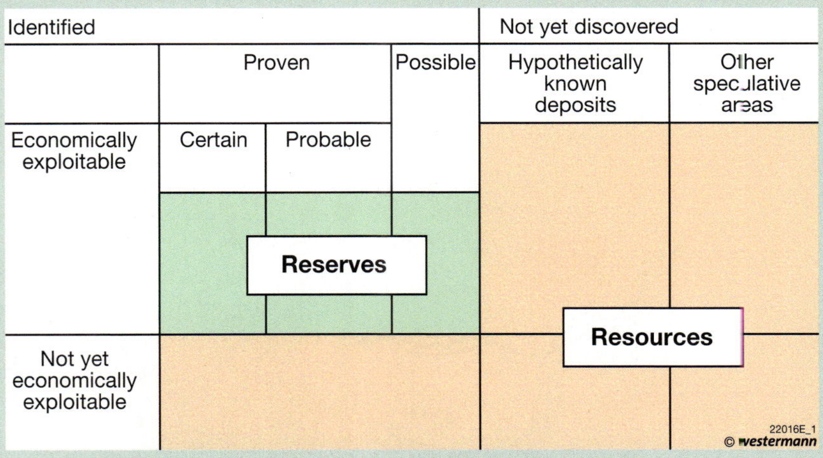

M7 Reserves and resources

1. Describe the development of global energy consumption since 1860 (M4/ M5).
2. Analyse the two scenarios for the development of the individual fuels in the future (M1/ M2/ M5). Formulate hypotheses about the situation of the future global energy economy.
3. Present the regional distribution of worldwide energy consumption. For this, make a simple sketch map (atlas, M4).
4. Explain the difference between resources and reserves (M7).
5. Find information about the energy consumption of your mobile phone and compare this with the energy consumption of other devices in the household.

Hard coal – a traditional energy resource material for over 100 years

Difficult hard coal mining with a problematic legacy in the Ruhr Area

"Deep in the west where the sun gathers dust", sang Herbert Grönemeyer in the 1980s. And even today, coal mining largely determines the image of the Ruhr Area.
Already in the 19th century, the Ruhr Area developed into one of the largest and most important hard coal mining regions in Europe. Mining made a decisive contribution to the rise of the region's economy. In the 1950s, nearly 500,000 people worked in coal mining in the Ruhr Area. Today the shafts and mines have mostly been closed, and problems have arisen that were not reckoned with.

1. Find information about the development of hard coal mining and present the changes in coal mining in Germany (M1 – M3, atlas, internet).
2. "German hard coal is not competitive!" Explain this statement (M3).
3. Ⓞ Explain the so-called northward migration of hard coal mining (M1/ M5) by:
   - **A** giving a precise description of the geological profile.
   - **B** depicting the profile in a simple sketch.
   - **C** preparing a short presentation on this subject.
4. The consequences of hard coal mining are particularly obvious in the Ruhr Area.
   - **a)** Describe the possible consequences of hard coal mining (M4/ M6).
   - **b)** Explain these phenomena (M4/ M6).
5. Ⓢ You are the editor of your school magazine. Draft an article on the subject of: "The legacy of hard coal mining – what does it mean for our future?"

Geological profile, Mine, Mining, Mining damage, Open-cast mine, Subsidence

Coal field	Number of employees	Extraction in mill. t (SKE)
Ruhr	13,795	8.4 million
Saar	2,604	2.0 million
Ibbenbüren	1,214	0.4 million
Share of German hard coal (in %)		
in primary energy consumption in Germany		3
in electricity generation in Germany		6
in hard coal consumption		20
in electricity generation from hard coal		29

Source: Gesamtverband deutscher Steinkohle e.V. 2013

M2 Data on hard coal mining areas in Germany

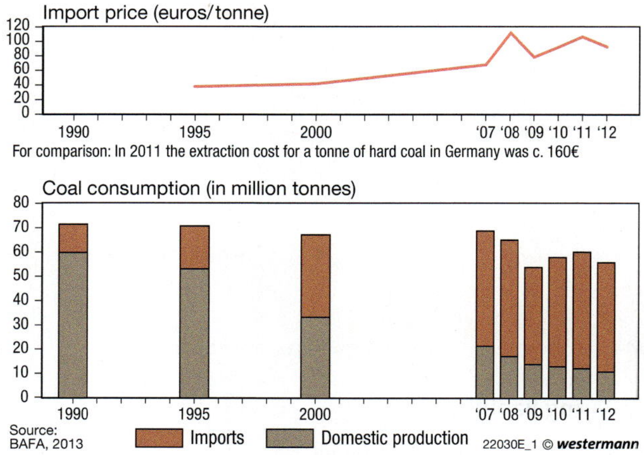

M3 Development of domestic production and imports of hard coal with price trends in Germany

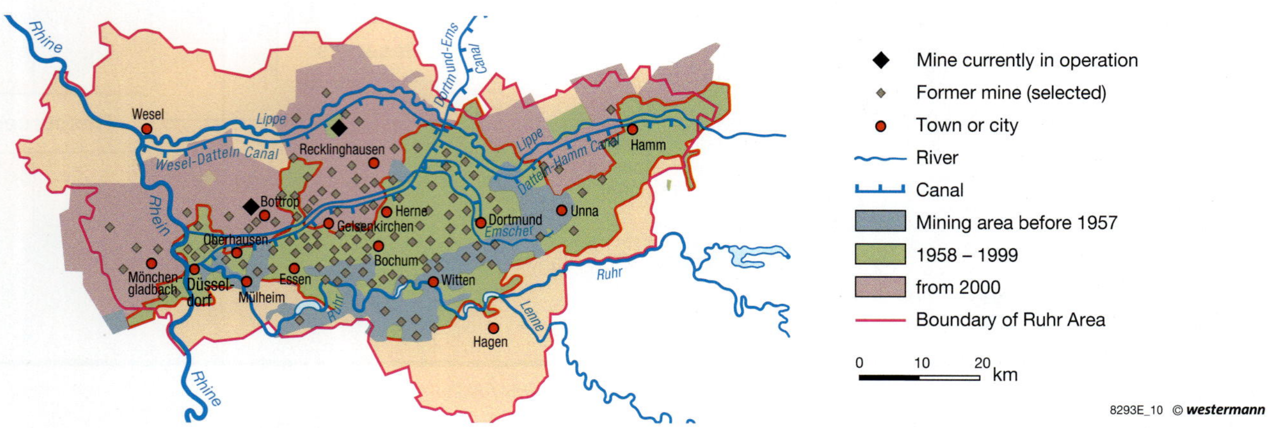

M1 Map of the Ruhr Area showing current and abandoned coal mines

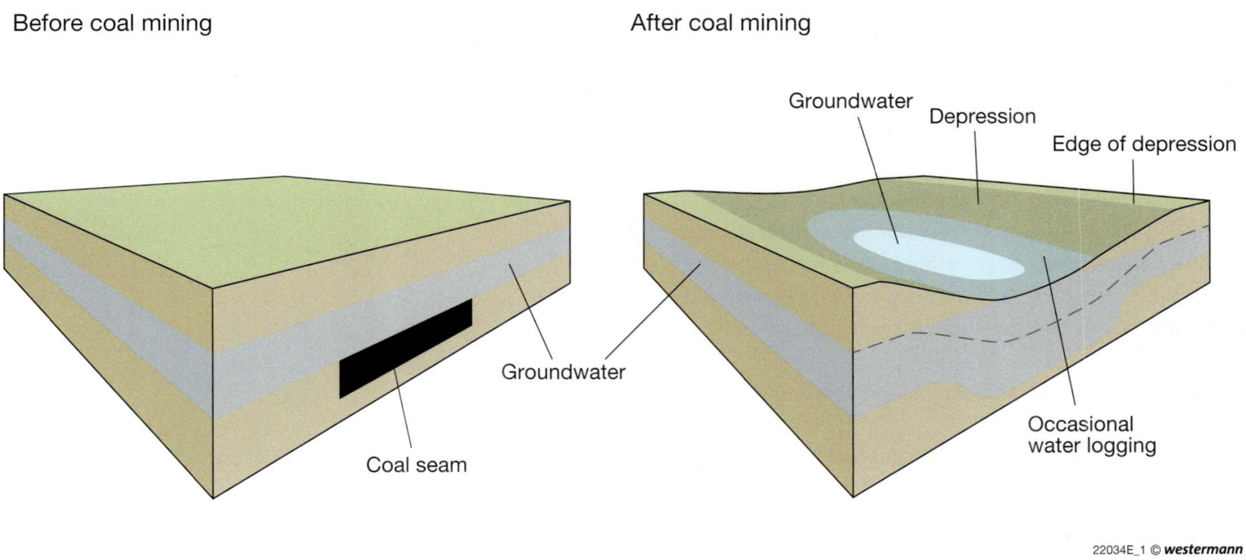

M4 Consequences of coal mining

Mining damage: whole sections of houses gape apart

Shift work in the mine shaft

Ingeborg Steensma grew up with coal mining. It dominated her everyday life and her family; her father was once a blaster in the coal mine. "When father died, it was clear that I would have to help," says Ingeborg Steensma. So she moved into her parents' home [...]. It was only some weeks after the move that she realised what she had let herself in for. There are centimetre-wide cracks in the basements, foundations and walls. Damp penetrates the walls and floor. Windows and doors are distorted and cannot be opened. The front part of the building tilts so much that her mother's rollator walker rolls away of its own accord if the brakes are not applied. Living in the house is a "daily nightmare," says the 64-year-old Steensma.

Based on: F. Dohmen, B. Schmid, www.spiegel.de, 18.02.2013

Enduring liabilities

Responsibility for the cost of damages in both hard and brown coal mining is in the hands of the mining firms such as RAG. According to RAG, the company bears costs amounting to some 300 million euros per year. When it comes to an end in 2018, coal mining will leave behind so-called enduring liabilities (Ewigkeitslasten) in the Ruhr Area. Water must be permanently pumped out – also to prevent water from collecting in the depressions left by mining subsidence and turning the Ruhr Area into a kind of lake district.

Based on: . T. Mader, U. Meinke, P. Wahl, Westfälische Rundschau, 22.11.2013

Mining subsidence – complete closure of A 43 near Witten in the direction of Wuppertal.
(WAZ, 23.01.2014)

Mining subsidence continues to halt trains at Essen central railway station. Galleries to be filled with concrete.
(WR, 22.11.2013)

M6 Newspaper reports

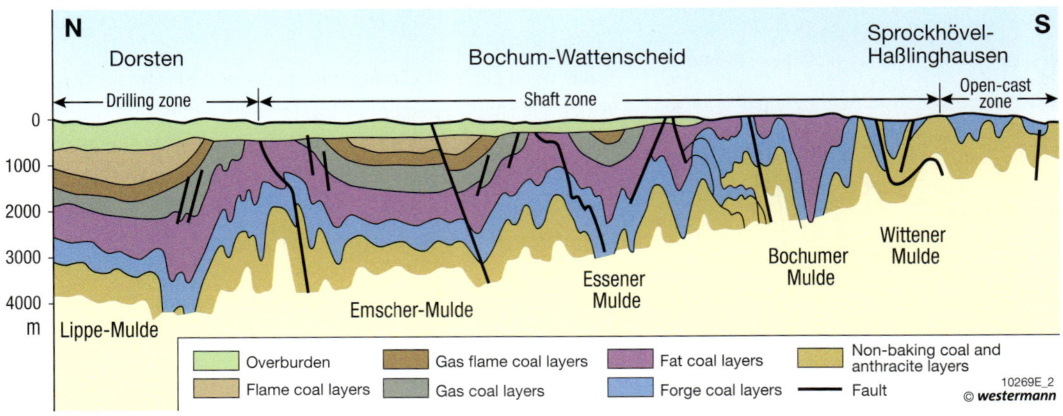

M5 Geological profile of the Ruhr Area from north to south

weblinks
- Federal Office for Economic Affairs and Export Control
- Ruhrkohle AG

Hard coal – a traditional energy resource material for over 100 years

Optimum conditions for hard coal extraction in Australia

Hard coal mining in Australia takes place in completely different conditions than in Germany. The immense hard coal reserves in Australia represent an important development factor for the economic future of the country. But at the same time there are increasingly strong protests from the population against any further extension of open-cast coal mining.

1. Describe the location and extraction of hard coal resources in Australia (M1/ M5/ M7).
2. Explain the development of the hard coal industry in Australia (M2–M4).
3. Ⓦ Characterise the importance of Australian hard coal for the country's economy (M2 – M4, atlas)
   - **A** based on a concept map.
   - **B** based on an entry in a travel guide book.
4. Give your opinion on the hard coal industry in Australia and explain possible future scenarios (M3–M8).
5. Ⓩ A high proportion of the hard coal that is used in power stations in Germany comes from Australia. Explain.

Land grabbing, Open-cast mine, Reserve

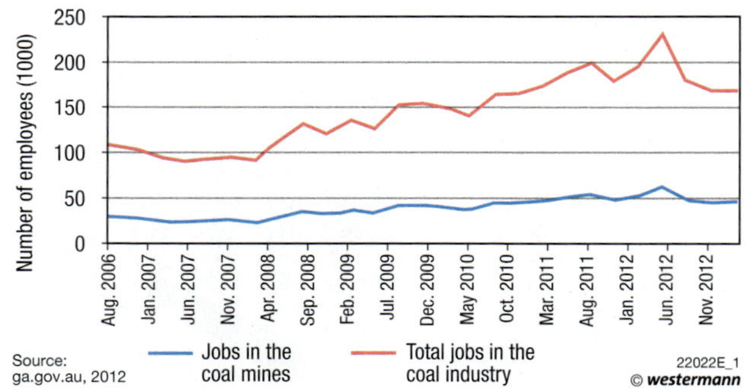

Source: www.ga.gov.au, 2012

M3 Development of employment in the Australian coal industry

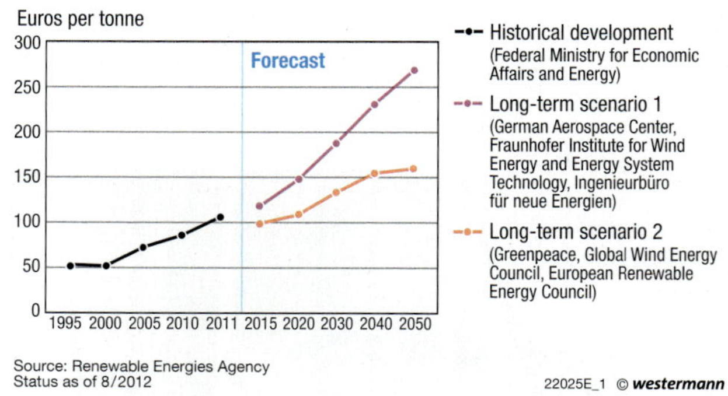

M4 Trend of coal prices on the world market

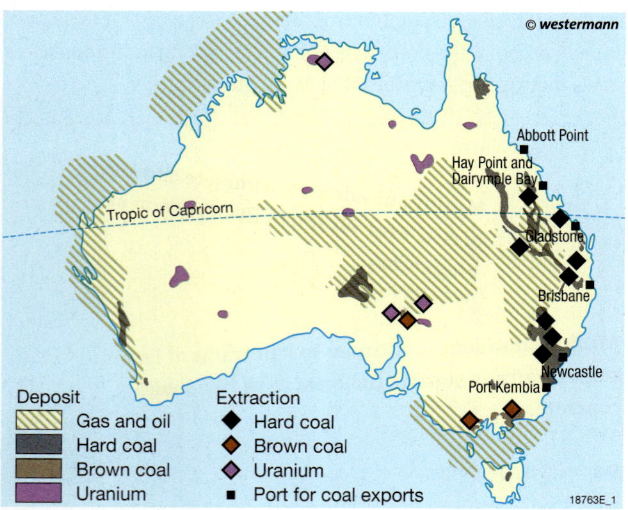

M1 Australia's energy resource materials

Michael Camden lives and works in Portland, a two-and-a-half hour drive from Sydney. Coal mining has left its mark on the landscape as far as the eye can see. But the long administration building, the recreation rooms and showers for the nearly two hundred workers at Michael's mine look tiny against the gigantic conveyor belts that operate all day and night. "Although raw material prices are not so great at the moment, our open-cast mining operation is generally in good shape," says Michael. And for his family there is much more money left over at the end of the month than in the past when, as an excavator operator with 700 dollars, he only earned half of his current wage.

The huge mining fields, which stretch to the horizon and which have been abandoned to nature after the cessation of coal extraction, are not particularly beautiful, Michael admits; he prefers to draw attention to the large number of jobs that they created.

Based on: M. Blümel, www.dradio.de, 29.03.2010

M5 A mine worker's viewpoint

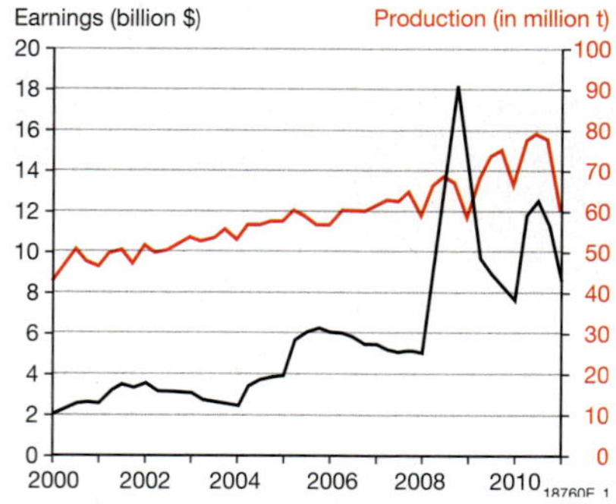

M2 Development of coal production and earnings in Australia

Today, most electricity in Australia is generated from fossil fuels. In 2011, more than 75 percent of total energy production was generated from coal and only 8.7 percent from renewable energy sources. The Australian government has therefore passed the "Renewable Energy Target" law, which is intended to ensure that by 2020 at least 20 percent of electricity generation will be covered by renewable resources.

Based on: Australian Government (publ.): Energy in Australia. 2012

M6 Changes in Australia's energy sector

M7 The coal terminal of Newcastle, New South Wales

Australia – a vast mining pit

[...] Ian Moore, a powerfully-built man with soft features, pulls his baseball cap down low over his face. He is the farmer here; his cattle are lazing in the shade of the eucalyptus. Behind them, hidden in the haze, are the vineyards. And the monstrous, grey-brown crater of the coal mine. In a few decades, the traditional Australian wine region of Hunter Valley, some 200 kilometres west of Sydney, has been turned into Australia's largest coal mining area. Untiringly, machines rip up the earth to meet China's hunger for raw materials. It has been like this for as long as the 63-year-old farmer can remember. The coal industry was already there when his grandfather was clattering through the idyllic meadows with his horse and cart. But in the last ten years, says Moore, the excavators have gone mad; working faster and faster, more and more is exported, and more and more often the coal industry digs up fertile farmland. Freight trains with up to 80 wagons thunder past. Now, gigantic cutters have also started to excavate coal underground; cracks have appeared in the highway and whole tracts of land with grape vines, homes and cattle meadows have suffered metres-deep subsidence. "The industry has gone completely crazy," says Moore with helpless rage in his voice. The farmer is not alone. Throughout Australia, mining now faces massive protests. Citizens from all parts of society are resisting the uncontrolled expansion of the coal and natural gas industry through which the country is ascending to the position of a global energy superpower. [...]

Since 1980, 42.5 percent of the farmland in Hunter Valley has fallen victim to coal mining, the national statistical office confirms. And now it is Ian Moore's turn. [...] "I am being treated like a second-class citizen," says Moore, his eyes staring dully into space. [...] "They want to destroy my livelihood! How can that be right?" The men from Nucoal Resources came last winter and explained their plans. Five valuable coal seams are located under the property, and they want to extract four of them. They want to dig underground galleries directly beneath the farmhouse and through to the neighbour who has already sold his fields to the mining people.

Moore balks at this. Against test drilling. Against selling up. He is afraid that his land too will suffer metre-deep subsidence and his house will collapse. But above all he is afraid that his precious groundwater will run dry. "Without water, my land would not be worth two dollars," he says. [...]

Over in the town, where one of Ian Moore's sons works as a boilermaker in the mining industry, the coal dust settles everywhere and makes the children cough chronically. Entry-level wages in the mining industry are 120,000 dollars per year. For many, the money quickly went to their heads, the farmer relates. [...]

Emotions have been further heated by the discovery by the Australia Institute, a left-oriented research group, that 83 percent of profits in the raw materials sector go into the pockets of overseas investors. Many speak openly of land-grabbing, of land robbery with the support of the politicians. Because according to the law, private individuals only own the upper layer of soil on their property. All the mineral resources beneath it belong to the Australian people. Anyone who denies access to a mining company with a valid mining licence is breaking the law. But many take the risk nevertheless. Everywhere in the country they hang yellow warning triangles on their gates with the fighting slogan "Lock the Gate", and refuse all cooperation with the giant mining concerns.

Based on: V. Eprothen, Zeit Online, 08.01.2012

M8 Newspaper report

Brown coal – a domestic energy resource

Opportunities and limits of brown coal mining – the Rhineland brown coal field

In the opinion of many experts, the reserves and resources of brown coal can continue to make an important contribution to global energy provision for many years. Germany is one of the largest producing countries, and brown coal also plays an important role in the country's energy industry. Following reunification, two large mining areas were added: the Central German and Lusatian coal fields. But today the Rhineland brown coal field still represents the country's most important deposit.

1. Compare the three largest brown coal mining regions in Germany (atlas, M2/ M3).
2. Describe electricity generation in Germany (atlas).
3. Characterise the situation of the Rhineland brown coal field and describe the development of the field (M8, atlas).
4. (O) Present the development of the individual open-cast brown coal mines in the Rhineland coal field (M7), by:
   - **A** putting the individual open-cast mines into a table.
   - **B** describing the development of selected open-cast mines in a text.
   - **C** entering the current volume of production in selected open-cast mines in a simple sketch map (atlas).
5. Evaluate the efficiency of brown coal as a fuel (M6/ M10).
6. (S) Draft an article for a 5th grade school textbook on the subject: "Brown coal – an important fuel in Germany".

Brown coal, CO_2 emissions, Open-cast mining, Power efficiency

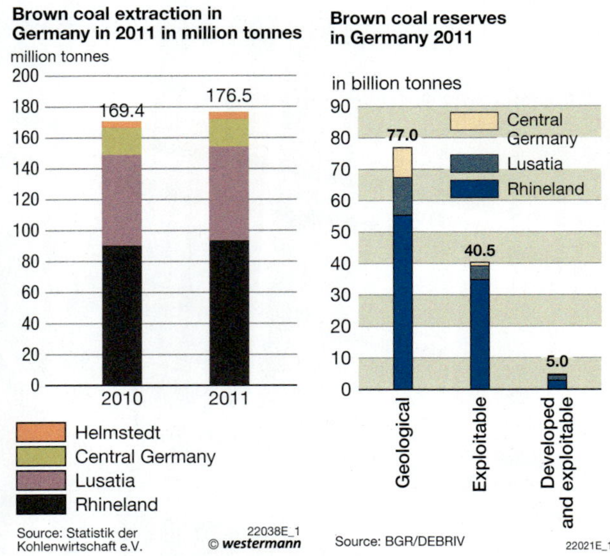

M2 Brown coal production and reserves

Coal field	Unit	Land use (total)	Operating areas	Renatured areas
Rhineland	ha	31,514	9,266	22,248
	percent	100	29.4	70.6
Lusatia	ha	86,135	31,455	54,680
	percent	100	36.5	63.5
Central Germany	ha	48,720	11,011	37,710
	percent	100	22.6	77.4

M3 Operating areas and areas returned to other uses

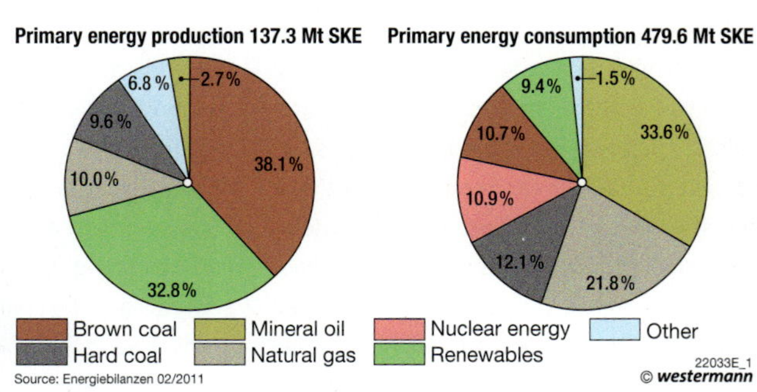

M4 Overview of Germany's energy industry, 2012

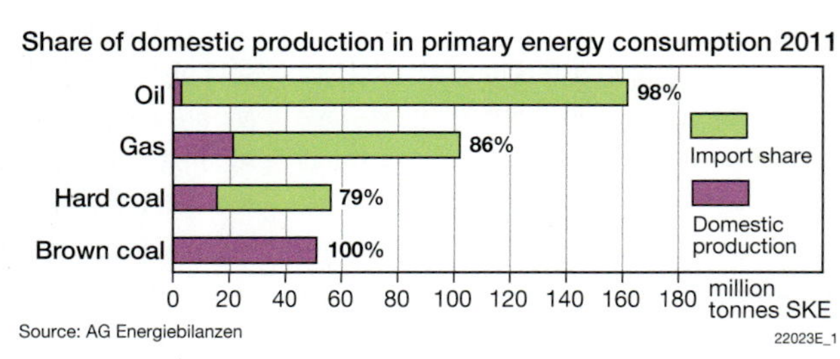

M5 Domestic production of primary energy sources

M1 Bucket wheel excavator in an open-cast brown coal mine

Brown coal: solutions for CO_2 emissions and efficiency

The German power station operators must use domestic fuels more efficiently. The greatest challenge they face is probably to find a solution for the greenhouse gas CO_2. On combustion, brown coal sometimes releases significantly greater quantities of this than the other fossil fuels. Moreover, progress in the efficiency of the power stations will remain limited as long as they are unable to make better use of waste heat. [...]
Technology in large brown coal power stations has improved significantly in recent years. RWE's brown coal power station at Neurath, and Vattenfall's new plant at Boxberg, can already achieve an electrical power efficiency of 43 percent. By comparison, the old brown coal plants that RWE will shut down when Neurath has been commissioned only use the fuel to 30 percent efficiency. The operators can further increase electrical power efficiency by drying the damp raw brown coal before burning it. Since October 2008, Vattenfall has been testing pressurised steam fluidised-bed drying at an experimental facility at Schwarze Pumpe. This uses steam from the power station process.
The technicians believe that a net electrical power efficiency of 50 percent would be achievable in brown coal power stations if coal drying were applied completely. This would then be comparable with hard coal power stations.
Based on: Stefan Schroeter, Ingenieur.de, 2012

M6 Newspaper report

Year	Production in mill. t
1970	369.0
1980	387.9
1990	356.5
2000	167.7
2007	180.4
2008	175.3
2009	169.9
2010	169.4
2011	176.5
2012	185.4

M9 Development of brown coal production in Germany

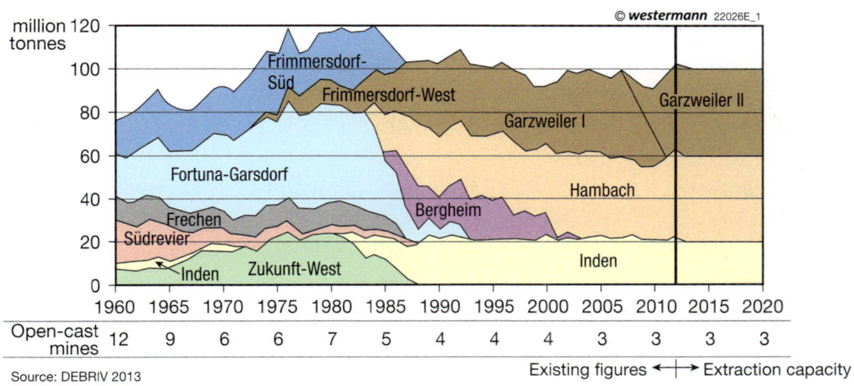

M7 Brown coal extraction in the Rhineland brown coal field by individual open-cast mines

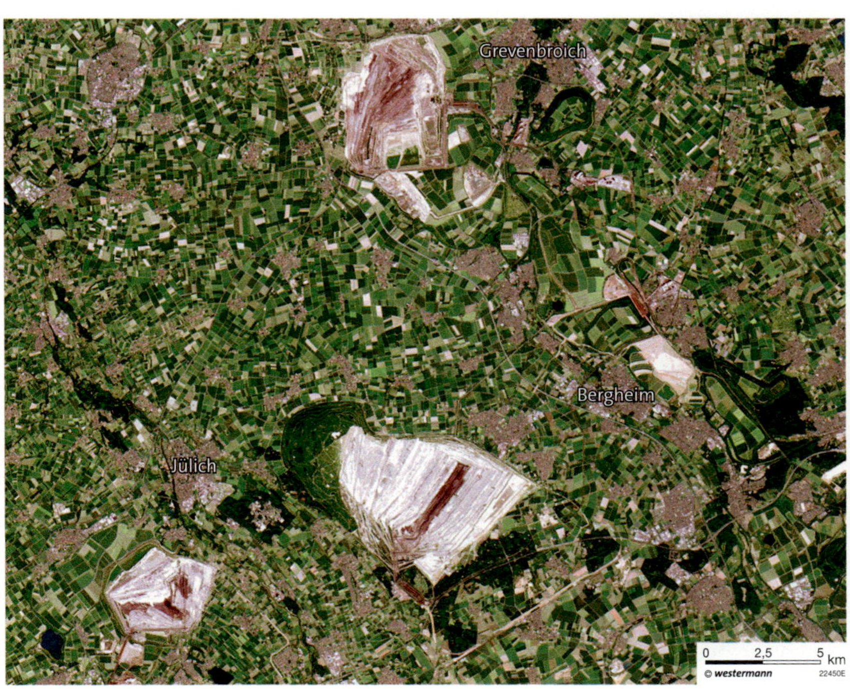

M8 Location of the Rhineland brown coal field

	Colour and lustre	Calorific value in k-calories/ k-Joules per kg
Peat	Brown, dull	1500 - 1000 6300 - 8400
Soft brown coal	Brown, dull	1800 - 3000 7500 - 12600
Hard brown coal	Brown to black, matt lustre	4000 - 7000 16800 - 29400
Flame to fat coal	Black, greasy lustre	7000 - 8000 29400 - 33600
Forge to non-baking coal	Black, greasy lustre	8000 - 8500 33600 - 35700
Anthracite	Black, brilliant lustre	8500 - 9000 35700 - 37800
Graphite	Black, semi-metallic lustre	Practically inflammable

10268E_2

M10 Coal types by energy content

Brown coal – a domestic energy resource

The Garzweiler II open-cast brown coal mine – between stimulus for growth and social and environmental resistance

The Garzweiler II open-cast mine in the Rhineland brown coal field is a frequent subject of debate between different interest groups. Questions about the future of Garzweiler II are many-sided and not easy to answer. Depending on the perspective, different answers are given concerning the future of open-cast mining.

1. Describe the appearance of open-cast brown coal mining and formulate the arguments of different interest groups for and against brown coal mining (M6).
2. Locate the Garzweiler II open-cast mine and explain possible positive and negative aspects of brown coal mining (M1/ M3, atlas).
3. Explain the German Mining Law in your own words and discuss the question of whether mining law should be given priority over individual rights (M2).
4. ⓞ Explain the terms renaturation and recultivation with reference to Garzweiler II (→ definitions, atlas).
   - **A** Produce a table with concrete measures.
   - **B** Make a simple sketch map, in which only recultivation measures are shown.
   - **C** Write a text on recultivation in the Rhineland brown coal field.
5. Work out arguments for key positions on open-cast brown coal mining in Garzweiler II (M5/ M4, internet) and carry out a debate for and against (e.g. in a role-play game).
6. Ⓢ You are a reporter for a regional daily newspaper. Draft a neutral article about the Garzweiler II open-cast brown coal mine.

CO_2, Fine dust, Mining law, Resettlement, Power efficiency, Spoil
→ Recultivation, Renaturation

German mining law originated as a customary law in the Middle Ages. Since 1982, the Federal Mining Law (BbergG) has been in force in Germany. The law is based on the principle of mining rights. This means that property rights can be withdrawn from all mineral deposits subject to mining rights as referred to in the law (metals, petroleum, natural gas, coal, salt, geothermal heat, etc.), so that the property owner only has rights to the mineral deposits on his own ground (sand, gravel, gypsum, clay, etc.).
But the ownership can be acquired by interested parties if a state-controlled award procedure has been gone through. In particular, this procedure clarifies whether the mining of the mineral deposit represents an overriding benefit for society.
If this is determined, the compensation of the affected property owner must be regulated in a further process. This can lead to considerable conflicts of interest, especially when it is associated with resettlement measures. The mining law also governs the way in which recultivation or renaturation must take place after mining has been completed.

M2 Germany's Mining Law

→ Renaturation

General "restoration" of landscapes or parts of landscapes, such as streams or woodland, to a natural state allowing the possibility of further natural growth.

→ Recultivation

This covers land management, water management, geotechnical, agricultural and forestry environmental measures to restore agricultural ecosystems that have been damaged or destroyed by human economic and technological activities and re-create the original or a newly-formed cultivated landscape.

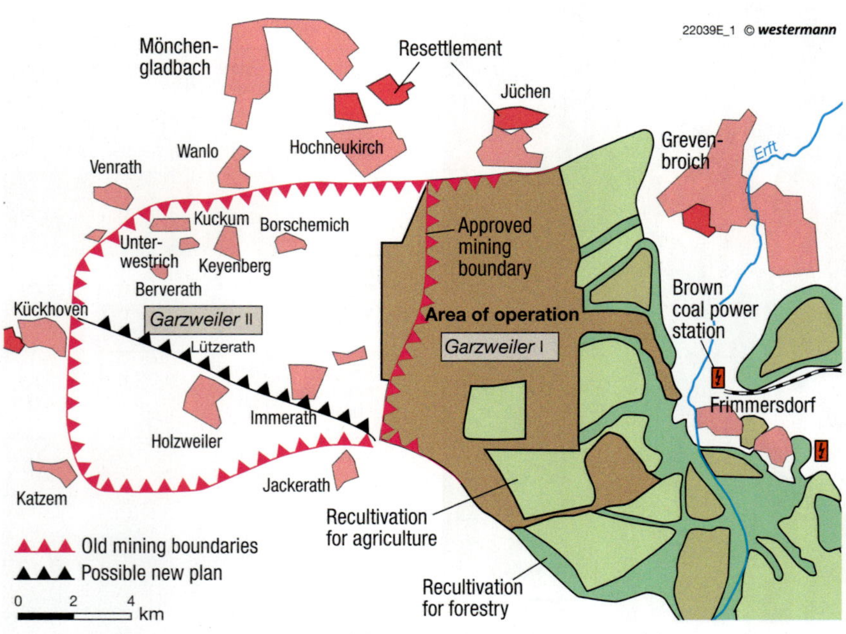

Location: Jüchen municipality, Rhein-Kreis Neuss
Size of mining area: 114 km²
Ratio of spoil to coal: c. 4 : 1
Number of employees: 1,725
Number of trainees: 125
Population of Jüchen: 22,379
Population of Rhein-Kreis Neuss: 439,225
Area of district: 576.52 km²

M1 Garzweiler II open-cast mine with brown coal power stations and data on open-cast mining and the region (2012)

Brown coal power station	Year of first commissioning	Capacity (in MW)	Power efficiency (in %)	CO_2 emissions (in 1,000 t) 2011[4]	Fine dust (in 1,000 t) 2011
Frimmersdorf	1959[1]	635	ca. 30	15,200	253
Neurath	1972[2]	4,400	30–43	19,600	283
Niederaußem	1965[3]	3,669	30–43	28,600	148

M3 Brown coal power stations south of Garzweiler II

1 In Frimmersdorf, many old power station blocks have already been abandoned and the last one will be abandoned in 2018.
2 The power station comprises a total of seven blocks, of which the two newest blocks (commissioned in 2012) produce half of the output and have a higher power efficiency of 43%.
3 With Block K, what was then the most modern brown coal power station block in the world was commissioned in 2002. The construction of an additional block at the Niederaußem site is currently under examination.
4 For comparison, today a modern small car produces c. 100 g of CO_2 per km.

Artificial landscapes instead of nature. In the opinion of Friends of the Earth, the severe intervention in nature and the landscape cannot be compensated. [...] It leads inevitably to a loss of natural soil diversity. [...] The loss of woodland also cannot be compensated.
Source: Friends of the Earth Germany (BUND), 2012

The recultivation in Rhineland is regarded throughout the world as exemplary, not only by specialists [...] The creation of expanses of water is one of the requirements of recultivation [...]. In many places there are more species of bird life than before the introduction of mining.
Source: RWE, 2012

M4 Opinions on recultivation in the Rhineland brown coal field

M6 Renatured areas on the fringes of the Hambach open-cast mine

Friends of the Earth Germany (BUND):

"Garzweiler II represents an irresponsible intervention in nature, the environment and social structures. RWE must no longer be allowed to hold those affected by open-cast mining as hostages for its retrograde corporate policy. The plans for brown coal must be quickly amended so that the 3,300 people affected are spared the planned forced resettlement. The Federal state government in North Rhine-Westphalia must also devise an exit scenario for brown coal that is in line with the provisions of the Federal state's climate protection law. If open-cast mining in NRW were to be continued until 2045 as currently intended, it would mean some 1.2 billion more tonnes of CO_2 for the atmosphere."
Based on: www.bund.net

Ilse Meyer

Ilse Meyer, now aged 85, has lived for decades in Immerath in the Erkelenz district. "We were independent, completely autonomous," she says. "We had a hospital, a baker, a butcher, a drug store. We didn't need to go anywhere. And when you went shopping you would always meet someone. I have lost my home place." Today, Mrs Meyer lives 10 km away in Neu-Immerath. "That is not a village, just a horrible housing estate," the widow says in despair. In particular the demolition of the church in Immerath, her church, saddens Mrs Meyer. "What has been done to the people here is a great injustice. We have even been forced to move the remains of our dead."
Interview with a former inhabitant of Immerath

RWE

RWE: Every fourth kilowatt-hour of electricity in Germany is generated from brown coal. This fuel has a share of some 25 percent in total German electricity generation. The Garzweiler open-cast mine currently stretches west from Grevenbroich to the southern part of the urban area of Erkelenz. Here, some 1.3 billion tonnes of brown coal are deposited at a depth of 210 m. Targeted measures reduce dust and noise from the open-cast mine and thus minimise the nuisance for the people living in the neighbourhood of the mine. The greatest importance is attached to the recultivation of exhausted open-cast mining areas. After the cessation of mining in 35 years, the Garzweiler mining area will largely be recultivated for agriculture. But its leisure value for people seeking recreation will also not be neglected.
Based on: RWE, 2013

NRW Federal State government

North Rhine-Westphalia possesses rich deposits of energy resources and mineral raw materials. For generations, mining has developed into one of the most important sectors of industry in our Federal state. [...] In North Rhine-Westphalia, coal mining is of special importance. [...] The brown coal open-cast mines in the Rhineland coal field have [...] a long-term perspective on account of their great importance for the energy industry. Even the wind is sometimes still, and the sun is hidden behind the Earth for many hours each day. This means that, until our engineers have developed large-scale storage, we will need gas- and coal-fired power stations. [...] This is the only way to ensure a reliable supply of electricity at an affordable price.
Based on: www.wirtschaft.nrw.de

M5 Four positions on brown coal in NRW and on Garzweiler

Petroleum – a guarantee of economic development?

The Middle East

The tallest buildings in the world, ski slopes and the FIFA football World Cup, the largest airports on the planet, artificial islands, magnificent palaces and Formula 1 racing circuits, immeasurable luxury and fabulous wealth – and all of this in the middle of the desert in the Middle East. Petroleum catapulted an entire region out of the Middle Ages and into the present. Until now, petroleum has been the guarantee for the region's economic development. But now the Gulf states are increasingly also using their oil income to equip themselves for the time after the oil runs out.
Petroleum oils the wheels of the global economy. Both the industrialised countries and the energy-hungry newly industrialising countries depend on it. They have a great interest in political stability in the Gulf region in order to secure their own oil supplies. The international community defends its supposed security by political and military means. Critics say that, in doing so, a blind eye is often turned to undemocratic and socially questionable conditions in the petroleum-producing countries.

1. Describe the regional and global importance of the petroleum resources in the Middle East (M4/ M5, atlas).

ⓞ 2. **A** Present the wealth that the Gulf states have earned through petroleum extraction (photos, atlas, internet).
   B The Gulf states are a classic example of how countries can develop through the extraction of mineral resources. Discuss this statement.

3. Using examples, explain the widely-spread investments by the Gulf states (photos, atlas).
4. Although the economies of the Gulf states are flourishing, there is growing international criticism of the economic and social conditions in these countries. Explain (M8/ M12).

ⓞ 5. **A** The Gulf region is often described as a "powder keg". Explain the international military presence in the region (M5/ M7/ M8).
   B Explain the measures that the big industrialised countries have taken to secure their access to the oil in the Gulf region (M7/ M8/ M11).

Ⓢ 6. Present the state of development of the Gulf states in comparison to their neighbouring countries (atlas).

Gulf region, Middle East, → Carter Doctrine, OPEC

→ OPEC

OPEC (the Organisation of Petroleum Exporting Countries) was founded in 1960 with the objective of protecting its member countries against a price collapse and reductions in their earnings from petroleum extraction. Its headquarters is in Vienna.

M3 The rapid development of a region: Dubai in 1970 and today

	GNP* 1978	GNP* 2012	GDP per capita** 1978	GDP per capita** 2012
Saudi Arabia	24.96	657	3,010	22,823
UAE	22	392.3	n.a.	45,880
Qatar	1.38	178.5 (2011)	8,320	97,967 (2011)
Bahrain	0.58	30.6	2,740	24,669
Kuwait	9.39	170.1	11,510	31,626

* in billion US dollars; ** in US dollars

Sources: auswaertiges-amt.de, Fischer Weltalmanach 1978, tradingeconomics.com

M4 Economic data for selected Gulf states

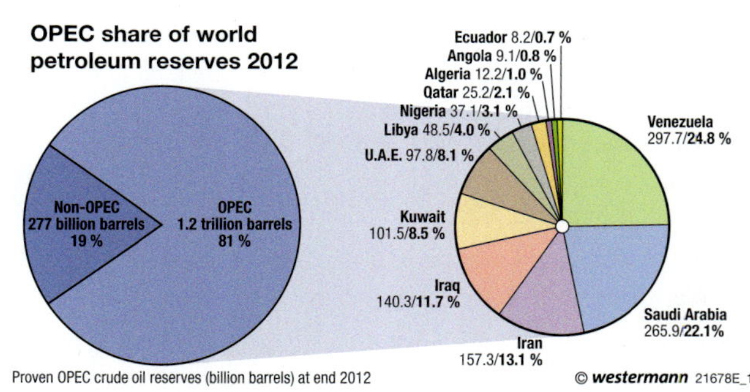

Source: OPEC Annual Statistical Bulletin 2013

M5 OPEC's share of world oil reserves, 2012

M1 Formula 1 racing circuit in Bahrain

M2 Royal palace in Riyadh

M6 Airbus A380 of Emirates Airline

◁ **M7** Conflict structures and tanker routes in the Gulf region

M11 The USS Abraham Lincoln is frequently stationed in the Persian Gulf

→ Carter Doctrine

Former US President Jimmy Carter expounded the so-called Carter Doctrine – which is still valid today – in 1980. It clearly defines the position of the USA: anybody who attempts to seize control of the Persian Gulf is attacking the "vital interests" of the USA and will be resisted by all possible means.

Tank deals with Saudi Arabia and Qatar

"According to a newspaper report, further contracts worth millions from the Gulf states of Saudi Arabia and Qatar are beckoning for the German armaments industry. In the next seven years, the Gulf emirate of Qatar [plans to] order a further 118 Leopard tanks as well as 16 Type 2000 self-propelled howitzers. The two weapon systems are manufactured by the German firms Krauss-Maffei Wegmann and Rheinmetall. Krauss-Maffei Wegmann recently confirmed that Qatar had initially ordered 62 Leopard tanks and 24 howitzers for a total of 1.89 billion euros. [...] According to the report, Saudi Arabia's interest in the purchase of 800 Leopard tanks has also become more concrete than was previously known. The acquisition of combat tanks is reported to be included in a list of planned armaments projects personally confirmed by King Abdullah. [...] On Friday however, the Handelsblatt newspaper reported that the sale of 270 Leopard tanks to Saudi Arabia is in danger of breaking down. [...] The export of armaments to Saudi Arabia and Qatar is deeply controversial in Germany. In the spring of 2011, both of these authoritarian countries participated with tanks and soldiers in the violent suppression of protests in Bahrain. In addition, Riyadh often responds violently to protests in Saudi Arabia itself. Critics also point out that the human rights situation in Saudi Arabia is highly problematic. [...]"

Source: fbo, AFP, www.n24.de, 14.07.2013

M8 News report

Forced labourers sweat for the football World Cup

"[...] In recent weeks [...] many migrants from Nepal have lost their lives while working in Qatar. The death rate [...] has now risen to one per day. Working conditions – measured according to the standards of the International Labour Organisation ILO – are akin to slavery. [...] Qatar will host the football World Cup in 2022.

According to the Guardian, the research showed that forced labourers are also employed in construction projects for the World Cup infrastructure.

Some workers have been denied their wages for months. Employers systematically confiscate the workers' passports and refuse to hand them out again.

This turns the workers into illegal immigrants. Workers reported to the Guardian that they are not allowed to drink water during their work in the great desert heat. In the sleeping quarters, twelve workers must share one room. [...]

It would appear that one of the richest nations in the world is exploiting one of the poorest in order to make itself fit for the World Cup. In Qatar, [...] more than 90 percent of the workers are migrants. The country is currently seeking 1.5 million more workers for the construction of stadiums, ports and hotels."

Source: T. Steffen, www.zeit.de, 20.09.2013

M12 Newspaper article

M9 The luxury Burj al Arab hotel

M10 Burj al Kalifa – the tallest structure in the world

M13 FIFA World Cup 2022 in Qatar – investments in the future in the Gulf region?

Petroleum – a guarantee of economic development?

Greed for oil knows no bounds – the case of Deepwater Horizon

The problem of the worldwide hunger for energy and the rapid increase in the extraction of fossil fuels is always drawn to our attention when men and machines fail and a disaster occurs. The case of the Deepwater Horizon oil platform shocked the world. It led to a renewed discussion about the extraction of energy resource materials from extreme deposits, such as in the deep ocean, and our dependence on petroleum for oiling the wheels of the global economy.

1. Make assumptions about what our everyday life would be like without petroleum (internet).
2. Locate the position of the Deepwater Horizon oil platform and describe the disaster (M1–M4).
3. Explain the economic and ecological consequences of the disaster and formulate possible questions about it (M6/ M7).
4. ⓞ Explain the extraction of petroleum from offshore deposits under the sea (M5),
   - **A** by preparing a report for a radio programme (duration: approx. 90 seconds).
   - **B** by making a simple sketch map.
5. "Stop petroleum extraction in the deep ocean!" This demand is made by many environmentalists. Evaluate it using the method for forming a judgement in geography (M5 – M9, internet).
6. Ⓢ Using the appropriate maps in the atlas, explain Germany's and Japan's dependence on petroleum.

Deep sea mining, Peak Oil, Petroleum

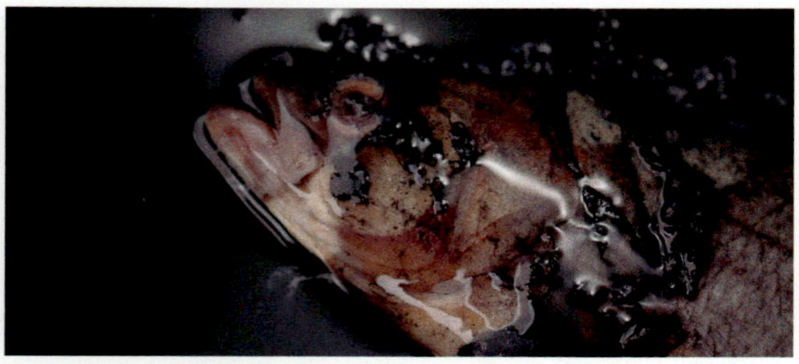

M1 Oil pollution – a disaster for flora and fauna

M2 The location of Deepwater Horizon and the spread of oil pollution in the Gulf of Mexico

"The first thing we noticed was that the machines in the winding tower suddenly ran hot." [...] Chief electrician Mike Williams reported in an interview how he experienced the accident. 114 of his colleagues were able to [...] reach safety; eleven died. [...] It was 20 April 2010. At 21:45 hrs the Deepwater Horizon oil platform, located some 80 km off the US coast, was shaken by a massive explosion. For two days, desperate attempts were made to put out the fires on the burning platform.
But the Deepwater Horizon sank, severing its connection to the borehole at a depth of 1,500 m. Oil flowed unhindered into the sea: an estimated 780 million litres – more than ten times as much as in the worst tanker accident in history when the Exxon Valdez hit a reef in 1989.

Source: D. Mc Connell, CNN, 2012

M3 Explosion on the Deepwater Horizon

M4 Deepwater Horizon after the explosion

One speaks of deep sea mining as a rule when the gap between the surface and floor of the ocean is greater than 1,000 m. Petroleum or natural gas can now be drilled at a depth of up to 3,000 m. The deeper the drilling, the harder it is to break through the rock, and the more the diameter of the drill must be reduced. The deposits are generally reached with a drill diameter of under 20 cm. The oil flows out through this opening.
Drilling may be in a horizontal direction. Deep sea drilling is very interesting for some countries because vast oil deposits are believed to lie there. Thus, for example, Brazil hopes to double its petroleum production by exploiting a gigantic offshore deposit – a huge opportunity for the country's economic development. But the Brazilians' hopes lie at a depth of 7,000 m beneath a kilometre-thick layer of rock and salt.

M5 Deep sea mining: top technical performance with high profits

According to calculations by the rating agency Moody's, the disaster has already cost BP 37.2 billion dollars. BP has set aside a further 14 billion dollars for civil actions.
In the worst case, the costs could rise to 90 billion dollars, a sum that could overtax even BP.

Source: M. Koch, Süddeutsche Zeitung, 26.02.2013

M6 BP fights for its future

According to its own account, the British company BP is now becoming active again in the Gulf of Mexico. The first test drillings started on 3 August in the so-called Tiber Oil Field, the company announced. The area is believed to hold gigantic deposits which could amount to up to three billion barrels of oil.
BP originally wanted to start test drilling in 2010, but was delayed by the sinking of the Deepwater Horizon.
At the same time, in the USA, BP is contesting the size of its compensation payments for the disaster. For the third quarter, a judge recently approved a budget of 130 million dollars in spite of protests from BP. In principle the compensation payments have no upper limit. Fearing costs of up to 21 billion dollars, BP has already sold facilities valued at five billion dollars.
Source: Reuters, 13.09.2013

M7 Protests against BP's plans

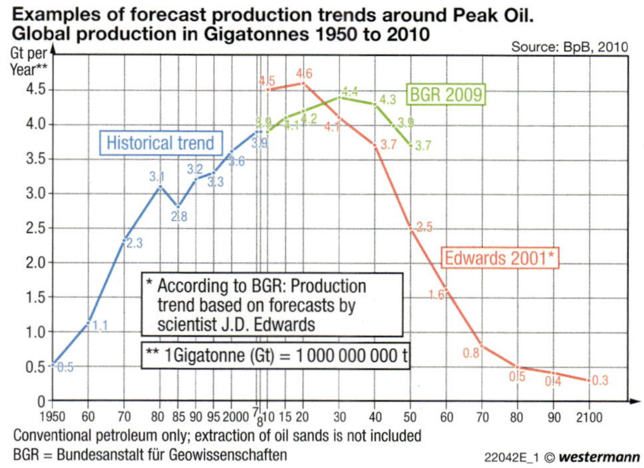

M8 Peak Oil (Fed. Institute for Geosciences and Natural Resources)

METHOD Forming a judgement in geography

In our increasingly complex world, the development of evaluation skills is of paramount importance for the ability to play an active part in social processes. You have become skilled at forming a judgement in geography when you possess the ability to evaluate geographical structures and processes in terms of the current and future realities of life. A differentiated evaluation must be based on well-founded factual information. It is only on the basis of facts that you can be in a position to form a true value judgement. In this chapter, many questions have already been raised that require judgements of this kind. The following steps, or the diagram shown here, can help you to reach such a judgement.

Six steps to developing evaluation skills in geography

STEP 1

→ Determine the problem requiring a judgement.

STEP 2

→ Describe the situation.
What are the facts of the situation?

STEP 3

→ Distinguish between different (possible) perspectives.
Who could want what and why?

STEP 4

→ Examine the consequences.
What results would different judgements have?

STEP 5

→ Analyse the associated norms and values (On what basis can a judgement be made? Use the triangle of sustainability, for example).

STEP 6

→ Form a reasoned judgement (perhaps with qualifications) by weighing up the advantages and disadvantages of the decision. What reasons can I give for my position?

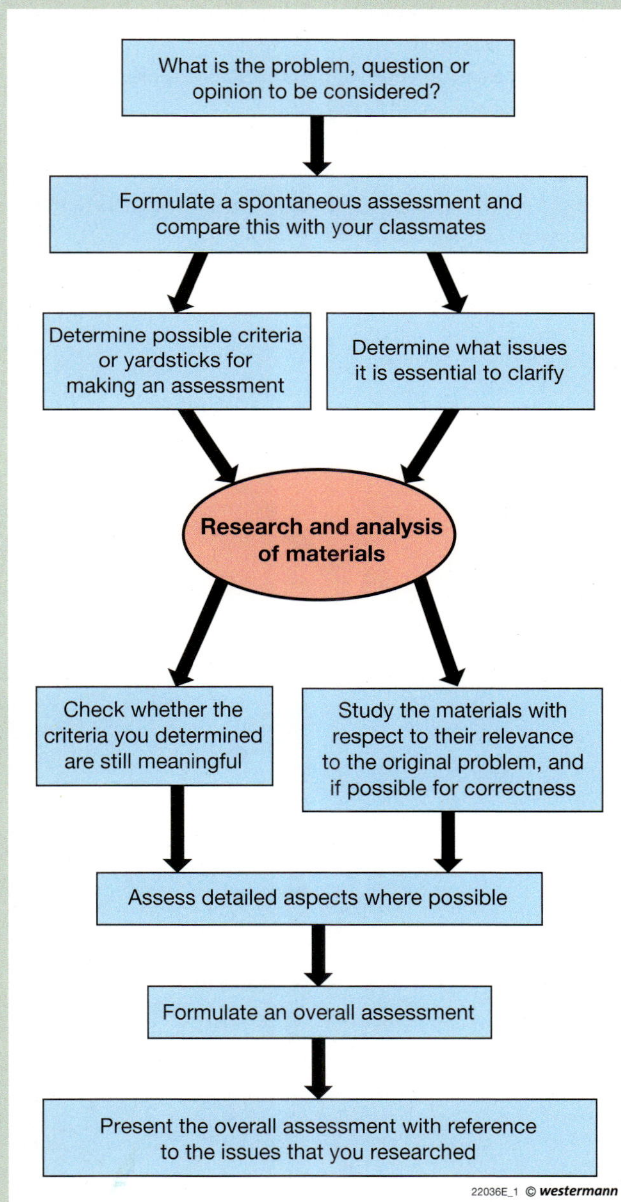

M9 Forming a judgement

Natural gas extraction – conflicts from unequal distribution

Russian gas for Europe's energy hunger

Natural gas is an important source of energy for Europe. Many countries in the European Union are making increasing use of natural gas. Most of the gas comes from deposits in Russia. Some individual countries such as Finland are completely dependent on Russian gas. But how does the natural gas come from the Russian deposits to the individual European countries? And what are the problems that can arise?

Ⓞ **1.** Describe the route taken by Russian gas from the extraction area to the purchaser (M1 – M3, atlas) by:
- **A** making a list of the individual points along the way.
- **B** writing a report.
- **C** preparing a short presentation.

2. Describe the gas imports into Europe from Russia (M4/ M5/ M9).

3. Explain the importance of natural gas for Russia's economy (M8, atlas, internet).

4. Analyse the cartoon (M7) and formulate possible problems of the European gas industry.

5. Give your opinion on the theme in the newspaper headline (M6).

Ⓢ **6.** Make a sketch for another cartoon that could illustrate the newspaper article.

Deposits, Natural gas, Pipeline, Reserves

M2 Gas extraction in the Zapolyarnoye gas field

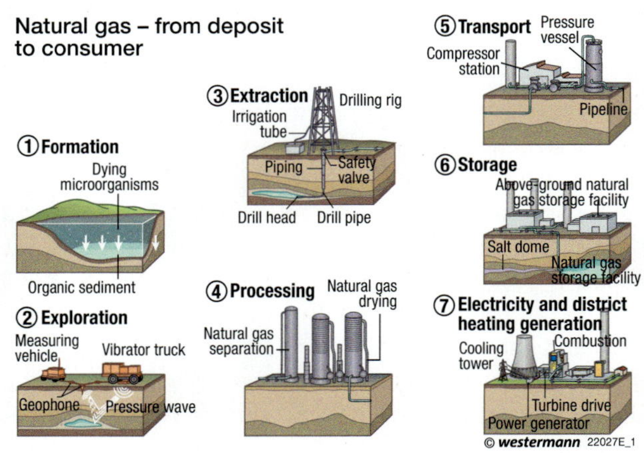

M3 Natural gas – from deposit to consumer

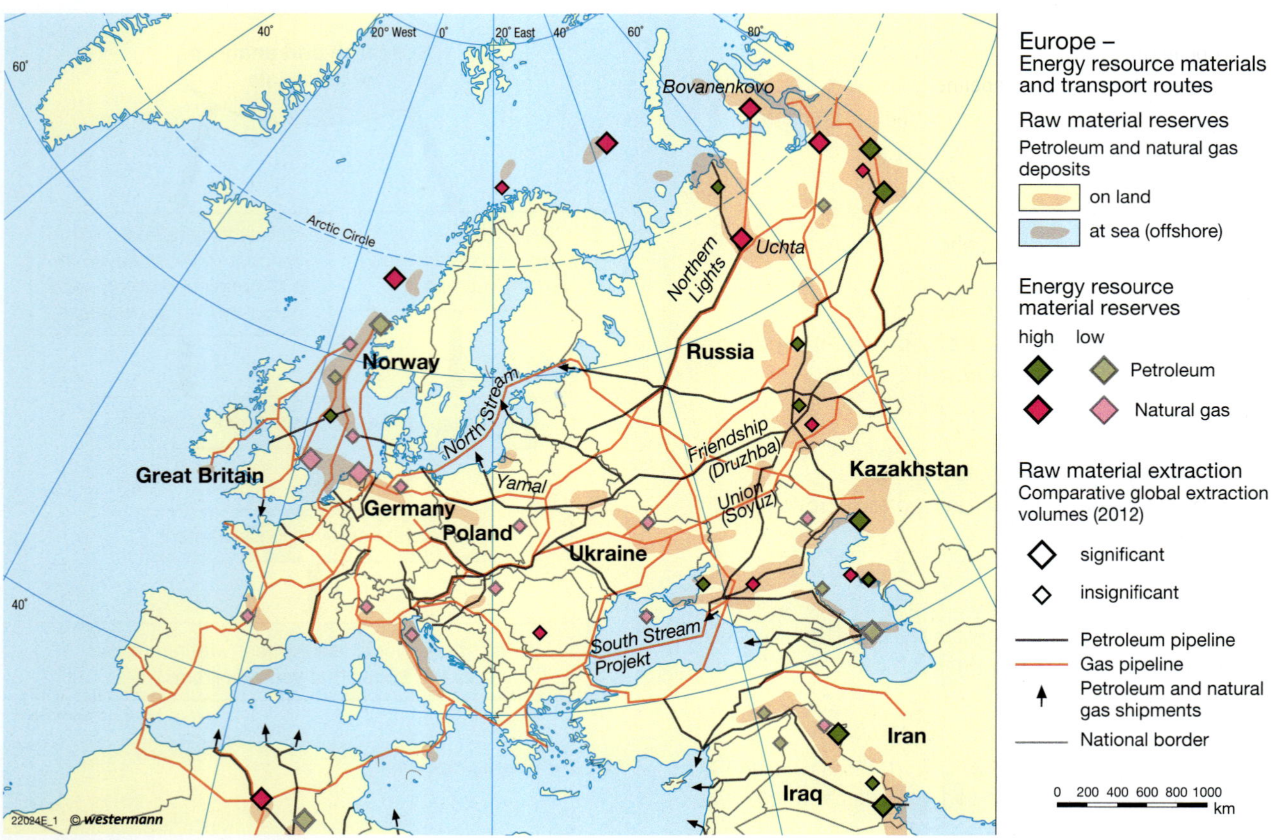

M1 Europe – energy resource materials and transport routes

M4 Gas sales by Gazprom

Year	Volume (in bill. m³)
1996	601
1998	591
2000	584
2002	595
2004	634
2006	641
2008	665
2010	649
2012	655

Sources: Russian Federal State Statistical Service
*Of the 655 billion m³ the state-owned energy giant Gazprom alone produced some 500 billion m³. Nearly one-third of the natural gas is exported

M8 Gas production in Russia

M5 A lever of power for the Kremlin: The Gazprom headquarters in Moscow

M7 Cartoon

Gas from Russia: power from the pipelines

The European Union and the Russian company Gazprom are waging a bitter war over pipelines and rules. From the Baltic to the Caucasus, the issue is whether gas can remain as a weapon in the hands of the Kremlin.
On New Year's Day 2009, the gas pressure suddenly dropped. Until New Year's Eve, Moscow and Kiev had been arguing about the price for which the Russian company Gazprom should deliver the gas.
When the new year began, Ukraine woke up without fuel. Soon the vast Ukrainian transit pipelines, through which the Russian gas flows to Central Europe, dried up. Hungary, Poland and the Czech Republic reported supply disruptions. Bulgaria had to close factories. Slovakia, where the heating broke down in the middle of winter, declared a state of emergency. Ukraine held out for 18 days, and then gave in. [...]
The gas had achieved what the Soviet Union had previously used tanks for – and Russia had demonstrated that it knows how to use its gas monopoly. Since then, not only Ukraine has discovered what this means. Estonia, Latvia and Lithuania too, which also belonged to the Soviet empire until 1991, are under pressure. Because all the gas pipelines to these countries still come from Russia, they are dependent on whatever prices Gazprom chooses to charge. According to a report by the European Commission in May, they pay 35 percent more for Russian gas than Germany for example. [...] The ambitions of the state-owned Gazprom are not limited to Ukraine and the Baltic. In southern Europe, its main strategic project is the South Stream gas pipeline which will connect southern Russia with the Balkans through the Black Sea. In the view of EU experts, this pipeline combines two political goals. Firstly, it would enable the main existing pipelines in Ukraine to be bypassed so that Moscow would be able to shut off supplies to Kiev in future without at the same time losing its delivery route to Europe. Secondly, South Stream should pre-empt a planned pipeline from Azerbaijan through Georgia and Turkey to the Balkans, the "southern corridor". If Europe could be supplied from Russia through South Stream, the calculation goes, gas from Azerbaijan would have no chance. Then, Russia would not have to share its role as the largest supplier to the EU [...] with other suppliers.

Source: K. Schuller, B. Triebe, Frankfurter Allgemeine Zeitung, 07.07.2013

M6 Newspaper report

Germany	35.6
Italy	23.8
Hungary	7.9
France	7.6
Czech Rep.	6.4
Poland	6
Austria	5.6
Finland	4.3

M9 Gas imports of selected European countries from Russia, 2012 (in billion m³)

Nuclear power – an acceptable risk?

Nuclear energy in Germany

Nuclear energy has a special position among the fossil fuels: it makes use of the fact that the splitting of uranium or plutonium, triggered by neutrons, releases energy and further neutrons. This form of electricity generation causes no CO_2 emissions. Electricity generation based on uranium is therefore less harmful to the climate than when it is based on coal, natural gas and petroleum. A further advantage for countries that are poor in energy resource materials is the fact that much less uranium needs to be imported for energy generation than coal for coal-fired power stations, for example. But the main problem is considered to be the safety of nuclear power stations, in view of the fact that radioactive radiation is extremely hazardous to health. The safe containment of radioactivity cannot be guaranteed in the event of either a nuclear accident or improper storage of nuclear waste.

In Japan in 2011, a tsunami resulted in multiple core meltdowns in the power station blocks at Fukushima, directly on the coast. This disaster put an end to the decades-long debate about nuclear energy in Germany: the German Federal Parliament (Bundestag) decided on a phasing-out of nuclear energy, which entered into legal force on 6 August 2011. Eight of the 17 German nuclear power stations in operation at that time were shut down, mainly for reasons of age. The last German nuclear power station will be shut down by 2022.

1. Describe the route taken by uranium from extraction to its conversion into electricity (M2/ M4, internet).

ⓞ 2. A Produce a sketch map showing the utilisation of nuclear energy in Germany and its neighbouring countries (atlas).
B Explain and compare the significance of nuclear energy using the examples of Germany and its neighbours (M5).

3. Discuss the phasing-out of nuclear energy in Germany.

ⓢ 4. Contrast the key arguments of the supporters and opponents of nuclear power.

Atomic energy, Meltdown, Nuclear energy, Residual risk, Uranium enrichment

In Germany, scientists and supporters speak of nuclear energy or nuclear power; its opponents use the terms atomic power or atomic energy. Why? In connection with the development of nuclear weapons after 1945, many Anglo-American expressions found their way into the German language; the word “atom” became dominant. Then, in the mid-1950s, the term “nuclear” came more and more to the fore as a result of the international fostering of research into the peaceful use of nuclear fission. The expression “nuclear energy” is more accurate in terms of physics, because the processes by which the energy is obtained take place in the nucleus of the atom. In the 1970s, the anti-nuclear movement popularised the word “atom” as opposed to “nuclear” which was seen as technical jargon. Thus, the conflicting positions were accompanied by a linguistic demarcation between supporters and opponents. In contrast to Germany, the majority of nuclear energy-using countries plan to hold on to it. Various countries are even planning substantial expansions in their nuclear power station capacity.

Based on: www.kernfragen.de

M1 Nuclear energy or atomic power?

Uranium is a natural component of the rocks in the Earth's crust, occurring in small quantities. From a geological perspective, even with any foreseeable increase in demand for uranium, no bottlenecks in the supply of fuel for nuclear energy generation are to be expected. But since the uranium content of the rock in uranium mines averages only 0.1 percent, its extraction produces vast quantities of rock as a waste product.

In addition, large quantities of radioactively contaminated wastewater are produced. A further problem of uranium mining – especially in poorer countries such as Niger in Africa – is the lack of adherence to safety regulations. In consequence, the workers are often subjected to massive health hazards.

In a uranium mill, two tonnes of extracted uranium ore can produce approximately one kilogram of yellowcake, which contains more than 80 percent of uranium oxide. In further processing steps, this precursor material is used to make the fuel elements for nuclear power stations. These contain the nuclear fuel and mostly consist of large numbers of thin rods which are immersed in coolant water in the reactor cores.

M2 From uranium ore to fuel element

Global warming: renowned climatologists call for a renaissance of nuclear power

Mankind needs more and more energy, the emission of greenhouse gases is increasing, climate change is proceeding apace. Four renowned climatologists are now calling for an expansion in nuclear power. This is the only way to make a reduction in CO_2 emissions realistic, they say.

(Source: H. Dambeck, Spiegel online, 04.11.2013)

M3 Headline in Spiegel online

M4 Emsland nuclear power station near Lingen

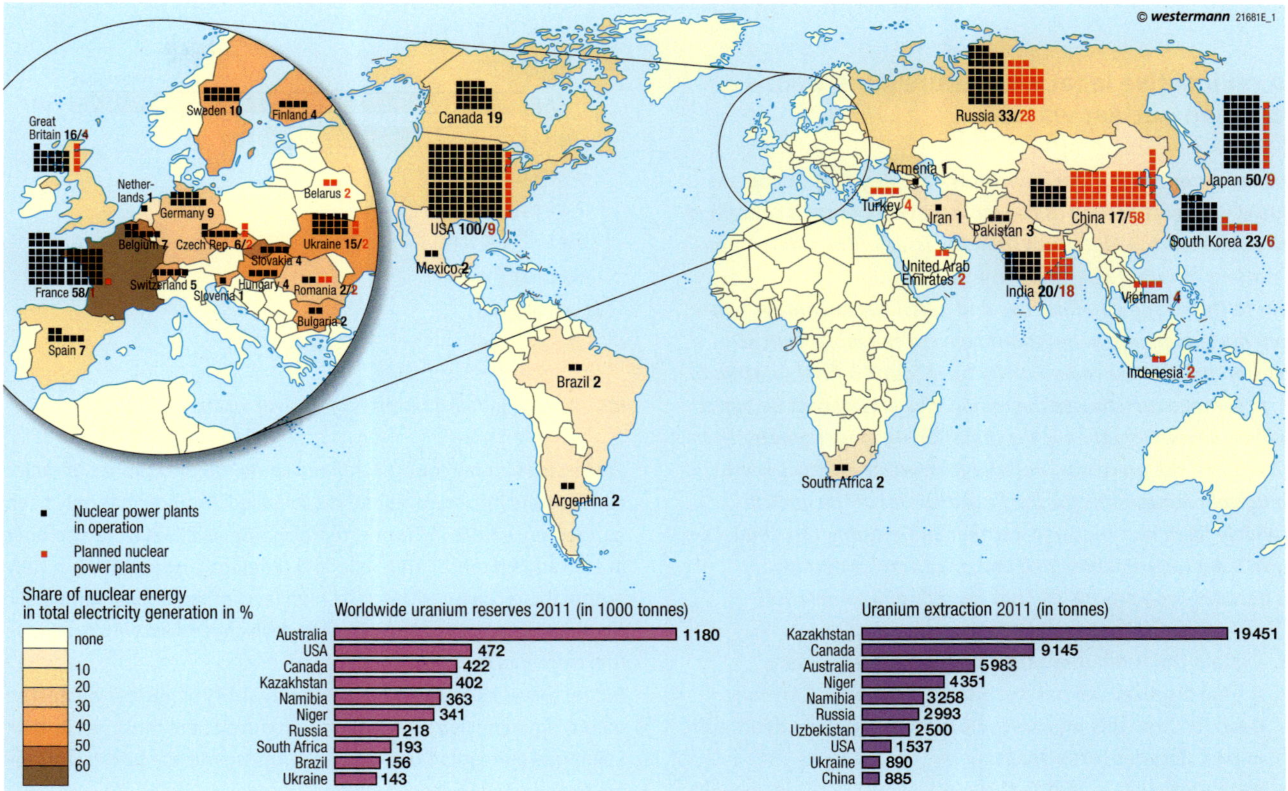

M5 Nuclear energy: Germany, Europe, global

Was the political decision to phase out nuclear energy in Germany too hasty in your view?

The nuclear power station operators in Germany have accepted the political decisions that were taken after the Fukushima accident and are implementing them. However, the policy of accelerated phase-out was determined by the Federal government very quickly at that time, even before conclusions about the cause of the accident had become available. This policy was then set in stone by legislation, even though the conclusions about the accident and a safety inspection of the nuclear power stations did not result in any safety-related justification.

Could an accident like in Fukushima also happen in Germany?

The accident was primarily the result of the failure to take precautions against a tsunami. The fact that, statistically speaking, tsunamis with a height of over 10 m hit the Japanese coasts every 30 years was an obvious risk that was ignored for decades. Such a failure in provision for natural dangers has not been discovered in any German plant [...]. The safety technology of the Fukushima power station was also not comparable with our German plants. There are very great differences in areas such as emergency power supply, emergency cooling or the management of major accidents.

What is your position on the residual risk and the manageability of operating nuclear power stations in Germany until at least 2022?

[...] The accident in Japan does not contradict the assumptions about residual risk and the manageability of a theoretical major accident in Germany, because the causes of the accident and its severe consequences do not apply to German power stations. In Germany we need to have safety precautions in accordance with the latest scientific and technical knowledge; in other words safety is being constantly improved.

M6 Interview with Dr. Ralf Güldner, President of the German Atomic Forum, on 20.01.2014

RESIDUAL RISK

The central question is: what can we, what can Germany, learn from the Fukushima disaster? From an event that would never have been thought possible and which nevertheless occurred. We learn that the probability of a nuclear disaster is nothing like one in a million, as the scientists and the industry have always told us. The risk is many times greater than previously assumed. In fact we must reckon with a severe nuclear accident with a substantial release of radioactivity approximately every 10 to 25 years.

MANAGEABILITY

We must also learn lessons from the helplessness of the Japanese state in the face of the disaster. The results show that nuclear power is not manageable!

Nuclear power does not forgive errors! [...]

EXTENT OF DAMAGES

[...] The consideration of the risks of nuclear power [...] must also focus on the consideration of the extent of the damages. This must include the question of acceptance: is society prepared to accept the consequences of a severe nuclear accident with hundreds of thousands of possible victims?

Are we also prepared to accept that hundreds of thousands of people would have to be resettled and that thousands of square kilometres of land would be uninhabitable for generations? [The extent of the damage from a nuclear disaster in Germany] could be much greater than in the events at Harrisburg [USA], Chernobyl [Ukraine] and Fukushima [Japan].

The German reactors are not the safest in the world, but they are among the most powerful. [...] Thus the German reactors possess a far greater radioactive inventory that could be released. [...]

Source: www.greenpeace.de

M7 Statement by the Greenpeace nuclear physicist H. Smital to the Ethics Commission for secure energy provision convened by Chancellor Merkel in Berlin on 28.04.2011

Nuclear power – an acceptable risk?

A radioactive legacy for future generations – German nuclear waste

Energy generation based on nuclear power has little negative impact on the environment – at least as long as the radioactive material is contained in the nuclear power station. But apart from the issue of the safety of nuclear power stations, another question arises:
What happens to the strongly radioactive and massively hazardous waste materials from nuclear energy generation? None of the more than 40 countries in the world that use radioactivity have found a final solution to this problem. In some circumstances the radioactive waste must be buried for up to a million years. Only then is the radiation no longer greater than that of uranium ore – the precursor material for nuclear energy generation. The geological prerequisites for the final disposal of the nuclear waste differ widely from one region to another. In Germany too, final disposal sites have been sought, examined and discussed for decades.
Why has this decades-long process still not been finally completed?

1. Explain the fundamental problem of nuclear waste.
2. A final disposal site has not yet been found in Germany. Describe how this unresolved situation has been dealt with in past decades (M1 – M12).
3. ⓞ With particular regard to this unresolved situation, explain the difficulties:
   A in the interim storage of German nuclear waste.
   B in the search for a final disposal site in Germany.
4. Ⓢ Set out your personal opinion on the use of nuclear energy in Germany.

Biosphere, Final disposal, Geology, Nuclear waste, Reprocessing

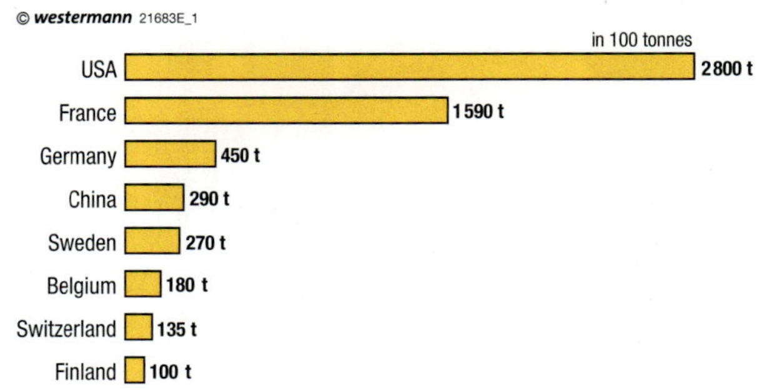

M3 Annual global quantities of nuclear waste

For decades, many countries tipped some 100,000 tonnes of radioactive waste into the oceans. This so-called dumping has only been banned since 1993. In Europe, reprocessing facilities were also built at La Hague (France) and Sellafield (England). Here, partly in fully automatic processes due to the high level of radiation, highly toxic weapons-grade plutonium is obtained which is mixed with expensive fuel elements for special reactors.
But reprocessing does not reduce the quantity of nuclear waste. The mostly liquid nuclear waste that is produced here must be vitrified. Spent fuel elements from various European nuclear power stations are transported to these facilities.
Since 2005, the transport of German fuel elements to these facilities has been forbidden. Transport by road and rail takes place in extremely resistant so-called Castor containers. As long as no final solution has been found, the nuclear waste is parked in so-called interim storage sites of which there are several in Germany. In the past, wastes with low-level radioactivity were also disposed of in abandoned mines.

M4 Dealing with nuclear waste, past and present

"Throughout the world there is no concept for the final disposal of nuclear waste. [...] The accidents in the nuclear waste dumps in the Asse and Morsleben salt domes with low and intermediate level wastes have disproved all the forecasts by official nuclear experts [...] after only 30 years. The final disposal of highly radioactive waste is even more [...] hazardous. [...] An ethical evaluation of atomic power must absolutely include the unsolved question of final disposal."

M1 From the statement by the Greenpeace nuclear physicist H. Smital to the Ethics Commission in Berlin on 28.04.2011

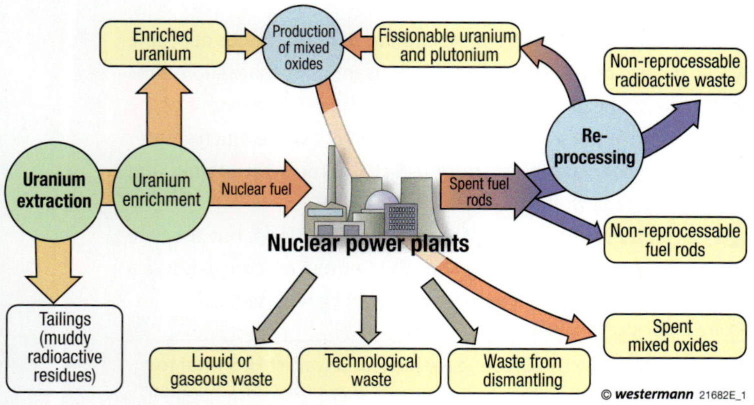

M2 Reprocessing

In 1995, the transport of some 30,000 tonnes of depleted uranium from [the uranium enrichment facility at] Gronau (NRW) to Russia for what was in effect its final disposal began. This export was only stopped in 2009 after massive protests from Russian, German and Dutch opponents of nuclear power.
But the pressure for disposal of uranium waste from Gronau is very great because the sheer quantities involved present insoluble problems for Germany's final disposal plans. Meanwhile, exports of uranium waste from Gronau to Russia have turned into a major environmental fiasco there. Russian environmentalists and the German media have frequently reported on improper storage in the Russian nuclear waste facilities as well as leaky drums of uranium waste in the open air. But the operating company of the uranium enrichment plant at Gronau, Urenco, and the German Federal Government, have so far refused to accept any responsibility for the nuclear waste from Gronau. In 2010, the export of highly radioactive nuclear waste from the interim storage site at Ahaus to Mayak in Russia, sponsored by the Federal Government, was stopped at the last second after strong protests by Russian and German environmentalists as well as by the NRW Federal State government. The region around the Mayak nuclear facility counts as one of the most radioactively contaminated in the world.
Source: www.contratom.de, 26.11.2013

M5 Final disposal in Russia

When delivered into interim storage, nuclear waste has a temperature of some 400°C. Further heat is released in the process of decay. Cooling to 200°C only occurs after 20 to 30 years. Only then can the drums be put into final disposal. Problems arise when the materials come into contact with groundwater in the final disposal site and are carried towards the biosphere. Therefore the nuclear waste must be stored in as dry a condition as possible for thousands of years, and should have practically no contact with the groundwater.
In principle, final disposal should be maintenance-free, for an unlimited period of time, and with no intention of retrieving the materials. The international search for the best geological formation for final disposal takes place in very different ways due to the geological circumstances in the different continents. Three so-called host rocks come into consideration: salt, clay and crystalline rocks (granite). In Finland for example, only crystalline rocks are available. In the year 2000, after 14 years of exploration, a final disposal site was selected at Olkiluoto, close to one of Finland's two nuclear power stations. The responsible local council supported the plan by 20 votes to 7. A survey among the local population resulted in 60 percent agreement. Parliament ratified this decision by 159:3. Operation of the final disposal site will start in 2020. It is planned to keep the final disposal site in operation until 2112 and then seal it in a long-time secure way by 2120.

M6 The search for a final disposal site

According to the Atomic Energy Act, the government is obliged to create a final disposal site for highly radioactive wastes by 2030. No final decision on a site has been taken. The current favourite is the exploratory salt mine at Gorleben in Lower Saxony. Exploration costs at the salt mine from 1977 to 2007 amounted to some 1.5 billion euros.

- **For Gorleben:** According to the nuclear industry, the current status of exploration confirms the suitability of the site. If the search for a final disposal site had to start again from the beginning, it would be very expensive for the nuclear industry.
- **Against Gorleben**: The suitability of the salt mine as a final disposal site for highly radioactive nuclear waste is contested by environmentalists and renowned geologists because of water ingress during the development of the exploratory salt mine. Over time, lethal radioactive isotopes could find their way into the biosphere via underground watercourses.

M7 For and against Gorleben as a final disposal site

M8 Citizens protest against Gorleben

M9 Protected Castor transport

The former salt mine at Asse is [in Lower Saxony]. [...] The experimental deposition of [...] radioactive wastes began in 1967. Various deposition techniques were demonstrated. These included for example stacking the drums in deposition chambers or tipping them [...], in order to minimise the workers' exposure to radiation. From 1971 onwards, a large proportion of Germany's low- and medium-level radioactive wastes were deposited in the Asse mine. By the time deposition ceased in 1978, a total of c. 126,000 [drums...] had been stored at a depth of c. 750 m. [...] During a routine inspection in 1988, an inflow of salt solution into the mine workings was discovered. The inflow currently amounts to c. 12 m³ per day. [...]
Source: www.endlagerung.de

M10 Experimental deposition in the former salt mine at Asse

The dilapidated salt mine at Asse should never have been used to store nuclear waste. This is the conclusion of Lower Saxony's committee of enquiry in its final report after more than three years of work. Politicians, the nuclear industry and scientists had pushed aside critical voices since the 1960s in order to solve the problem of final disposal and prove the suitability of salt domes. The dangerous penetration of water in Asse was known but ignored. [...] The government's disposal commission doubts that retrieval and transport of the drums of nuclear material is possible, because the mine is threatening to collapse and flood with water. "Just as today at Gorleben, it was asserted then at Asse that flooding was impossible." In the current search for final disposal sites, it should be ensured that wastes can be retrieved, said [the SPD Member of Parliament] Tanke.
Source: www.zeit.de, 18.10.2012

M11 Committee: Asse should never have been used

M12 Tipping of nuclear waste in an underground mine

The essentials in brief

M1 Power stations use energy sources to generate electricity

In our highly-developed technological world, good energy provision is of vital importance. In large parts of Europe, people take it for granted that they will have secure energy supplies for heating and electricity. For this, so-called fossil fuels such as coal, petroleum, natural gas and uranium still play a very important role. For example, our main means of transport are fuelled principally by petroleum products. The great importance of these finite energy sources will not change in the foreseeable future.
The opportunities to use fossil fuels are very unequally distributed in the world because some countries possess practically no deposits of such energy resources. In many countries in the Southern Hemisphere, universal electricity supply is not taken for granted. In particular, developing countries without their own fossil fuel deposits cannot fully supply their own needs. The leading industrialised nations, on the other hand, can secure their access to deposits by military means if necessary.

Fossil fuels as a motor for development

But for individual countries or regions, available reserves can provide a decisive stimulus for development. The geographical examples presented in this chapter, the Middle East – here especially the Gulf region – and Russia, make this clear. High incomes and employment make fossil fuel deposits a godsend for the producing regions. To protect these developments, many oil-producing countries have joined together in OPEC.

No extraction without risks

But the extraction of fossil fuels is not unproblematic. It is frequently associated with grave ecological and social consequences for the region. Particularly in Germany, certain negative consequences of hard coal and brown coal mining are frequently the subject of public debate. Even today, large-scale mining of hard coal in the Ruhr Area results in massive subsidence and severe mining damage. Numerous villages have already been displaced by open-cast brown coal mining in the Rhineland coal field and the people have been forced to resettle elsewhere. Petroleum extraction is also often met with resistance from the population. More and more examples can be seen in the daily media. There have been detailed reports from Nigeria for example, and in Ecuador there is also resistance by the indigenous peoples.

Extraction and utilisation at any price?

Especially when disasters occur, questions are asked about the acceptability of extracting fossil fuels in extreme locations. In previous years, such sites as oil platforms in the deep ocean and the mining of available deposits there were not considered. But now, constantly rising prices for energy resources make their extraction at such locations profitable for the energy companies.
For nuclear energy, the issues are different. Its use is particularly controversial in Germany because of the fear of nuclear accidents. This discussion was intensified by the disaster at Fukushima in Japan, where a tsunami resulted in a meltdown in the nuclear power station. However safe nuclear energy may be, there will always be a residual risk, and in spite of reprocessing the question of the final disposal of nuclear waste has still not been adequately answered today. The investigation of the geological conditions at potential final disposal sites is continuing. The primary objective is to protect the biosphere from contamination for thousands of years.
Not only the extraction but also the processing and utilisation of fossil fuels involve a conflict between economic and environmental or social goals. In this context, in Germany the large brown coal and hard coal power stations have been a particular focus of attention in recent years.

Energy wealth = welfare for all?

It remains to be said that the extraction and utilisation of energy sources can represent both a blessing and a curse for a region or a country. The great earnings potential must be set against the sometimes incalculable consequences and risks for the environment. Furthermore, there is no guarantee that the income from the available energy sources will reach the whole population. But one thing is certain: eventually even the last remaining resources of each of the fossil fuels will be exhausted.

Skill Check

All the skills that you were able to acquire in this chapter are listed here.
You can judge for yourself if you have mastered each skill: ***securely, moderately or barely.***

Explanatory skills

Can I...		Unsure? Look it up on these pages
1.	analyse the development of global energy consumption from a regional perspective and explain the development of the individual energy sources?	108/109
2.	analyse the high energy consumption of industrialised nations from the point of view of sustainability?	108/109
3.	present the availability of fossil fuels, depending on the conditions of their geological deposition, as an important factor in economic development by describing the economic opportunities for individual countries arising from large uranium, petroleum or natural gas deposits?	118–123
4.	explain the economic, ecological and social impacts of fossil fuels extraction by discussing for example the consequences of hard coal and brown coal mining?	110–117
5.	explain the connections between worldwide demand for energy resource materials, the stimulus for development in the producing regions, and its domestic and international conflict potential, by presenting for example the dispute over natural gas between Russia and European countries?	122/123

Methodological skills

Can I...		
6.	orient myself with the aid of physical and thematic maps, e.g. by locating Europe's energy resource materials?	122
7.	using suitable search strategies in libraries and the internet, research information and analyse it in an issue-oriented way, for example by finding information on the latest developments in the dispute around the Garzweiler II open-cast mine?	116/117
8.	substantiate written and oral statements with correct and appropriate references to materials and citations, for example by setting out my own arguments in a debate, supported by the materials?	114–117

Evaluation skills

Can I...		
9.	evaluate geographical facts, issues and measures on the basis of factual criteria?	120/121
10.	evaluate the importance of fossil fuels for the development of regions from an economic and ecological perspective?	118–121
11.	assess facts, issues and measures relating to geographical areas with explicit reference to and application of the underlying standards, values and norms?	121

Action skills

Can I...		
12.	present in class the results of work on issues relating to the extraction and utilisation of fossil fuels using correct terminology, in a way that is factual, problem-oriented and appropriate to the target group?	114–117, 124–127
13.	adopt different positions on conflicts over land use and argue for them, for example on the question of deep sea mining or brown coal extraction?	116/117, 120/121

V New Production Technologies

Prolonging the age of fossil fuels with calculable risks?

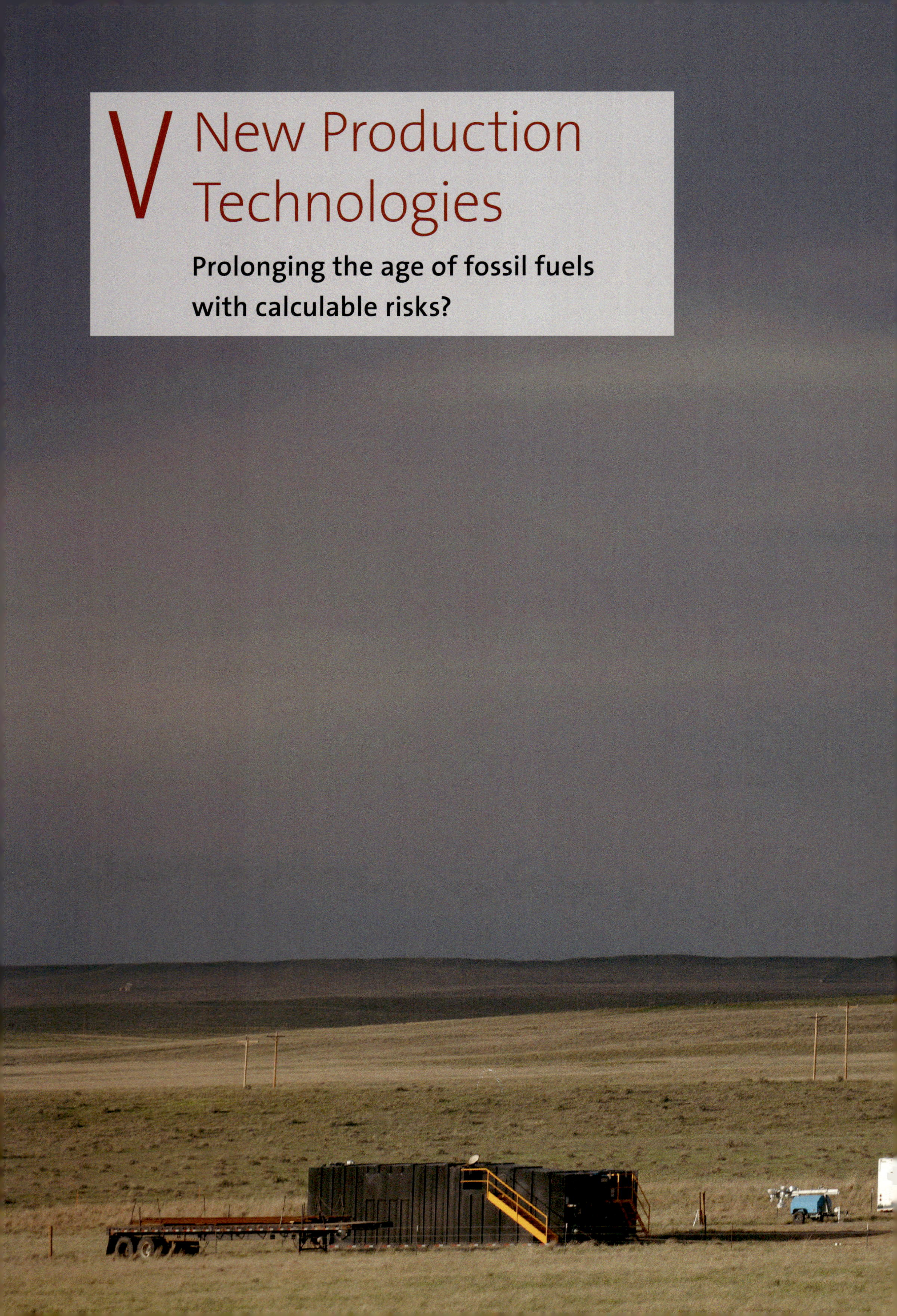

Fracking operation in the USA

New production technologies – answer to the energy shortage?

Conventional production
No specialised drilling techniques are required in this production method because the natural pressure in the reservoir is so high that the gas or oil flows to the surface by itself after drilling.

Unconventional production
Specialised drilling techniques and processes are required to extract oil or gas, e.g. oil sand production, fracking and gas hydrate production.

Will the supply situation become a global issue?

The population of our planet is constantly growing. The economic development of highly populated nations such as China, India and other emerging, newly industrialised countries is so rapid that the already growing demand for energy has been further significantly accelerated. It is disputed as to whether we are just before "Peak Oil" - maximum global oil production - or whether that point has already been passed. But one thing is clear: in many production areas, the quantity of petroleum and natural gas produced by conventional means is already in decline.
The looming energy shortage poses a huge problem. The shortfall in supply and growing demand inevitably results in increasing raw material prices. The price of one barrel of oil (approx. 159 litres) has long since passed the 100-dollar mark. As such high prices make very high investments in fuel production profitable, natural mineral deposits where production seemed unthinkable only a few years ago, e.g. the arctic regions or kilometres below the ocean floor, are now being exploited with new technologies. In other regions "fracking" is carried out. With this method, rock far below the Earth's surface is fractured by means of great pressure to extract oil and gas. In large areas in the middle of the taiga, oil is produced from oil sand. Even gas hydrates, until now a largely unknown fuel resource, are drilled for at great ocean depths. Regions that are literally in the global periphery and previously subject to very little attention have suddenly become the focus of political disputes due to the growing worldwide demand for energy. There are an increasing number of conflicts between nations over the distribution of existing reserves of raw materials. In the race for finding more and more new methods of producing raw materials the question of sustainability is often side-lined.
However, can new production technologies satisfy our hunger for energy?
Can they delay Peak Oil once more?
What risks are associated with these new technologies? Do new production technologies provide an alternative for bridging the energy gap?

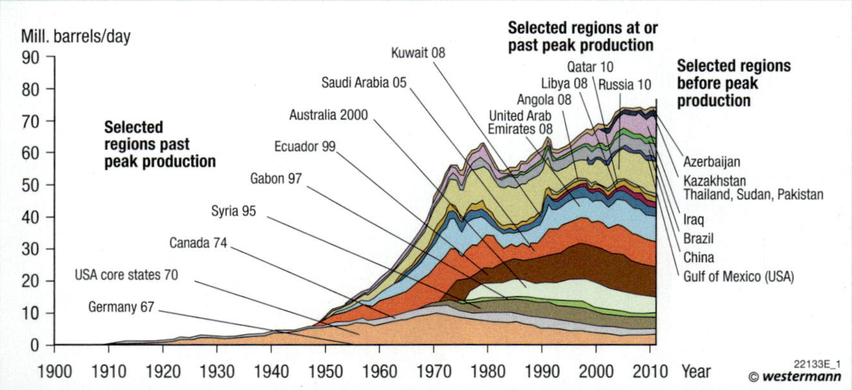

◁ **M1** World oil production 1900 – 2011

Fracking, Oil sand, Peak Oil, Reserves, Resource

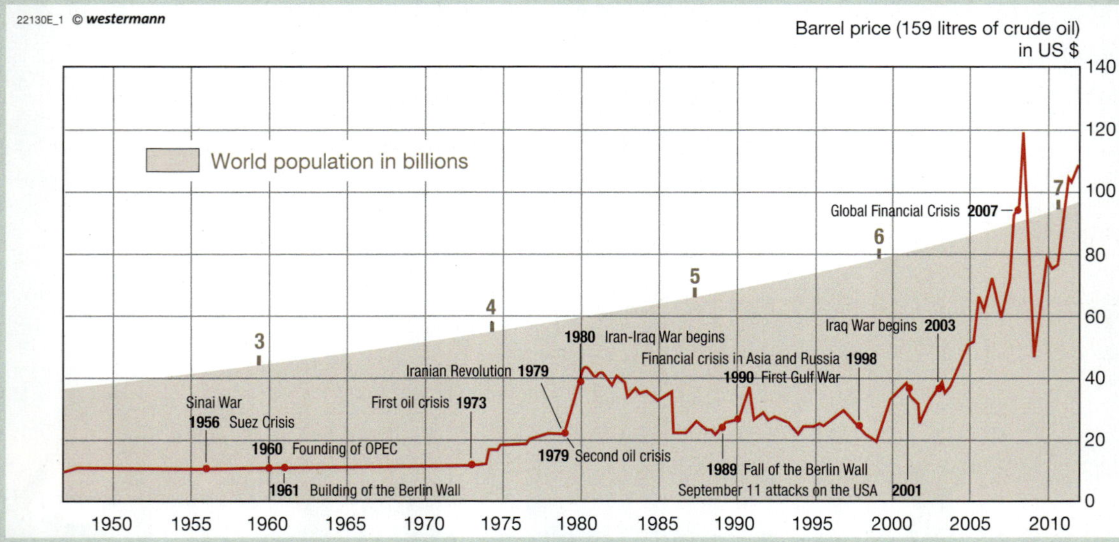

M2 The development of crude oil prices and world events that influenced the price. In the past, even relatively short-term events caused crude oil prices to skyrocket. But how will prices develop in the case of a long-term shortage?

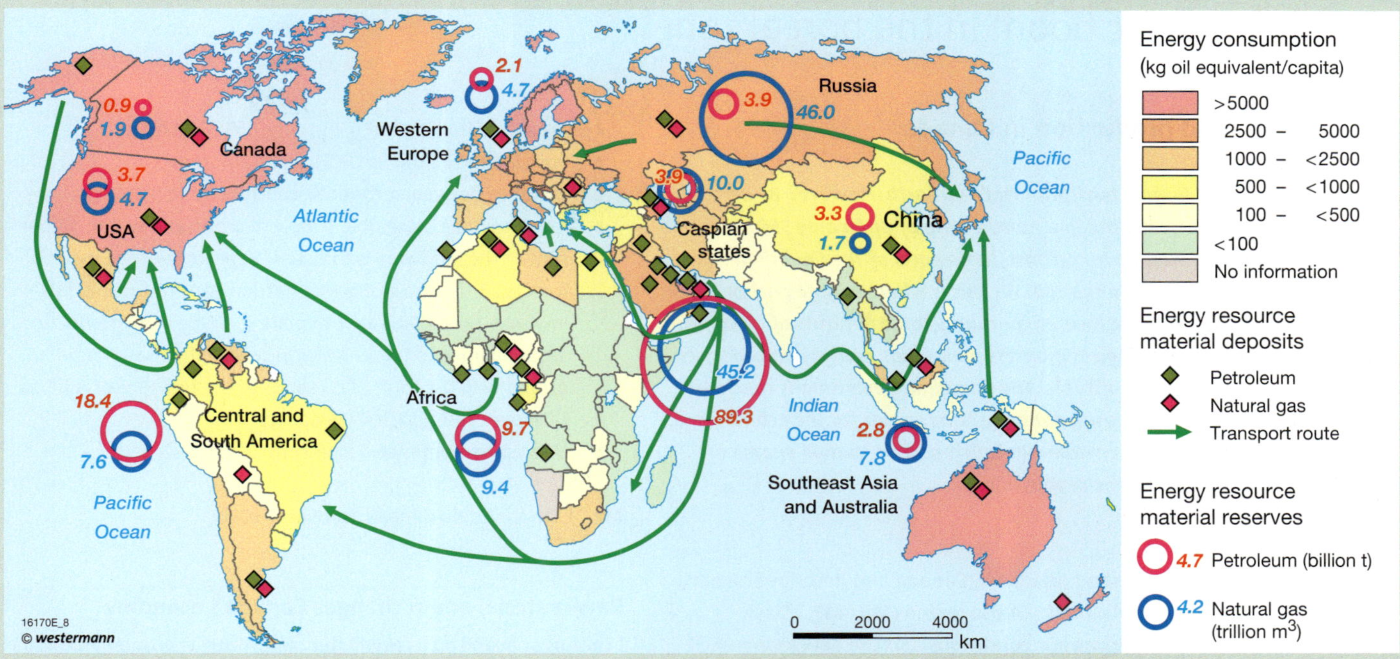

M3 Energy resources and energy consumption

M4 Protests against fracking in the USA

M5 One way street?

M6 Shortage

1. Explain the relationships illustrated in M1, M2 and M3.
2. Describe the main transport routes for oil and gas between the producing countries and importing nations (M3, atlas).
3. ◎ Sketch possible future scenarios based on:
   A The images M4, M5 und M6.
   B The diagram on p. 109, M5.
4. After working through this chapter, find an answer to the following question: Do new production technologies provide an answer to the imminent energy shortage?

Economic boom in the boreal zone

Oil sand production in Canada

Canada is the second largest territorial state on Earth. However, 85 percent of the workforce is concentrated in a narrow strip of up to only 350 km in width north of the US border. In contrast, large parts of the Canadian subpolar region are virtually uninhabited. Yet this is where a large part of the country's mineral deposits are located, primarily large deposits of oil sands. Rising oil prices and technological progress mean that production processes are becoming increasingly profitable. The Canadian government considers oil sand production to be a significant and sound motor for the economic development of the country. Environmentalists and the native people are however up in arms against the production.

1. Locate the oil sand regions in Canada and describe the natural habitat and climate of the region (M6/ M7, atlas).
2. Describe the economic conditions and the development of oil sand production in Canada (M2 – M5, p. 132 M2, atlas).
3. ⓞ Explain the conflicts of interest with regard to oil sand production (M3, M5 – M7):
   A from the perspective of the indigenous population.
   B from the perspective of the Canadian government.
4. Explain the environmental impact of oil sand mining (M5/ M6, atlas).
5. ⓢ Discuss oil sand production in Canada in view of the conflict potential.

Economic migrant, Indigenous people, Oil sand, Surface mining

- **Area**: 10.0 million km² (Germany: 357,000 km²)
- **Population**: 3.2 million, 3.5 inhabitants/km².
- **Distance between economic centres:** Up to 5,000 km.
- An almost inexhaustible supply of wood and water: 46 percent of the land area is forested (10 percent of the world's forests). 3rd place globally in natural gas production, following USA and Russia; 6th place globally in oil production. Second largest oil reserves worldwide, after Saudi Arabia, primarily in the form of oil sands in the province of Alberta.
- **GDP per capita:** 52,364 US dollars
- **Unemployment rate**: 7.2 percent (2012)

M2 Economic data Canada, from May 2013

Raw material boom changes Canada's economy

"The huge oil sand projects will stimulate Alberta's economy for years", says head economist Craig Wright from Canada's leading bank, the Royal Bank of Canada. [...] The long-term resource boom has permanently transformed Canada's economy. Provinces that were once underdeveloped, such as Saskatchewan and Alberta, are now growth drivers and not only in regard to increases in the GDP. Wages are rising rapidly and are attracting tens of thousands of economic migrants from other provinces and Europe. [...] Workers in the oil sand regions of Alberta annually earn 8,000 dollars more than Canadians with comparable jobs in other provinces. [...]

Source: M. Gärtner, www.manager-magazin.de, 03.01.2012

M3 Newspaper article

Low net energy

By 2025, Canada aims to extract 3.5 million barrels of oil from oil sand daily. [...] 70 dollars per barrel – that is what the oil has to cost for oil sand mining to be profitable [...].
At the moment the US blend WTI costs around 95 dollars, the North Sea blend Brent costs 115 dollars. One problem with unconventional production is that much more energy is required for oil sand extraction than in conventional oil fields. Experts therefore speak of low net energy.

Source: Wirtschaftswoche, 19.07.2011

M4 Newspaper article

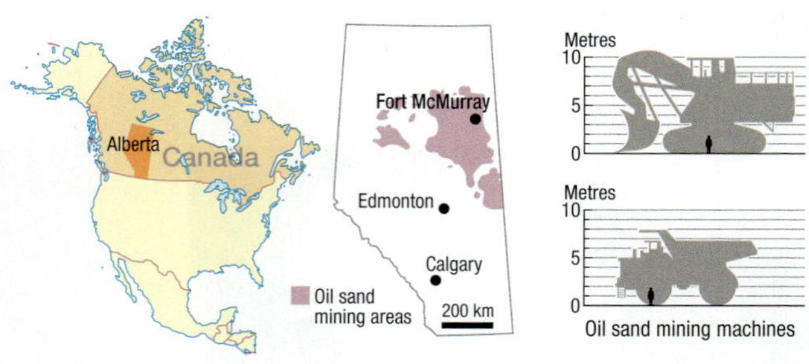

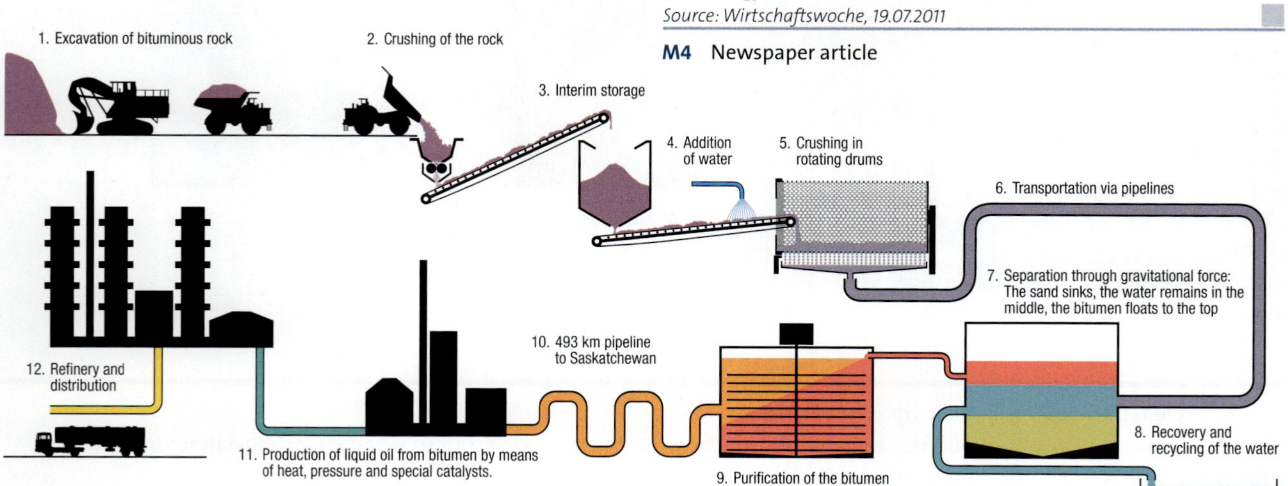

Oil sand is a mixture of sand, water and viscous petroleum or bitumen. In regions where deposits of oil sand are covered by a sediment layer of less than 75 metres it is extracted by open-cast mining. When extracted underground the bitumen in the oil sand is heated with steam, liquefied and pumped to the surface through horizontal pipes.

22131E_2 © westermann

M1 Process of oil sand extraction. Large-scale oil sand mining has been carried out in Canada since 2003.

M5 Oil sand production in Canada

4.5 barrels of water are required to extract one barrel of oil from oil sand. To convert bitumen into petroleum around one sixth of the energy of the petroleum produced is expended. Three to five times more environmentally harmful gas is emitted per barrel of extracted oil than in the production of conventional oil. Additionally, billions of tonnes of carbon dioxide are stored in Canada's coniferous forests, which have to make way for the oil sand mining. This will be released when the forests are cleared. The flora and fauna have adapted to the extreme climatic conditions of the polar and subpolar zones. The dominant landscape type is moor. Moor covers about 60 percent of the total area and also stores gigantic amounts of carbon dioxide. This will be released when the peat is removed in the course of mining, thus creating additional climate issues.

The stench of money

[...] Celina Harpe holds the map up in the air like a bill of indictment. "The oil companies have now advanced as far as Moose Lake", she says and taps the paper angrily. The workers have already begun surveying the land. "When I heard that I cried", says the tribal elder of the "Cree First Nation" from Fort MacKay in the Canadian province of Alberta, "I was born there". [...]
Only a stone's throw away, the mighty Athabasca River flows sluggishly. "We can't drink the water anymore", says Harpe, aged 72. [...] "And they didn't even ask whether they could take the land." [...]
The epicentre of the oil sand operation is the region around the small town of Fort McMurray [...] in the northeast of Alberta. [...] Crumbled oil sand looks like coffee grounds and stinks like diesel. It is the stench of big money. About 40 km from Fort McMurray the smell hangs in the air night and day.
The journey takes us north on Highway 63 into a seemingly endless coniferous forest. But the forest soon opens up again providing an unobstructed view over the smokestacks of a gigantic industrial plant in the middle of a seemingly apocalyptic landscape. [...]
The plant is the Mildred Lake Mine belonging to the company Syncrude. 300,000 barrels of oil are produced here every day. [...] On average, oil sands are composed of ten percent bitumen. [...] Over 90 percent of the bitumen can be extracted from the oil sands.
The last few percent of the mass are problematic. Mixed with water, sand and clay they land in huge settling basins that meanwhile cover a land area of around 170 km² in Alberta. [...] The brew contains poisonous heavy metals and chemicals. Increased concentrations of lead, cadmium and mercury have been measured in the nearby Athabasca River. The indigenous people tell of malformed fish and complain of rare forms of cancer. It is not yet clear whether this is connected to the oil production.

Source: P. Bethge, Der Spiegel, 10.10.2011

M6 Newspaper article

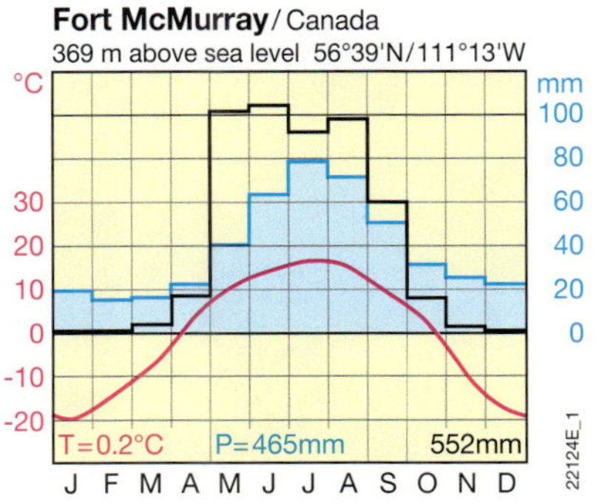

M7 Climate diagram for Fort McMurray

Fracking – a controversial production technology

Fracking in the USA

Most industrial nations on Earth have to satisfy a large percentage of their huge demand for fossil fuels through imports. Access to petroleum and natural gas far from their own national borders is therefore often a cause for political – and sometimes military – disputes.
The USA has the largest economy on Earth and at the same time it is the country with the highest energy consumption. The USA pins its hopes on fracking. In the process of fracking – a so-called unconventional production method – gas deposits are extracted by means of artificial fractures forced into the bedrock to release the gas. Using this innovative production technology, the idea is to tap into further resources in their own country, namely natural gas and petroleum stored in rock.
What economic opportunities and environmental impacts could result from this? What consequences in terms of foreign policy could the increase of production in their own country have for the superpower's global military involvement?

1. Name reasons for the current fracking boom in the USA (M1/ M3 – M6).
2. Explain the political background behind the US government's decision to expand natural gas production by means of fracking (M3/ M6/ M8).
3. Characterise the potential for domestic conflict caused by the fracking boom in the USA (M7/ M9, internet).
ⓞ **4. A** Analyse possible consequences for foreign policy that could result from the fracking boom in the USA (M6/ M8, atlas).
B In particular, explain possible consequences for the USA's military involvement in the Middle East (M6/ M8, p. 118).
ⓢ **5.** Produce a causal diagram to show the diverse potential impacts of fracking in the USA.

Fracking, Shale gas, Shale oil

President Obama on the topic of fracking:
„We have created 600,000 new jobs and have energy for several hundreds of years right beneath our feet."
(www.zeit.de, 04.12.2012)

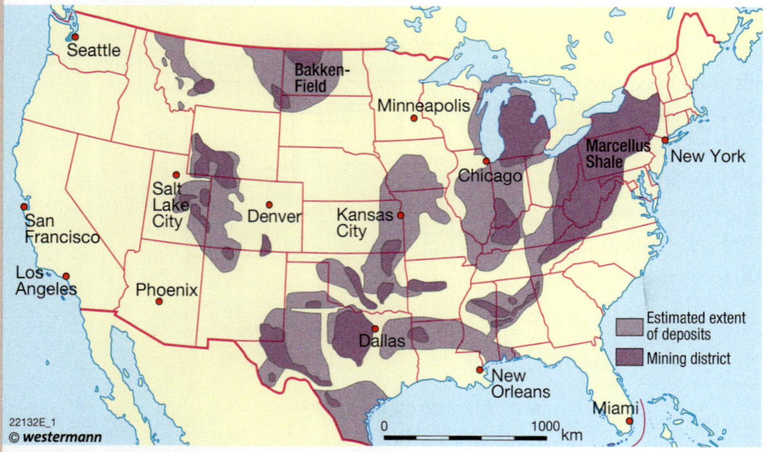

M1 Shale gas extraction by means of fracking in the USA

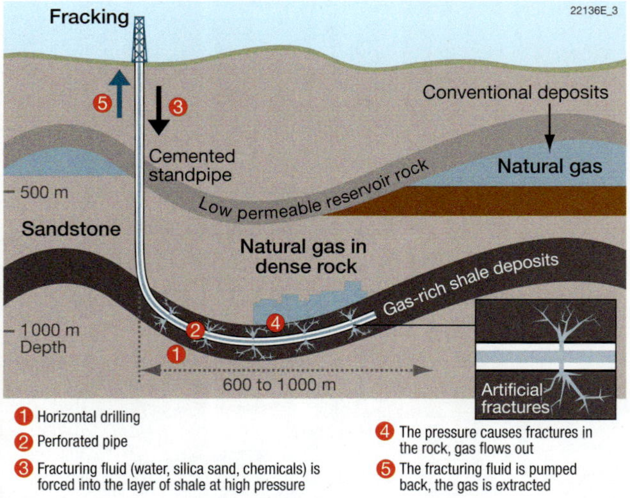

M2 Fracking Technology

Oil rush

This corner of North Dakota used to be no man's land. You turned on cruise control and read the newspaper behind the wheel. The greatest danger was running into a bison that had strayed onto the road. [...] Today, more than 12,000 trucks thunder along the main connecting roads in the northwest of the state. [...] The reason: In the Bakken region of North Dakota an unprecedented gold rush is underway. The gold of the 21st Century is called shale oil and shale gas, whereby oil is of greater significance. There is light crude oil located under the prairie that is as soft as shoe cream at 20°C and easy to process. The fact that there is oil and gas beneath the peacefully grazing bison has been known since 1951. But up until five years ago this treasure wasn't very valuable, because the technology to extract it was lacking. But then along came "fracking" – the new magic word. Fracking creates jobs. [...] Alone here in the Bakken region 65,000 jobs have recently been created. North Dakota is proud of the lowest unemployment rate in the USA and a budget surplus of 3.5 billion dollars. Many come from far away [today]: California, Alabama, Texas. The men leave their families behind because there is no accommodation for them, at least nothing affordable. Rents and house prices have increased fivefold in only a short space of time. The new heroes can afford to pay: they earn over 100,000 dollars annually. That is a lot in a country where the minimum wage is $ 7.25 an hour.
Source: Neue Züricher Zeitung, 19.05.2013

M3 Newspaper article

Development impetus

"The additional output resulting from the shale gas and oil revolution in the United States will contribute an additional two to three percent to real GDP growth by 2020. [...] The International Energy Agency [has] established that low gas and energy prices give the American industry a competitive edge. Low energy prices have led to a "re-industrialisation" because the manufacturing sector is investing hundreds of millions of US dollars in US chemical, fertiliser, steel, aluminium and plastics factories and export figures for these products are skyrocketing."
Source: blogs.usembassy.gov, 06.02.2013

M4 Excerpt from a speech by the US Ambassador Philip D. Murphy on 23.01.2013 in Berlin

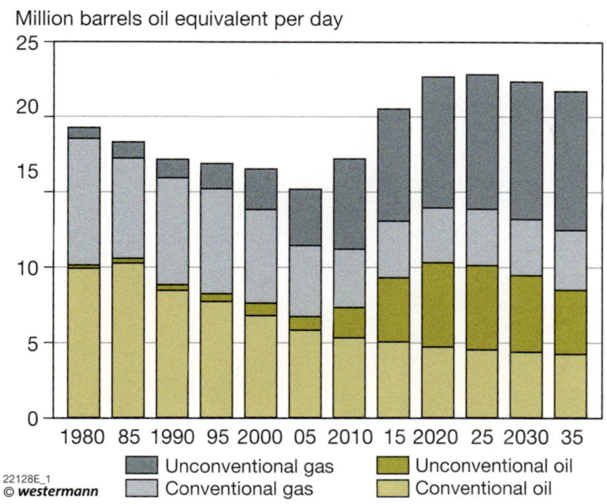

M5 Predicted development of oil and gas production in the USA, based on OECD/IEA 2012

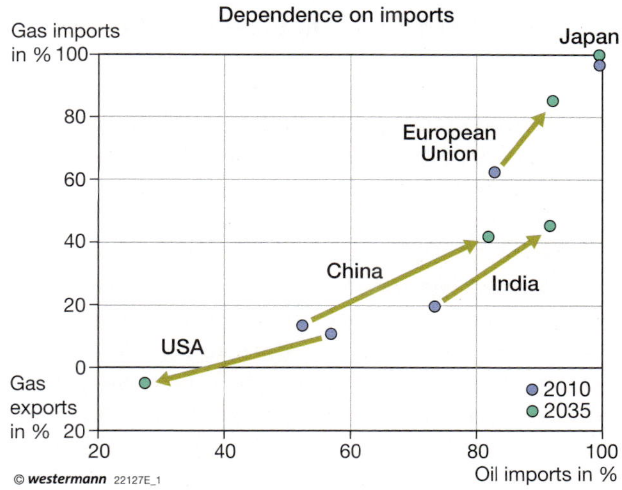

M8 Net oil and gas import dependency in selected countries, based on OECD/IEA 2012

Oil glut in the USA will change the Middle East

The new abundance of gas and oil in the USA will significantly change politics in the Middle East according to estimations of intelligence experts. [...]

The Gulf region will be most affected. The USA has previously been so heavily involved in the Middle East, both politically and militarily, because of its dependency on the region's fuel supplies [see also p. 119]. However, the USA could soon completely forego supplies from the region, [a confidential study of the Federal Intelligence Service] predicts. This would significantly increase "political leeway in terms of foreign policy and security" for the government in Washington. [...] Until now, according to the authors, the USA has ensured the security and independence of trade routes by investing billions of dollars in its globally operating naval fleet.

China, in particular, benefited from this. As interest in the security of raw material supplies e.g. from Saudi Arabia wanes, the willingness in Washington to continue investing billions in the military capacity of the region will be likely to drop.

Source: Reuters/sara, www.welt.de, 17.01.2013

M6 Newspaper article

The scenario projections in the WEO 2012 by the International Energy Agency stating that around 2020 - 2025 unconventional oil and gas production will make the USA independent from imports rest (1) on the assumption that demand declines considerably and (2) on the speculation that huge possible resources will be transformed into proven reserves and then be put into production. [...]

Actually, we believe that shale gas production [in the USA] is very close to its peak. This is due to rising development costs, much too low gas prices well below production costs, rising environmental side effects and conflicts with the population, and – most important – shrinking gas production volumes of new wells and shrinking access to new well sites. [...]

The deficit caused by a lack of conventional production is so great that the decline of shale gas production could lead to a collapse of gas supplies in the USA if not compensated for through increased import capacities early enough.

Source: Energy Watch Group, March 2013

M9 Results of a study

Fracking forces gas into drinking water

Researchers examined wells in the fracking region of Pennsylvania and claim to have proved the contamination of drinking water for the first time. A study has found indications that unconventional gas production [...] contaminates drinking water with gases. Some wells near fracking operations were contaminated with methane, ethane and propane, reported a group of researchers from Duke University in the US state of North Carolina. The researchers took samples from 141 private wells in the Marcellus Shale area in the northeast of the state of Pennsylvania [...]. According to the study, the methane contamination was six times higher in wells within a one-kilometre radius of fracking operations than in other wells. The concentration of ethane was as much as 23 times higher [...]. There were no robust findings about adverse health effects caused by the gases.

Source: boj/AFP, www.spiegel.de, 25.06.2013

M7 Newspaper article

M10 Fracking in Pennsylvania

Fracking – a controversial production technology

Fracking in Germany – Should we start or not?

In Germany there are regions that essentially display the geological prerequisites for the formation of shale gas. Until now though it has not been produced like it has in the USA. Territorial licences have however already been granted for the exploration of unconventional natural gas deposits. In contrast to conventional gas production, the deposit pressure in unconventional production is not high enough for the gas to flow to the surface on its own. The industry sees opportunities for profitable production. As yet however, there is no verified information about the potential of fracking in Germany. At present, Germany finds itself in a state of legal limbo that is unsatisfactory for all parties. Companies eager to begin production, concerned citizen's initiatives and the responsible regional authorities are unsure of what the future holds for them. The topic of fracking is a controversial issue in Germany.

1. Locate regions in Germany that are assumed to have potential for fracking (M2).
2. Ⓞ **A** Compare the assumed deposits in Germany with those in the USA (M2, p. 136 M1).
   B Compare the size and location of the fracking regions (M2, S. 136 M1).
3. Explain how the population density in Germany differs from that of the fracking regions in the USA and the resulting conflicts over land use in Germany (M2/ M3, pp. 132 – 133, atlas).
4. Discuss the different positions represented within the general public, politics and science on the topic of fracking (M4 – M9, internet).
5. Ⓢ Design a poster to present your opinion on the topic of "Fracking in Germany".

Citizen's action group, Environmental Impact Assessment (EIA), Exploration, Fracking, Shale gas

M2 Shale gas regions in Germany

According to expert estimations, conventional natural gas reserves in Germany will be depleted in about 10 years at the current rate of production. The industry and business sectors are therefore turning their attention towards unconventional deposits of natural gas [...]. To date, shale gas is not produced in Germany. However, fracking technologies have been employed for around 60 years – for producing other gases and in the production of petroleum and geothermal energy. Based on an estimation made by the Federal Institute for Geosciences and Natural Resources [...] the total quantity of shale gas potentially available in Germany amounts to 13 billion cubic metres. It is however assumed that only 10 percent can be technically exploited [...]. This amount could secure gas consumption in Germany for about 13 years [...].
The Clausthal scientist Professor Kurt Reinicke considers the risks of fracking to be assessable. "We have years of experience employing the technology and during that time nothing has gone wrong", he says. Reinicke worked at the Institute of Petroleum Engineering at the Clausthal University of Technology until he retired one year ago. "The technology could and can be improved – for instance with regard to the quantity of required additives and the securing of drilling sites."
Source: www.braunschweiger-zeitung.de, 03.04.2013

M1 Newspaper article

Fracking doesn't pose a big problem in the often sparsely populated regions of the USA. The US state of North Dakota, for example, is half the size of Germany but has a population of only 640,000.
When developing a new shale gas deposit, new wellbores have to be sunk repeatedly. Experts calculate an average of six wellbores per square metre. An area of one hectare is usually required for one drilling site and this gives access to about two to four square kilometres underground.
In this context, federal mining authorities in Germany are examining technical questions relevant for exploration and production as well as diverse environmental aspects including emissions, water consumption and disposal, waste management and conservation. The search for economically significant mineral resources in Germany – including shale gas – is subject to the regulations set out in the Federal Mining Act, in short mining law.
Minerals that are referred to as "free for mining" – almost all ores and fossil fuels – are not the property of the landowner. Instead, mining rights granted by the mining authorities are required for shale gas exploration and production.

M3 The issue of area requirement

According to existing studies on ecological relevance, the implementation of fracking technologies for unconventional natural gas extraction [...] is a technology with significant risk potential. The effects for people, nature and the environment have not yet been adequately scientifically established. Drinking water and health have absolute priority.
We reject the use of environmentally toxic substances in fracking processes implemented for the exploration and extraction of unconventional natural gas deposits. Decisions on applications for approval can only be met when the information base necessary for evaluation is available and it is established beyond doubt that it will not be detrimental to the water quality. [...]
The coalition will determine in a joint process with companies, and taking the federal states and science into account, the exact information that exploratory drilling must deliver in order to eliminate knowledge deficits and establish an adequate basis for possible steps to follow. This is to be a transparent process. [...]
In the short term, the coalition will submit changes in the Water Resources Act aimed at the improved protection of drinking water and regulations regarding environmental impact assessment (EIA) of mining proposals. This will make provisions for a compulsory EIA and public involvement before approval can be granted for the exploration and extraction of unconventional natural gas deposits by means of fracking.

M4 Extract from the coalition agreement between the CDU, CSU and SPD published on 27.11.2013

Planned fracking regulations in coalition agreement

GREENPEACE

There is neither an intended ban [...] nor mention of a moratorium. [...] Instead the commencement of industrial fracking is to ensue with exploratory drilling, the filling of knowledge gaps and by establishing a substance database. This will not lead to a future ban on fracking. If the coalition decides that sufficient data is available, there will be nothing to stand in the way of fracking. [...]
Source: www.greenpeace-magazin.de, 11.11.2013

M5 Statement by Greenpeace

BÜNDNIS 90 DIE GRÜNEN

Due to unforeseeable health and environmental risks we reject the production of unconventional natural gas in particular through the implementation of toxic chemicals. Production [...] obstructs climate protection and the energy revolution. We therefore demand clear statutory regulations that rule out the implementation of this technology. [It must] be ensured that it is no longer permissible to inject toxic reservoir water from oil and gas production underground. In addition, a [...] reform of German mining law is required. There must be a much greater focus on conservation and the 'polluter pays' principle.
Source: www.gruene.de

M6 Statement on fracking by the Green Party

EU Commissioner Oettinger (CDU) demands test drilling in Germany

Test drilling in one's own country would give an edge in knowledge over other countries. Especially an engineering country like Germany should react quickly – among other things in order to better gauge the costs of energy extraction. [...] One should not neglect the opportunities that natural gas deposits below one's own soil signify for energy supplies, the German newspaper Die Welt quoted the EU commissioner. Furthermore, fracking offers great potential and therefore a legal basis for demonstration projects and practical testing should be established. If we approve test drilling we will be a lot wiser in a few years and better understand the costs involved. An engineering country like Germany is strongly advised to do this.
Source: www.deutsche-wirtschafts-nachrichten.de, 21.05.13

M7 From the German Economic News

BGR

The Federal Institute for Geosciences and Natural Resources (BGR), the Helmholtz Centre Potsdam GFZ German Research Centre for Geosciences and the Helmholtz Centre for Environmental Research have now published their common positions on "environmentally friendly fracking" in the so-called "Hanover statement". In the "Hanover statement" the three institutions draw their conclusions from the congress "Umweltverträgliches Fracking?" (Clean Fracking?) [...] at which more than 200 experts from politics, economics and science exchanged views on the topics of shale gas and hydraulic fracturing (fracking). Here is a short summary of the most important points:

1. Shale gas extraction can contribute to stabilising the decline in local petroleum production.
2. The implementation of fracking technologies for shale gas extraction in Germany requires an environmentally friendly procedure. The protection of drinking water is of utmost priority.
3. Whether fracking can be carried out in an environmentally friendly way is to be tested on a case-to-case basis according to the geological conditions and accompanied by suitable monitoring measures. To this end an environmental impact assessment must be carried out in each case.
4. The implementation and development of technologies for shale gas extraction in Germany requires a transparent and step-by-step approach.

Source: www.bgr.bund.de, 06/2013

M8 BGR press release on the conference "Clean Fracking?"

Citizen's initiatives and nature conservation associations demand that the production of unconventional gas deposits must be prohibited. This is a high-risk technology, the consequences of which are uncontrollable, irreversible and irreparable. All methods trialled to date are too high-risk for the environment, people and resources. In the event of an accident there are no countermeasures that can be implemented.
Source: www.gegen-gasbohren.de

M9 Statement by organised fracking opponents

The search for new raw materials – a delicate issue

In the Arctic – are climate change and innovative production methods creating a new global hot spot?

A new era has begun in the Arctic. As a result of climate change the polar ice is melting more and more quickly. As a consequence thereof, deposits of energy resources that were previously not exploitable are becoming the focus of attention. Many nations have designs on these deposits, for it is estimated that 25 percent of the world's natural gas and petroleum reserves are in the Arctic alone and they could, for the moment, help to satisfy the global hunger for energy – if it proves possible to develop these resources with innovative technologies. On the other hand, the further development of this region harbours considerable potential for ecological, economical and geostrategic conflict. Environmentalists fear e.g. devastating consequences for the fragile ecosystem in the event of an accident.

Ⓞ **1.** Demarcation criteria in the Arctic region vary and are partially disputed – a greater or lesser number of countries are entitled to make claims depending on how the boundaries in the Arctic are defined.
   A Describe which countries have asserted territorial claims on regions in the Arctic and which demarcation definitions these are based on (M4/ M6).
   B Produce a table of countries that border the Arctic. Differentiate between the varying definitions of the Arctic (M4/ M6).

2. The extreme Arctic region places high demands on the technology used for the production of energy resources.
   a) Describe the prevailing climate in the region (atlas).
   b) Explain the two possible variants for extracting the energy resources (M1/ M3).

3. Two potential conflicts have arisen at once in the Arctic: Politically and economically motivated. Analyse M2/ M4 – M8 under these aspects.

Ⓢ **4.** Antarctica is also increasingly becoming the focus of attention. Find out what mineral resources are on this continent and compare it with the Arctic (atlas).

Climate change, Geostrategy

M2 Protest action against the production of raw materials in the Arctic

Underwater factories

In search of oil and gas the Norwegian energy company Statoil wants to anchor entire production operations on the Arctic seabed. Technically speaking it is a brilliant idea and economically attractive. But can it be justified?

Observed from above it appears as if the environmentalists have prevailed in the Arctic. But below the water surface it looks like a futuristic industrial complex. Dozens of fully automatic drilling and tapping sites are connected to distribution stations, to separators for separating oil and gas, to pumps and compressors the size of a football field. Gigantic machine parks have sprung up on the seabed. In complete darkness they extract oil and gas from the bottom of the Arctic Sea, which is then transported to land via long pipelines. Because no human could survive here, the underwater factories are maintained exclusively by robots.

This underwater scenario is still a simulation. But it is in the process of becoming reality. The Norwegian energy company Statoil aims to build the first production plant completely underwater by 2020 [...]. The news agency Reuters once referred to underwater production as the "holy grail" of the oil and gas industry. Using this technology, corporations could tap into hard-to-access Arctic regions where until now conventional drilling platforms would have been threatened by strong storms and drifting icebergs the size of small towns.

Source: F. Rohrbeck, Die Zeit, 02.10.13

M3 Newspaper article

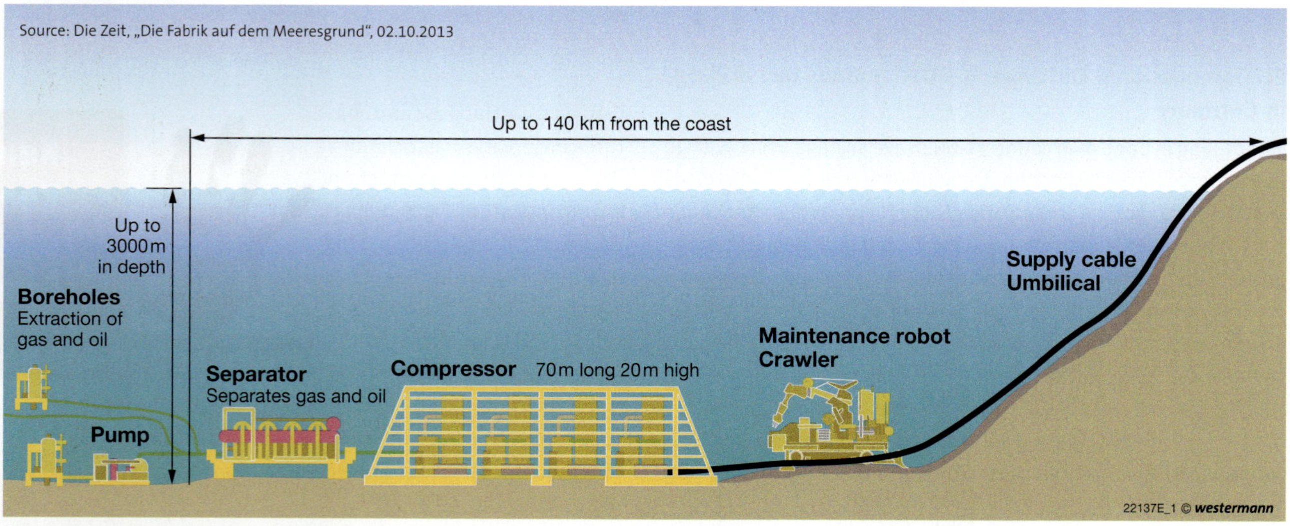

M1 Concept for underwater production of oil and gas in extreme regions – without platforms on the surface

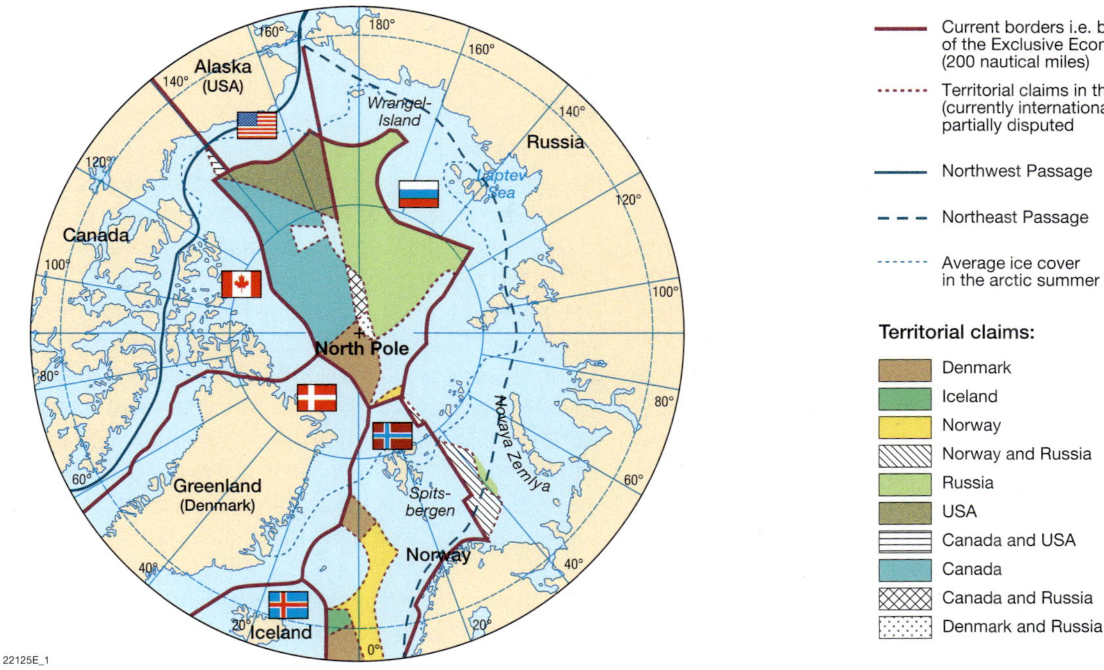

M4 The different territorial claims in the Arctic region

The Arctic Council

Members: The Arctic Council was established in Ottawa in 1996. The following countries are members: Canada, USA, Russia, Denmark (Greenland), Finland, Sweden, Norway, Iceland. Furthermore, representatives of the indigenous people also have seats in the Council.
Function: The Council aims to facilitate reconciliation of interests between the neighbouring states – keyword: geostrategic territorial claims – and the indigenous people, deal with the impacts of climate change and assist security in the region. Among others, the six observer states include Germany, France and Great Britain.
The EU, China, Italy, Turkey, South Korea and Japan have applied for observer status.

M5 An international committee is called upon to coordinate differing interests

Dispute over resources: Russia's claim to leadership

As the natural barriers of ice melt, Russia is strengthening its border troops in the Arctic region. At the international Arctic Conference the country acts like the leader of the region – and reinforces its claims to the massive deposits of natural resources in the sea. The other neighbouring countries also assert their claims. At the two-day ministerial conference of the Arctic Council in Kiruna on 13 May 2013 the Russian Minister of Foreign Affairs Sergey Lavrov referred to this forum as being historical, because of its tremendous ramifications. Russia will attempt to convince the Council of the inappropriateness of Arctic claims made by "non-regional protagonists" such as the European Union, China and India as well as a number of other countries and organisations.

M7 The protagonists attempt to stake their claims – politics motivated by geostrategic considerations

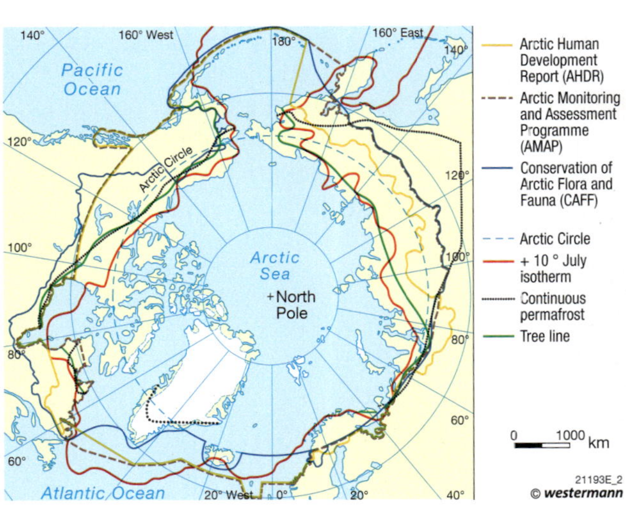

M6 Different demarcation definitions of the Arctic region – with considerable potential for conflict

M8 Greenpeace activists in action: An attempt to enter a Russian oil platform

The essentials in brief

Because the world population is growing and the demand for energy is also increasing in developing and newly industrialised countries, global, conventionally extractable energy reserves are becoming scarcer. Many deposits are already either completely depleted or productivity is declining. Diverse studies have come to the conclusion that we have passed "Peak Oil" and that energy shortages are to be expected. The consequences can already be seen today in sharply rising fuel prices. The higher prices allow for the production of petroleum and natural gas in regions where, only a few years ago, production was not economically viable. Thus newly developed, unconventional production technologies are increasingly being put into practice. The extraction of oil sands and the implementation of fracking technologies, for example, are on the advance worldwide. The production of gas hydrates is under consideration. These innovative technologies are now transforming energy resources that were previously not extractable for technical reasons or because their production was uneconomic into energy reserves and consequently into utilisable potential. However there are grave ecological concerns regarding their production.

Oil sands and fracking

The mining of oil sands in sparsely populated regions of Canada is a motor for the country's economy and attracts vast numbers of both domestic and international economic migrants. Large quantities of water and energy are used to extract liquid oil from bituminous sands. The net energy (see p. 134, M4) of this oil is however low. The indigenous people, in particular, protest against the destruction of their homelands and the contamination of flora and fauna caused by oil sand mining. These protests have, to date, had little effect.

Large industrial nations, such as the USA, are still highly dependent on imports of natural energy resources from producing countries often located in distant regions of the world. The economic and military expense for securing access to these energy reserves is huge. Huge expansion in the production of shale gas and shale oil by means of fracking could reduce this dependency and necessary expense in the future. Diverse studies present future scenarios that predict the possibility of the USA being able to dispense with energy imports from the Gulf region altogether. Other sources however describe shale gas production in the USA as a short-term boom that could quickly be followed by a collapse of gas supplies in the USA in the absence of imports.

The implementation of fracking technology has however already greatly stimulated the US economy. Previously underdeveloped regions such as North Dakota now find themselves in nothing less than a "gas rush". The state now draws large numbers of job seekers from all over the USA with the lowest unemployment rate in the country and well-paid jobs. However, due to possible contamination of drinking water opposition against fracking is also growing in the USA.

The situation in Germany is completely different. The expanded use of fracking technologies was recently the subject of extensive discussion. For a long time no agreement could be reached between the political parties and even within the parties. Citizen's initiatives were and still are up in arms about fracking. Meanwhile, there is political consensus that in the future a compulsory environmental impact assessment must be carried out for each proposed operation.

Exploration of the Arctic

New production technologies are even turning the Arctic into a new hot spot in regard to climate change. Particularly the geostrategic location of the bordering states, but also the inconsistent demarcation of this region results in territorial claims being asserted by many states. This alone harbours much potential for conflict that is further exacerbated by environmental and social aspects. Production glitches can be devastating for the fragile ecosystem; the protection of the indigenous people living there has not yet been the focus of attention.

High hopes are also pinned on a fossil energy resource that has until now received little attention: gas hydrates, also known as "burning ice". Its energy potential is estimated to be more than double the known reservoirs of petroleum, natural gas and coal. Gas hydrate production is however technically very complicated and extremely cost-intensive.

Ultimately, experts differ on how long the era of fossil fuel can be extended by implementing unconventional production methods.

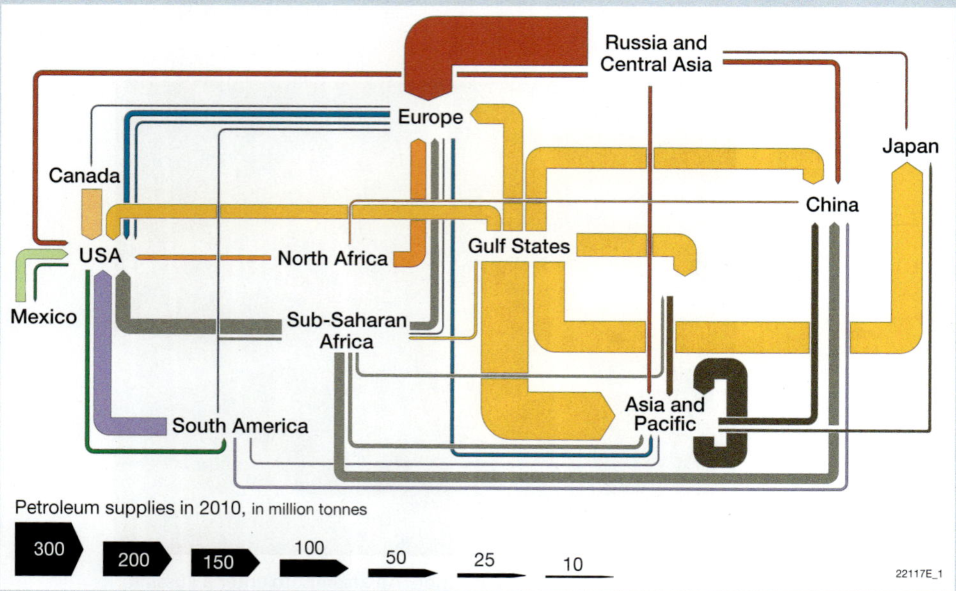

M1 Global movement of petroleum

Skill check

All the skills that you were able to acquire in this chapter are listed here.
You can judge for yourself if you have mastered each skill: ***securely, moderately or barely.***

Explanatory skills

Can I ...		Unsure? Look it up on these pages
1.	explain the effects and consequences of human intervention in the network of geofactors?	134–141
2.	describe economic and political factors influenced by regional development processes, e.g. caused by fracking in the USA?	132–141
3.	describe land use claims and conflicts as well as approaches for possible solutions, e.g. based on the Arctic region?	134–141
4.	organise structures and processes in a spatial raster at different scale levels, e.g. the global interdependence of the USA in terms of energy policies and the consequences for the state of North Dakota?	136/137
5.	present the availability of fossil fuels and their dependency on geological stratification as the most important locational factor for economic development, e.g. based on the Canadian province of Alberta or the US state of North Dakota?	134–137
6.	explain the economic, ecological and social effects of fossil fuel production, e.g. based on the Canadian province of Alberta?	134–137
7.	explain the connections between worldwide demand for energy resources, development impetus in the production regions and domestic and international conflict potential?	140/141

Methodological skills

Can I ...		
8.	orient myself using physical and thematic maps?	132–141
9.	identify problem-related geographic facts and develop corresponding questions?	130–141
10.	analyse different kinds of presentation and working materials to answer geographic questions?	130–141
11.	use appropriate search strategies to research information in libraries and on the internet and analyse it in an issue-oriented way?	136–139
12.	use correct terminology to present geographic issues orally and in writing in a way that is problem-oriented, factually and logically structured, and relevant to the assignment, operators and materials?	132–141

Evaluation skills

Can I ...		
13.	evaluate geographic facts, issues and measures on the basis of factual criteria?	132–141
14.	discuss problems which arise from contradictions and probabilities when evaluating geographic facts?	134–139
15.	critically assess results of my own work in regard to the underlying issues and my working methods?	132–141
16.	assess the suitability of economic and settlement areas based on different geofactors, e.g. by using the potential implementation of fracking technologies in Germany as an example?	138/139
17.	assess measures for overcoming natural boundaries in ecological and economic terms?	134–141

Action skills

Can I ...		
18.	present in class the results of work on geographic issues in a way that is factual, problem-oriented and appropriate to the target group as well as using correct terminology?	132–141
19.	adopt different positions on conflicts over land use and argue for them?	138–141

VI Renewable Energies

Realistic alternatives for the world's energy hunger?

Bathers in the "Blue Lagoon" in Iceland, a natural depression in the lava filled with two-thirds salt water and one-third fresh water. A geothermal power station is in the background.

Renewable energies on the advance

Sun, wind and heat – what are the possibilities?

With economic growth, energy consumption also increases. In future, energy consumption in developing and newly industrialising countries will also significantly increase. But fossil fuels will only be available for a limited time, and their use fosters climate change. This is why renewable energies are becoming increasingly important.

Germany has decided to phase-out nuclear energy and start moving towards the greater use of renewable energies. This process is called the energy transition (Energiewende). It could show the way for other countries. But the question arises of whether the use of renewable energies can make a contribution to the sustainable conservation of resources and the environment. To evaluate this, the possibilities and limits of renewable energy use and its globally very varied energetic potential must be considered.

The purpose of this chapter is to analyse and assess this potential, i.e. what renewable energies can offer in terms of energy supply. The geographical preconditions for, and the effects of, the use of renewable energies will be discussed. This will create the foundations for forming a factually-based opinion on the future importance of renewable energies for energy supply. This will take into account both economic interests and the demands of climate protection.

M2 Biogas plant

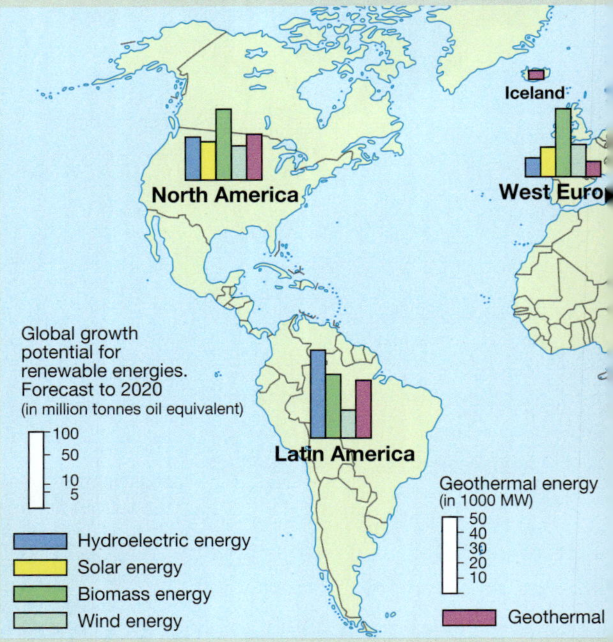

M3 There are wide variations in energy potential around the world

M1 Wind turbines (3D model)

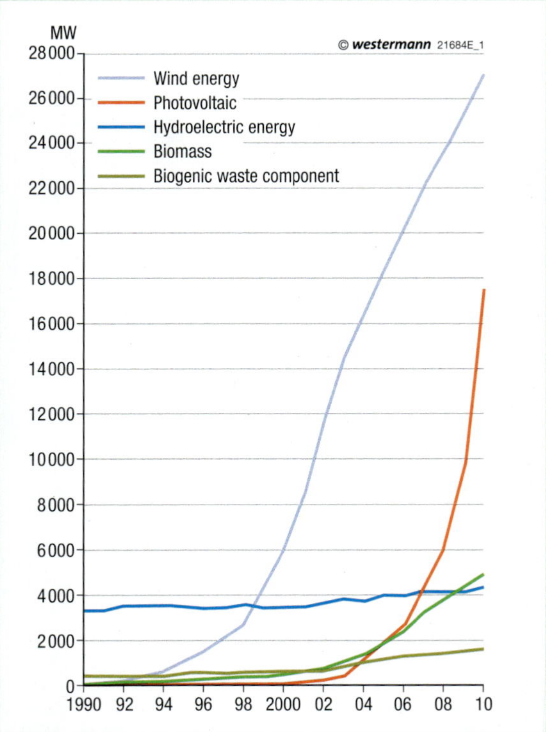

M4 Installed capacity for electricity generation from renewable energies in Germany since 1990

M5 Solar plant

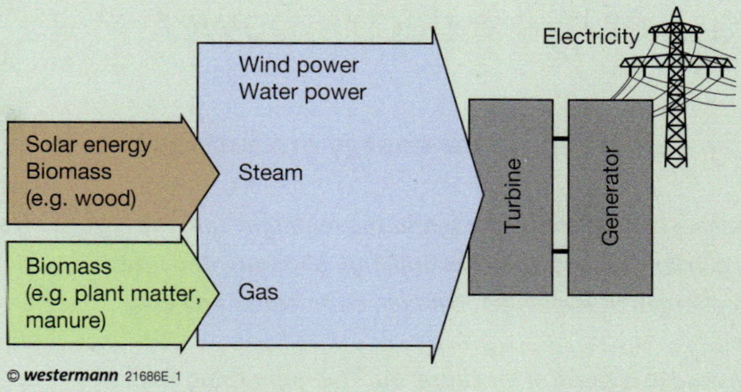

M7 Energy conversion

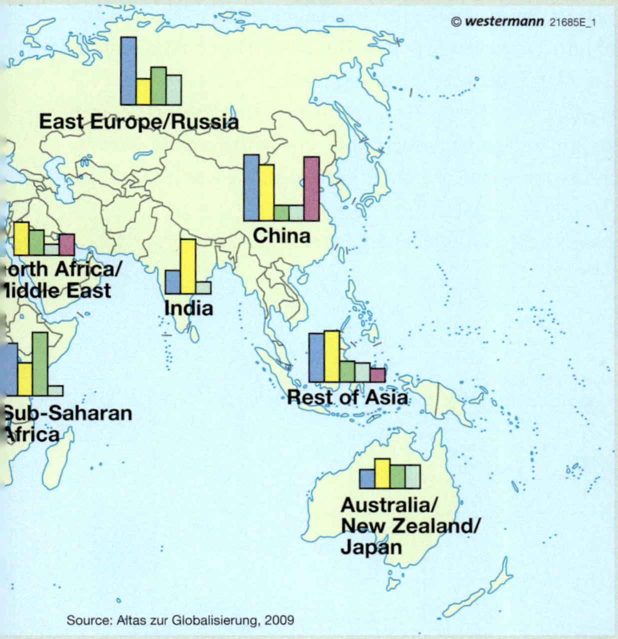

M8 Geothermal power station

M6 Dam (Möhne Reservoir, NRW)

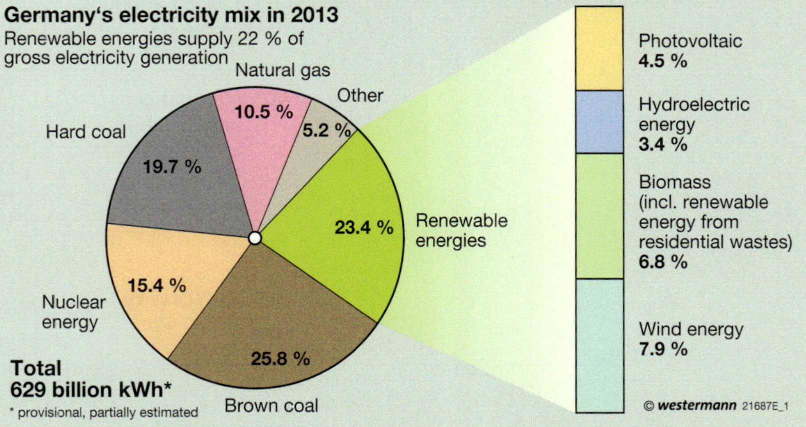

M9 Share of renewable energies in electricity generation in Germany

1. A subject can be structured with the aid of a mind map. Using the materials, produce a mind map on the subject of renewable energies (M1/ M2/ M5 – M8).
   Produce sub-branches for each of the five main branches (wind, geothermal, hydroelectric, solar, biomass) and use the mind map as a structural aid for working through the chapter.
2. Note your assumptions about the geographical preconditions for the examples of renewable energies named in the mind map.
3. Analyse the development and current importance of renewable energies for energy supply in Germany (M4/ M9).
4. Put together in a table the renewable energies that are used in the vicinity of your school and in the vicinity of your home.

The use of hydroelectric power

Norway – ideal region for energy production?

"Norway, the fascinating land of trolls, fjords and the midnight sun, the country of inspiring glaciers, mountain peaks and high plateaus, and interesting cities like Oslo, Bergen or Stavanger. Pure air, clean water and endless nature await you."
This is what you can read in a tourism brochure. But the main thing that energy experts see when they look at Norway is the potential that the landscape offers for energy production from hydroelectric power.

1. Explain energy generation in pumped storage hydroelectric power stations and draw from this the necessary preconditions for the use of hydroelectric power from this technology (M5/ M7, → definitions).
2. ⓞ Analyse the potential for the use of hydroelectric power in Norway (M1 – M4/ M7).
   A Summarise your results in a report.
   B Present the results in graphical form.
3. Discuss the possibilities and limits of the use of hydroelectric power in Norway for Europe's energy supply (M6).
4. Assess whether hydroelectric power can represent a solution for Europe's energy hunger. Consider which locations in Europe apart from Norway could be suitable for pumped storage hydroelectric power stations.
5. ⓢ The total installed capacity of hydroelectric power stations in Germany is some 4,720 megawatt-hours, while in Switzerland it is around 35,830 gigawatt-hours. Give reasons for the difference (atlas).
6. Research the use of hydroelectric power in run-of-the-river or tidal energy hydroelectric power stations and prepare a lecture.

Fjord
→ Pumped storage hydroelectric power station, Relief energy

Norway has hundreds of hydroelectric power stations. The country produces 98 percent of its electricity from hydroelectric power and can also export energy. [...]
The organisation German Environment Aid assumes that, even today, Norway's potential for renewable power is sufficient to fully supply Germany's needs from renewable energies. And Norway's water can do even more. Theoretically, Norway's hydroelectric power could replace the electricity from 60 European nuclear power stations. [...] This mainly requires cables through the North Sea and access to the electricity grid. [...]
And an additional benefit: with the aid of undersea cables, surplus electricity from German wind turbines could be fed to Norway and converted into hydroelectric power. This electricity could then flow back to Germany whenever it was required.
[...] The development of energy storage systems is necessary to absorb peak loads and bring energy supply into line with demand. Norway could thus become a kind of storage battery for the whole of Europe.
Source: G. Rehsche, ee-news.ch, 11.10.2010

M3 Norway's potential as an energy supplier to Germany and Europe

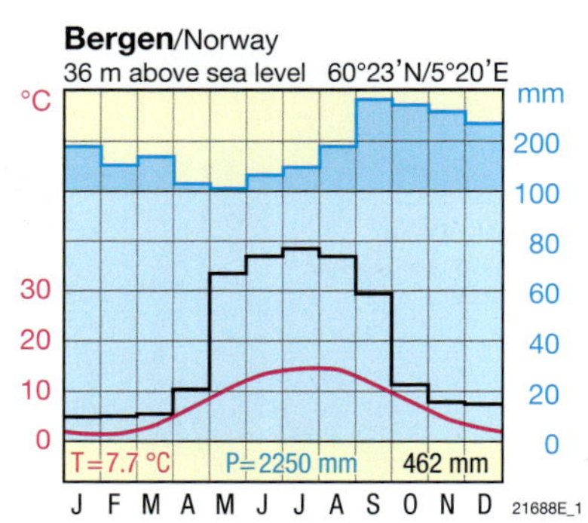

◁ **M4** Climate diagram of Bergen

M2 Information on population and energy supply in Norway ▷

Norway

Population (2013)	5,063,709
Population density	13 inh./ km²
Hydroelectric power supply	98%

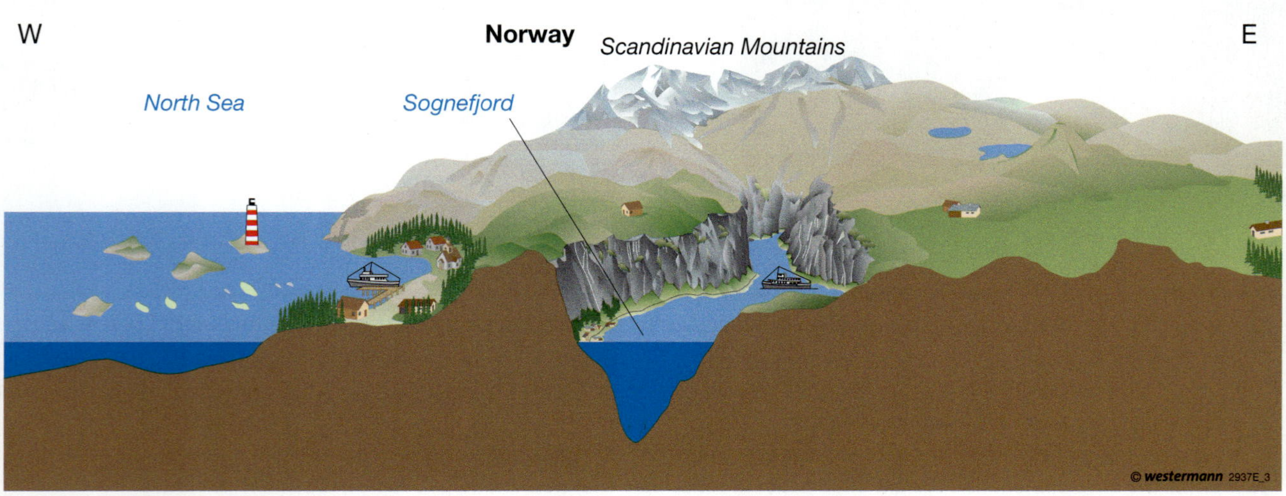

M1 West-to-east profile through Norway

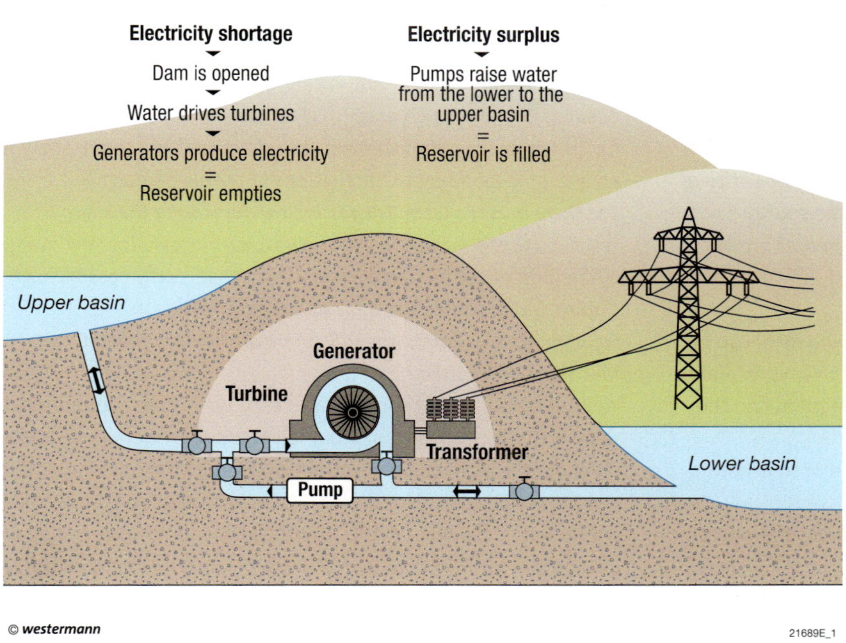

M5 Diagram of a pumped storage hydroelectric power station

→ Pumped storage hydroelectric power station

Pumped storage hydroelectric power stations are hydroelectric stations with two storage reservoirs at different levels. The force of the falling water is used for electricity generation. Part of the electricity thus obtained is used to pump the water back into the upper reservoir. An advantage is that the water is stored in the upper reservoir and is available whenever needed. But the pumped storage power station can only deliver electricity until the upper basin is empty. Since releasing the water and pumping it back up again leads to water losses, a regular inflow of water from rivers is necessary.

→ Relief energy

The output of a pumped storage hydroelectric power station depends on relief energy, i.e. the difference between the highest and lowest point and the angle of slope. The greater the relief energy, the greater the force of the falling water.

Nord.Link cable project: the North Sea electricity highway is coming

Electricity grids of all countries, unite! From 2018, a 1,400 megawatt cable will connect Germany and Norway.

The two countries have agreed on this project, which will cost billions of euros. It will turn hydroelectric power stations in the far north into buffer storage for German solar and wind energy. [...]

Whenever the planned offshore wind farms in the North Sea produce too much energy for the German grid, the surplus energy could flow to Norway – and drive electrical devices there or act as a buffer in pumped storage. Surplus solar energy could also be used in this beneficial way.

The only things that were missing for this bold vision were the technical prerequisites. Now it seems that the first step has been taken. The governments of the two countries [Norway and Germany] have agreed in principle to the laying of a cable on the floor of the North Sea. [...]

The plan is to lay the 600 kilometre connection between southern Norway and Schleswig-Holstein, called Nord.Link, by 2018. With high-voltage direct current transmission, 1,400 megawatts will flow in one direction at a time. [...] But if the North Sea electricity highway is really to fulfil the high expectations, power transmission networks on land must also be expanded. And that is equally difficult in both Germany and Norway. [...]

Norway has them too: protesters. In the area around Hardanger Fjord, the company was subjected to massive protests from inhabitants and environmentalists. These campaigners had little interest in seeing ugly new power lines in the landscape – and certainly not with the objective of possibly storing surplus energy from mainland Europe.

Source: C. Seidler, www.spiegel.de, 21.06.2012

M6 Newspaper article

M7 Dam of a hydroelectric power station in Norway

- Flooding of ecologically significant areas by damming the water
- Absence of springtime floods in snow melting periods and a seasonal shifting of the maximum water level into the winter months for the stretches of river below the dam
- Silting processes in the reservoir
- Influence on winter icing on and around the dammed lakes and runoff watercourses
- Changes in water temperature and air humidity

M8 Ecological consequences

Energy from the sun – solar energy

The power plant on the motorway and on the roof

The greatest source of energy in our solar system is the sun. The energy it delivers could solve all the energy supply problems in the world, because – in spite of energy losses on the way to the Earth's surface – the quantity of energy is more than five thousand times greater than all the energy needs of mankind throughout the world. Why do we not make greater use of this energy source?

1. Solar power plants like the one in Hardenberg (M8) can be seen more and more often, also along motorways. Explain how these power plants differ from "power plants on the roof" (M2/ M4/ M5/ M8).
2. Describe and give reasons for the distribution of solar plants (photovoltaic) in Germany (M6, atlas).
3. Give reasons for the choice of location for the solar power plant in Brandenburg (M6/ M8).
4. ⓞ Discuss the possibilities for the greater use of solar energy in Germany for future energy supply (M3/ M9/ M10).
   A Argue from the perspective of the German Federal Government.
   B Present your own opinion.
5. ⓢ With the aid of a map in the atlas, investigate the suitability of your school's location as a site for a solar power plant.
6. Research the possibilities and limits of solar updraft power plants for German's energy supply (internet).

Photovoltaic, Solar thermal energy
→ Renewable Energy Act

The Federal Office of Economics and Export Control offers financial subsidies for the installation of solar plants on buildings that are already built and in use, and, in the framework of innovation support, also on new buildings.
www.bafa.de

M1 Subsidies for solar plants

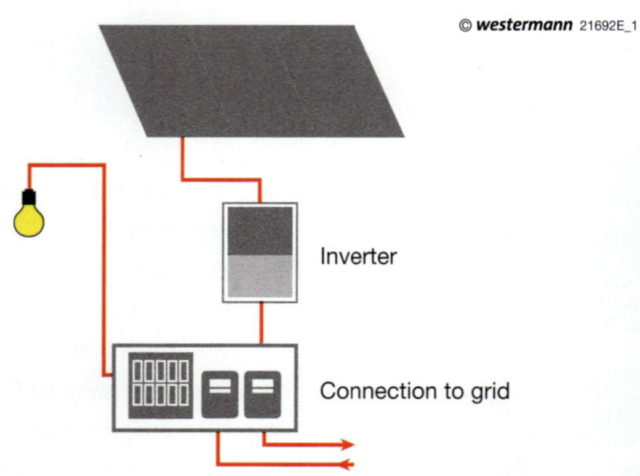

M2 How a photovoltaic plant works. Direct current is converted to alternating current in the inverter. The plant also works when the sky is cloudy.

The sun delivers limitless quantities of energy. Why is greater use not made of this energy source?
Solar energy can only be used when the sun's rays come into contact with the solar plant, i.e. not during the night. The efficiency also depends on how much energy is lost on the way to the ground. A further problem is that the solar energy cannot be stored. There is no way of responding to fluctuations in energy demand.
But there must be technical possibilities for storing the energy?
Yes, if you use the solar energy to boil a kettle you can store the energy in the form of hot water, so to speak, but that is only possible up to a point and of course there are also energy losses.
Solar plants are climate-friendly; there are no harmful emissions. Isn't that very positive?
You must also bear in mind that energy, water and chemicals are required for the production of the solar modules. I would also like to point out the high cost of electricity generation. The consumer must pay the added costs. This is also regulated through the Renewable Energy Act. If you want to increase the efficiency of solar energy by installing large plants, you need large areas of land. This can lead to land-use conflicts.

M3 Fictitious interview on the advantages and disadvantages of solar energy usage

→ Renewable Energy Act

The Renewable Energy Act (Erneuerbare-Energien-Gesetz, EEG) says that electricity from renewable energies can be fed in to the grid at favourable rates. It guarantees fixed feed-in tariffs for electricity producers.
Source: www.bafa.de, 2012

	Primary energy	Converted by	Secondary energy	Main uses
Heat	Thermal radiation	Solar thermal energy Solar collector	Hot water	Heating, service water
	Light	Photovoltaic Photo cell	Electric current	Heating, light, mechanical energy

© westermann 21693E_1

M4 Solar energy: capture and use

M5 Solar modules (photovoltaic) below and solar collectors (solar thermal energy) above

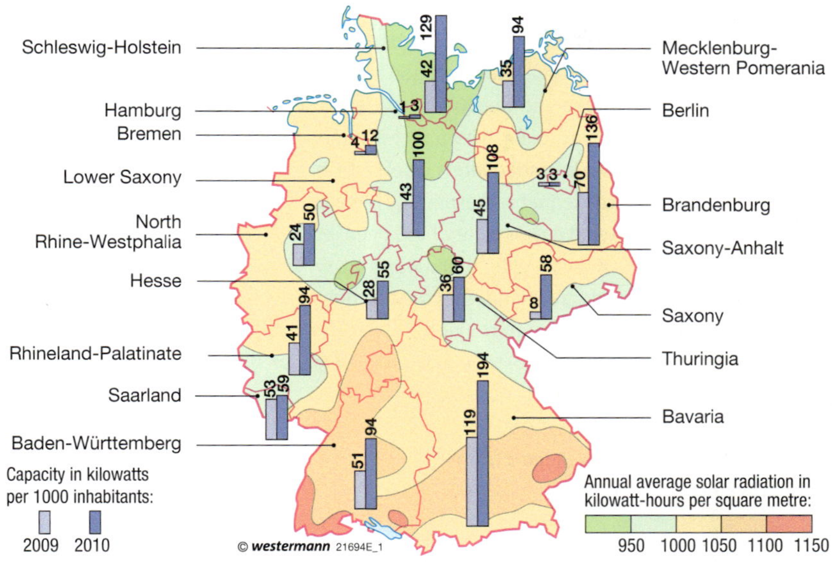

M6 Photovoltaic plants and solar radiation in Germany

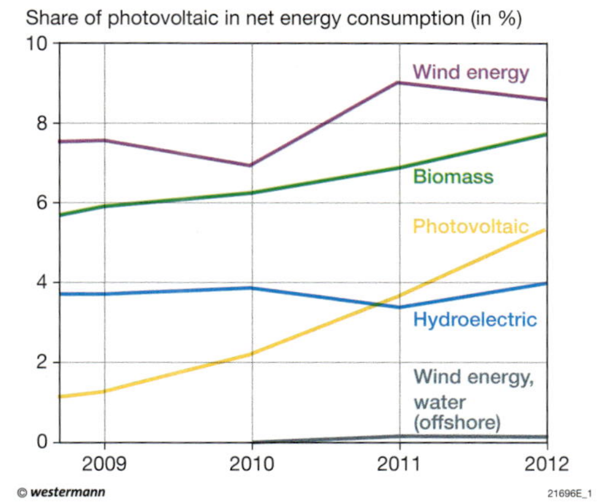

M9 Share of photovoltaic in electricity consumption in Germany

Brandenburg: The right position in Europe?

Electricity for 48,000 households: at Neuhardenberg airfield in eastern Brandenburg, a solar farm with a maximum capacity of 145 megawatts was installed in a record time of only five weeks. The electricity generated is sufficient to supply 48,000 households at peak times. Despite the photovoltaic power plant, the location is still used as an airfield.

Energy generation is based on the photovoltaic principle. In bad weather and during the night, the quantities of energy actually generated fall drastically to almost zero.

It will be more difficult to implement such gigantic projects in Germany in future, because projects with a capacity exceeding ten megawatts will no longer be subsidised. Comparable projects in Central and Northern Europe can only be operated with the aid of intensive support, because the energy yield is rarely efficient.

M8 Hardenberg solar power plant in Brandenburg

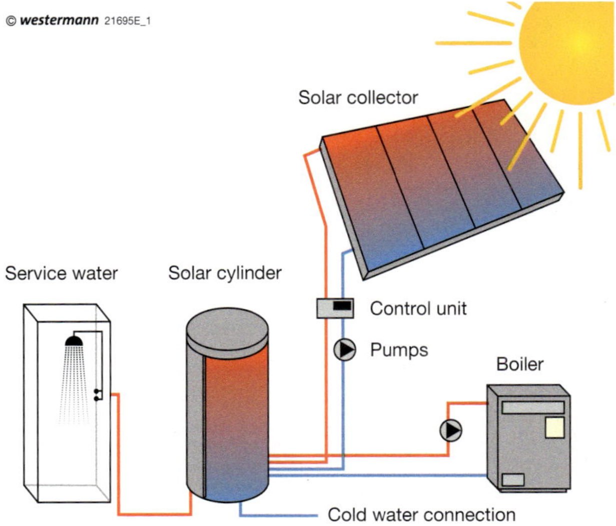

M7 How a solar collector works

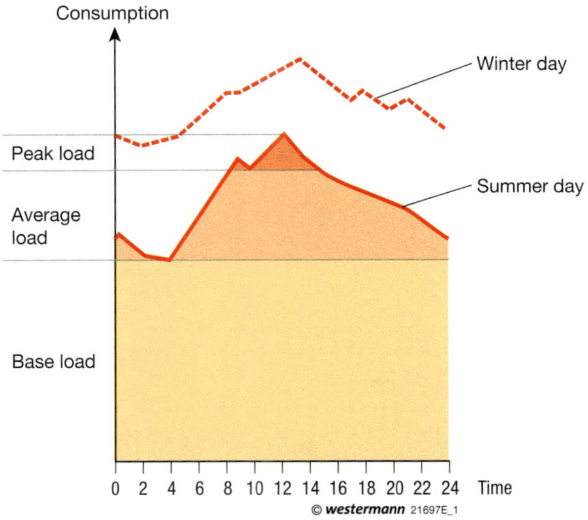

M10 Electricity consumption in Germany on summer/winter days

Not just a windmill – energy from wind turbines

Offshore wind farms under discussion

Wind energy is considered to be inexhaustible. The best sites for wind turbines are where the wind is strongest and where it is constantly blowing. These sites are in the sea, off the coast. Offshore wind power plants could therefore make an even greater contribution to our energy supply, but they are controversial.

1. Describe how the electricity from offshore wind power plants is fed to the consumer (M1/ M9).

2. Explain the problems arising from the location of the offshore wind power plants and the location of the electricity consumers (M4/ M8/ M9).

Ⓞ **3.** The installation of electricity transmission lines through the area where you live is discussed at a public meeting. Give reasons for your viewpoint.
   - **A** You represent the view that the high-voltage transmission line should be accepted.
   - **B** You protest against the route of the high-voltage transmission line.

4. Evaluate the importance of offshore wind power plants, taking into account their supply potential as well as their significance for sustainable conservation of resources and the environment (M2/ M5/ M6).

Ⓢ **5.** In a sketch map, present the transmission of electricity from the Alpha Ventus offshore wind farm to Stuttgart. Mark existing power lines and new route extensions (M9).

Ecological rucksack,
Offshore wind farm

weblinks

- Protest against power lines
- Alpha Ventus
- Bard offshore

Recently there have been heated debates about the environmental compatibility of wind energy. This discussion has many facets, ranging from the destruction of the landscape to the negative effects on fauna and the quality of life of the inhabitants.
In particular, the impact of large offshore wind farms on the animal world (underwater noise) is still being researched. The environmental consequences, especially the danger of collisions between birds and wind power plants, have been researched in many studies. The result of the investigations was that, while such a danger is comparatively minor on land, the construction of offshore plants must pay special attention to major bird migration routes and resting places. There are also possible safety problems for ships that have strayed from their proper course.

Source: www.vde.com

M2 Report from VDE

M3 Alpha Ventus offshore wind farm

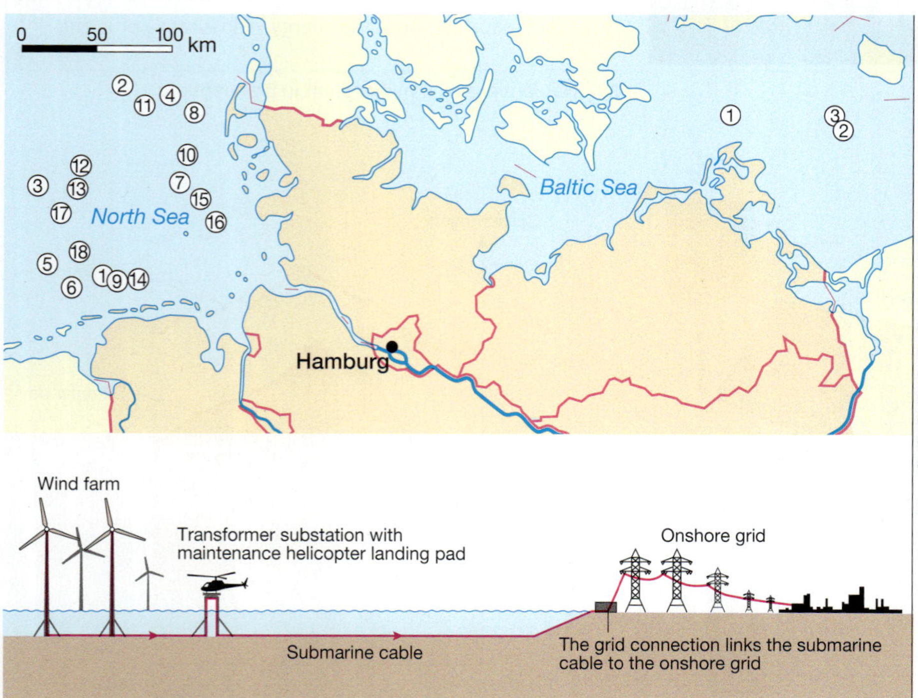

18 wind farms in the North Sea

① Alpha Ventus (formerly Borkum West)	12/208	60
② Sandbank 24	80/980	400
③ Bard Offshore I	80/320	400
④ Dan Tysk	80/300	400
⑤ Borkum Riffgrund West	80/458	400
⑥ Borkum Riffgrund	77/180	231
⑦ Nordsee Ost	80/250	400
⑧ Butendiek	80/80	240
⑨ OWP Delta Nordsee I (formerly: ENOVA)	48/251	216
⑩ Amrumbank West (& research platform)	80/80	400
⑪ Nördlicher Grund	80/402	400
⑫ GlobalTech I	80/320	400
⑬ Hochsee Windpark Nordsee	80/508	400
⑭ Gode Wind	80/224	400
⑮ Meerwind Ost	40/k.A.	200
⑯ Meerwind Süd	40/k.A.	200
⑰ Hochsee Windpark „He dreiht“	80/80	400
⑱ Borkum West II	80/k.A.	400

3 wind farms in the Baltic Sea

① Baltic II	80/80	300
② Arkona Becken Südost (& research platform)	80/210	400
③ Wikinger	80/200	400

Number of wind turbines: Pilot phase/planned final expansion (pink)
Total capacity in the pilot phase in megawatts (purple)

21702E_1

M1 Electricity from offshore wind farms

M4 High-voltage transmission lines in the landscape

M8 Citizens protest against the expansion of the transmission line network

Wind power cannot claim to be environment-friendly. Wind power is an expensive alternative energy. [...] The ecological rucksack of wind power needs to be calculated: how much energy is consumed until the rotors are in place, from the workers' journeys to the factory, through the energy for steel and concrete, to the electricity that [the] turbines themselves consume. Land use for building roads, for transmission lines, damage from stationary and moving machinery, [must also be] included in the calculation.
An environmental compatibility study is needed. Among other things it should also mention the burdens on people and their property, the noise footprint for homes and residential areas, nuisance in recreational areas, dangers for air and road traffic, natural damage to fauna and flora. Naturally [wind power] is also not uncontroversial and has its ecological price. The ecological rucksack for wind power is nevertheless significantly smaller in comparison to other energy sources.

Source: B. Dahlbender, Landesvorsitzende des BUND, www.windkraftgegner-schwarzwald.de

M5 Opinion on the use of wind energy

The ecological rucksack is a model developed by scientists as a way of presenting the pressure placed on the Earth by mankind's lifestyles and consumption. Every product is allocated a virtual weight based on the processes of extracting the resources through to the manufacture of the final product. The more nature is "used up", the heavier the ecological rucksack becomes.

M6 The ecological rucksack

The crisis in the offshore industry continues. The Bard Group has ceased its operations. The pioneer project Bard Offshore 1, the largest German offshore wind farm, will be taken over by a new company. The project with 80 units about 100 kilometres north of Borkum was opened at the end of August 2013. Its capacity of 400 megawatts theoretically corresponds to the annual electricity requirements of more than 400,000 households. But with investment costs of well over two billion euros, it was considerably more expensive than planned. Bard was also faced with technical construction difficulties and bad weather. The opening was delayed by several years.

Source: Handelsblatt, 13.12.2013

M7 The offshore industry

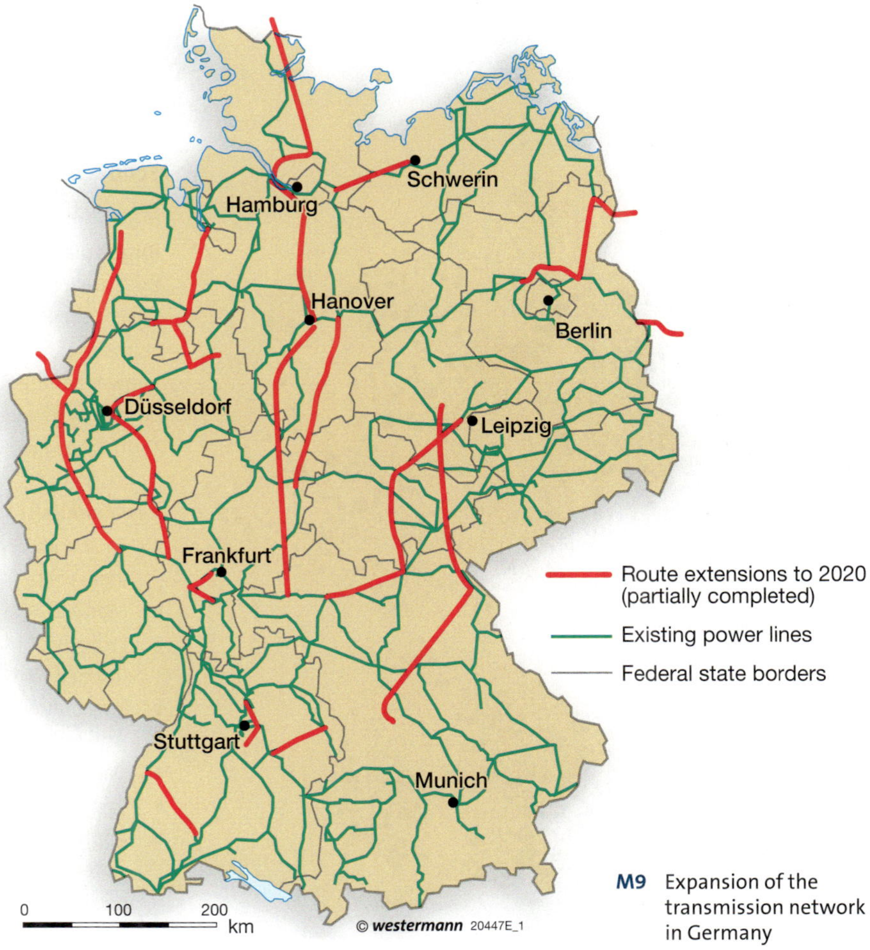

M9 Expansion of the transmission network in Germany

Biomass as an energy supplier

Biogas production under discussion

Biogas plants are shooting out of the ground like mushrooms. Their numbers in Germany have grown enormously. For many farmers, biogas production has become a second mainstay. The required biomass is made up of renewable raw materials such as maize and cereals, as well as manure and food leftovers. Currently however (2011), biogas only accounts for six percent of Germany's electricity supply, and the future role of biogas in sustainable energy use is controversial.

1. Describe the possibilities for the use of biomass for energy generation as well as its structural development over time (M5/ M8/ M6/ M11).
2. Compare biogas production at Schöpcherhof with biogas production at Bioenergie-Park Penkun in terms of capacity, input materials, supply potential and impact on the area (M3/ M4/ M7/ M10).
3. ⓞ Present the geographical preconditions for biogas production and the geographical impact of biogas plants
   A in a causal network (M1/ M8/ M9).
   B in a table (M1/ M8/ M9).
4. Assess biogas as an energy source in terms of its contribution to sustainable economic activity (M3/ M4/ M7/ M9/ M10).
5. Ⓢ With the aid of a map on agricultural land use in the atlas, analyse which regions in Germany are suitable locations for biogas plants (M1).

Biogas, Biomass, Renewable Energy Act, Sustainability

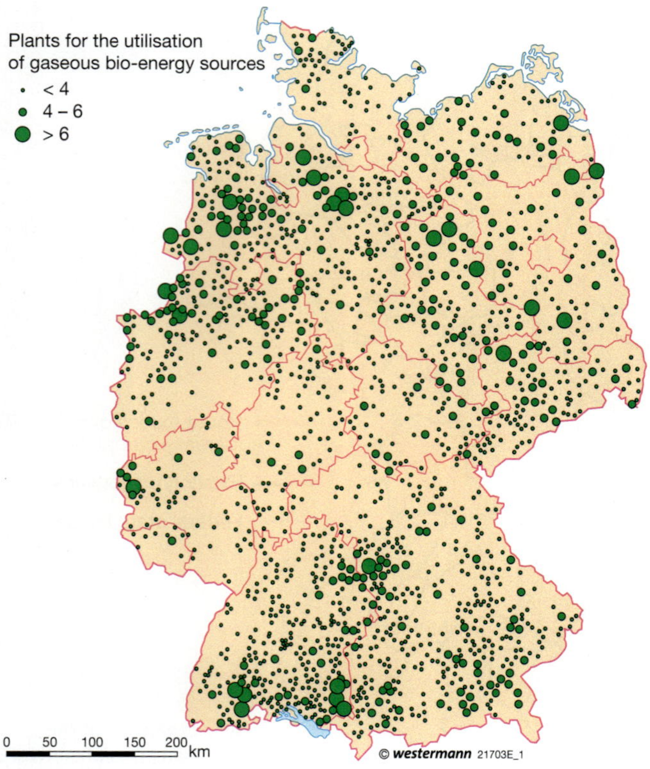

M1 Locations of biogas plants in Germany

M2 Biogas plant at Schöpcherhof in Lohmar

Were there no protests in the village against the biogas plant?
Dr. Lüpschen: I promised our neighbours that the nuisance from noise and odours would be reduced to a minimum. We live here too – we are the first to be affected. Our neighbours have given us their support.
But didn't you have problems in getting approval?
Yes, initially, because there were no biogas plants here in the Cologne district. The land use plan had to be amended.
How is the cooperation with your business partner Achim Schlehecker managed?
Mr. Schlehecker's company undertakes the disposal of the fermentation residues, i.e. the liquid and solid residues produced by the fermentation. He is also responsible for operating part of the shared plant.
Source: N. Kreuzberger, 2013

M3 Interview with the managing director of Schöpcherhof

Schöpcherhof Lohmar

Grassland operation with 200 dairy cows, free stall barns

Biogas plant:
Built area: 0.5 ha
Cost: 4.5 million euros
Number of fermenters: 2
1 power plant with 800 kW unit:
Electricity generation: 800 kilowatts per hour
Tariff for in-feed to electricity grid: 9.5 cents/kWh

Raw materials:
- Food and meal wastes as well as agricultural residues: 50% (delivery of food wastes from three businesses in Cologne)
- Manure: 50%
- Delivery of manure from farmers in the surrounding area (up to 10 km distant)
- Raw material deliveries per day: 60 t
- Raw material from own farm: 10 t
- Daily processing of biomass: 70 t
- Application of residues: fertiliser
- Application of waste heat: own use and supply of hot water and heating to an industrial company and a guest house in the village

M4 Profile of the biogas plant at Schöpcherhof, Lohmar

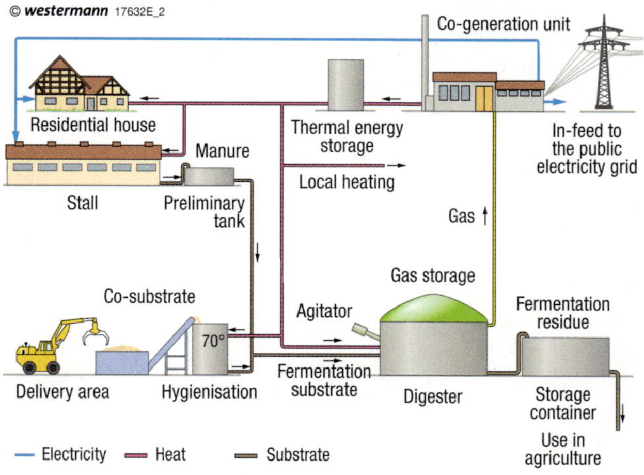

M5 Schematic diagram of a biogas plant

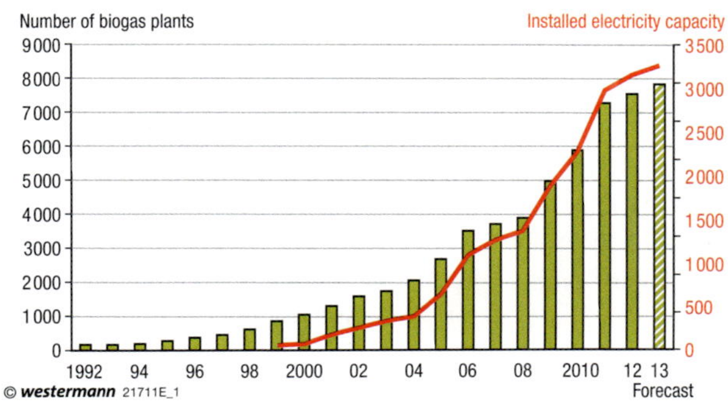

M6 Development of the number of biogas plants and installed capacity in Germany

Bioenergie-Park Penkun (Mecklenburg-Western Pomerania)

Limited company,
the world's largest biogas plant

- Built area: 15 ha
- Cost: 78 million euros
- Number of fermenters: 40
- Power plants, each with 500 kW units: 40
- Electricity generation: 20 megawatts per hour (equivalent to electricity supply for 40,000 households)
- Income per plant per year: 600,000 to 650,000 euros

- Total biomass capacity: 92,000 m^3
- Maize, cereals, grass processed per year: 460,000 t
- Manure processed per year: 84,000 t
- Delivery of raw materials from farmers in Poland, Brandenburg, Mecklenburg-Western Pomerania
- Application of residues: fertiliser
- Application of waste heat: heating, drying, steam generation, cooling, processing of residues in fertiliser factory

M7 Profile of Bioenergie-Park Penkun

When vegetable matter rots, fermentation gases are given off. Important for the use of these fermentation gases is methane (CH_4), a flammable, energy-rich gas, which however is also a greenhouse gas. It is produced by methane bacteria which convert the energy in the biomass to gas by fermentation. In biogas plants, vegetable matter is loaded and converted in a fermenter (digester) with methane bacteria with the exclusion of air. The gases are cleaned and fed into a co-generation unit for energy generation, while the liquid residues are processed into fertiliser. The waste heat can also be used (power-heat combination).
Maize, cut grass, cereals and sunflowers, but also manure and organic wastes, can be used for biogas production.
The starting materials, manure and organic wastes, are permanently available and not dependent on weather conditions. Local processing of the manure avoids the need for long-distance transport. The most frequently used bio-energy crop is maize, because it has a high energy yield and a high yield per unit area. For germination, maize needs a temperature of 7 – 9°C.

M8 Functioning and location requirements of biogas plants

- Biogas plants require industrialised, inappropriate animal husbandry in order to obtain manure.
- The landscape is changed by the large-scale cultivation of maize.
- Growing of maize in monoculture also burdens the groundwater.

M9 Criticism by an expert from Friends of the Earth (BUND)

You have installed 40 biogas units here, so it is a real bio-energy park, as the name indicates.

Why did you choose Penkun as the site?

Dr. Schramm: We found outstanding conditions in Penkun.
In the immediate vicinity of the bio-energy park, 30 to 60 percent of the land is used for agriculture, that is more than 230,000 hectares of farmland.
The park's supply of agricultural raw materials (especially silage maize) from the surrounding farms is secured by long-term contracts. [...]
In addition, the electricity can be easily in-fed into the grid through existing and nearby in-feed points.

Source: Forum Biogas spezial, EnviTec Biogas

M10 Interview with Dr. Balthasar Schramm, chairman of the operating company

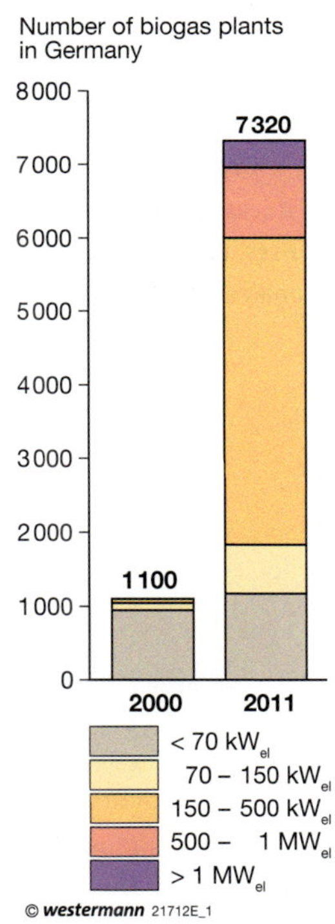

M11 Biogas plants by size (kWel) 2000 and 2011

Biomass as an energy supplier

In the tank or on the plate? Food crisis in Mexico

"What is celebrated on one side of the border as a breakthrough in environmental technology is a cause of fears about everyday survival on the other side." This report appeared in Tagesspiegel on 2 February 2007. It was about the so-called "tortilla crisis". For Mexicans, the tortilla is part of their daily diet. To produce tortillas, maize is required. But maize is also the basic raw material for the production of ethanol, a bio-fuel.

1. Explain the reasons for the tortilla crisis (M1/ M3/ M5/ M9).
2. Analyse the cartoon (M10).
3. Summarise the advantages and disadvantages of ethanol production in the USA in a table (M1 – M3/ M7 – M9).
4. ⓞ Give your opinion on the statement that the use of bio-energy crops for the production of bio-fuel makes a contribution to sustainable economic activity.
   - **A** Argue from the perspective of a supporter of this thesis.
   - **B** Argue from the perspective of an opponent of this thesis.
5. ⓢ Discuss the possibility of covering seven percent of petrol consumption in Germany with bio-fuels from renewable raw materials (M9).

M4 Citizens protest in Mexico City on 31 January 2007 against rising prices for maize, the basic foodstuff for Mexico's lower income groups

Bio-fuel

The price of maize on the world market depends on supply and demand. When global maize stocks are low, the price rises. When demand for maize is high, prices are also high.

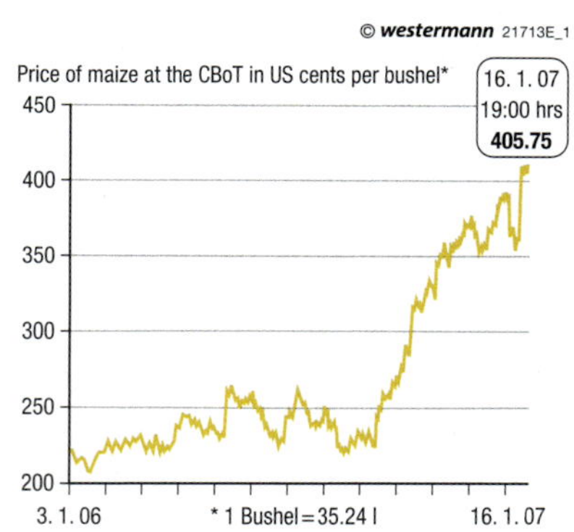

M2 Maize price on the Chicago contract market, 2007

Ethanol production in the USA	
2000	8 billion litres
2007	18 billion litres
2011	53 billion litres

M1 Ethanol production in the USA

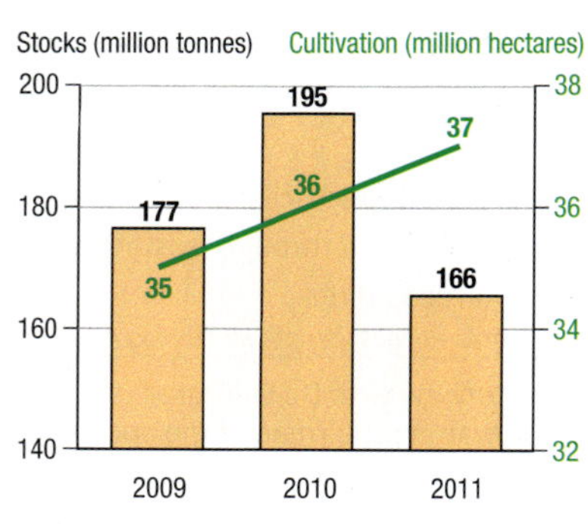

M3 Maize cultivation and stocks in the USA

Tortilla crisis in Mexico

It is an object lesson in so-called free trade. After the accession of Mexico to the Nafta free trade agreement, many farmers could no longer compete with the highly-subsidised maize from the USA. As a result, the country became more and more dependent on its rich neighbour for food. But with increasing demand in the USA for maize for the production of bio-alcohol for cars, many Mexicans could no longer afford to buy the basic ingredient for the production of tortillas. [...]

Mexicans obtain nearly 50 percent of their total calorie intake from the tortilla, which is also known as "Vitamin T". With an average income of only 200 pesos per day, it is easy to imagine what rising prices mean.

But 20 million Mexicans are already classified as extremely poor and have to live on 20 pesos per day, about 1.5 euros. Last year, this was enough to buy over three kilos of tortilla for a family. But now a kilo is already traded for 10 to 11 pesos in Mexico City, while in some parts of the country the price is already pushing towards the 20 peso mark. [...]

Many other products, such as chicken and pork, are also affected by the rising prices because the animals are fattened on maize.[...]

The competition for the crop between the 800 million car drivers in the world, who wish to maintain their mobility, and two billion of the poorest people in the world who just want to survive, is developing into a dramatic problem.

In dozens of low-income countries which are dependent on grain imports, like Indonesia, Algeria, Nigeria and Mexico, exploding food prices could lead to food riots in the cities. [...]

Various organisations complain that the tortilla crisis has been used to promote massive imports of genetically modified maize into Mexico. The environmental organisation Greenpeace declared that certain multinational companies that have patented GM maize are also using the crisis to "back up their demand to open up the country for the cultivation of this crop."

It is anticipated that permission for this will already be granted this year. As a first step, between half a million and one million hectares of land will then be sown with GM maize.

Source: R. Streck, www.heise.de, 29.01.2007

M5 Cause and effects of the tortilla crisis in Mexico

M6 Maize cultivation in the USA

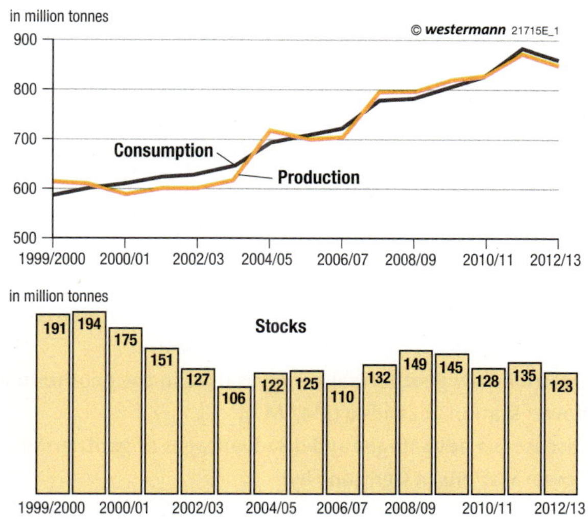

M8 Maize: production, consumption and stocks worldwide

Protest against production of bio-fuels in the USA

The Hamburg-based environmental organisation Rainforest Rescue has called for a protest action against the promotion of so-called bio-fuels by the US government. The production of ethanol and other bio-fuels competes directly with the cultivation of food and feedstuffs, the organisation writes. [...]

"Politicians still present the system of highly-industrialised agriculture and the production of bio-fuels as a great success," Rainforest Rescue complains. They ignore the ecological, social and health consequences: "Because this form of agriculture is directly reliant on the enormous use of fossil fuels, herbicides, pesticides and artificial fertilisers. The chemicals used cause long-lasting pollution of the rivers and soil."

The worst drought for 50 years has already hit 61 percent of the land area of the USA. Emergencies have been declared in more than 1,000 rural districts. No rain is expected in the foreseeable future – instead there will be further increases in temperatures and aridity. A few days ago, the US Department of Agriculture (USDA) quantified harvest losses for maize alone at 46 million tonnes; losses that could double or triple if the drought continues until the end of August. "Prices for maize and cereals on the commodity exchanges are already shooting up to record heights," the NGO [non-governmental organisation] writes.

In its current study, "Bio-energy – Chances and Limits", the National Academy of Sciences Leopoldina comes to the conclusion that, in Germany too, bio-energy as a sustainable energy source can make a contribution to the energy transition, the protest call continues. In Latin American countries, the use of food crops for fuel production has met increasing resistance in recent years.

Source: www.Portal amerika21.de, 31.07.2012

M7 Position statement on the production of bio-fuels

The USA puts more focus on environmental protection in transport, so tortilla prices explode in Mexico

The United States is by far the largest maize producer in the world. It accounts for 70 percent of world exports. The State of Iowa alone produces more than the whole of Canada. Therefore, whatever happens in the USA has worldwide consequences. And here, demand has shifted dramatically. [...]

"The attempt to solve one problem, [namely] the growing dependence of the United States on oil imports, creates a much greater problem," says Lester Brown [head of the Earth Policy Institute in Washington]. Instead of building alcohol refineries, it would be much cheaper to produce fuel-saving cars.

But the subject of "food for cars" is not confined to the USA and Mexico. In Germany too, researchers see the possibility of covering seven percent of petrol consumption with renewable raw materials (especially oilseed rape); the worldwide figure could reach 27 percent.

Source: N. Piper, Süddeutsche Zeitung, 19.05.2010

M9 Newspaper article on "food for cars"

M10 Cartoon

Using the Earth's heat – geothermal energy

Energy from the depths

The Earth's heat is available in unlimited quantities. In contrast to other renewable energy sources such as the sun, wind and water, geothermal energy is independent of the season or time of day, wind or weather. Because it is always available, no energy storage is needed.
Geothermal energy is nevertheless only used in Germany to a very limited extent.

1. Analyse the prerequisites for the location of geothermal power stations in Germany (M3/ M6, atlas).
2. Describe how geothermal energy is used in the geothermal power station at Landau (M4/ M7).
3. ⓞ Discuss the advantages and disadvantages of geothermal power stations in Germany by:
   A writing a letter from the perspective of a member of a citizens' action group (M5/M8).
   B writing an article for the school magazine (M5/ M8).
4. ⓞ Hypothesis: In future, geothermal energy will become more important for the supply of energy from renewable resources in Germany.
   A Assess the possibilities and limits of the use of geothermal energy in Germany.
   B Give your opinion on this hypothesis.
5. Ⓢ Investigate what areas around the world are suitable for the use of geothermal energy. For each continent, note an area that is particularly suitable (atlas).

Earthquake, Geothermal energy, Hypocentre

weblink
- Planet Wissen Erdwärme

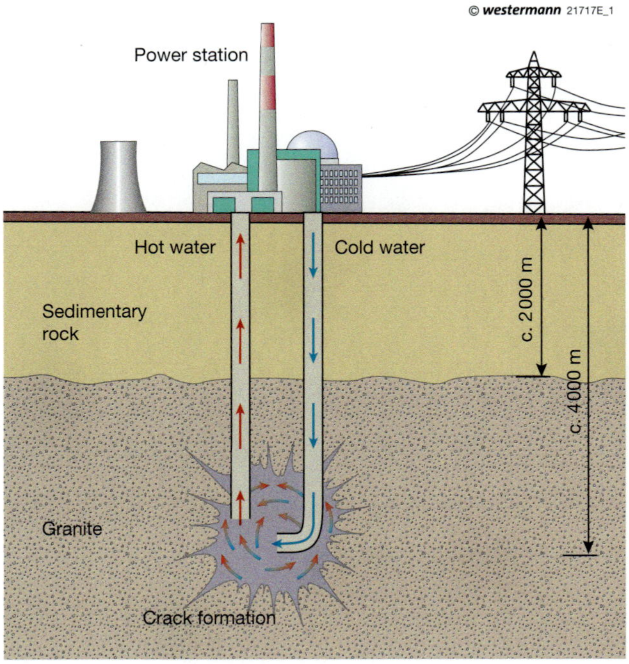

M2 The Hot Dry Rock process

Thermal springs (springs with a water temperature of at least 20°C) are produced by heat from the Earth's interior: geothermal energy. At geological fault zones, the heat energy can easily rise to the Earth's surface. This heats the water that is present in the rock layers.
Geothermal power stations use this energy. There are now several hundred geothermal power stations around the world, generating some 8,400 MW of electricity. In Germany, more attention has been paid to the use of geothermal energy since the beginning of the 21st century. The first geothermal power station was built at Neustadt-Glewe in Mecklenburg-Western Pomerania in 2003.
A distinction is made between shallow and deep geothermal energy. Shallow geothermal energy (up to a depth of 400 m) is used directly for the heating of buildings or for hot water supply by means of e.g. geothermal heat probes.
With deep geothermal energy, the thermal water is raised to the surface from a depth of more than 400 m and transferred into a water circuit by means of a heat-exchanger. This can be used either to generate electricity or to feed-in heat into a district heating grid. The cooled water is fed back into the earth.
The Earth's heat can also be used when no thermal water is present in the deep layers. This is done by the Hot Dry Rock process, in which water is forced underground where it heats up, thus creating "artificial thermal water".

Based on: S. Bodmer and others, Thermalwasser – Ein heißer Schatz aus der Tiefe. In: Praxis Geographie, Heft 4/2010

M3 Use of geothermal energy

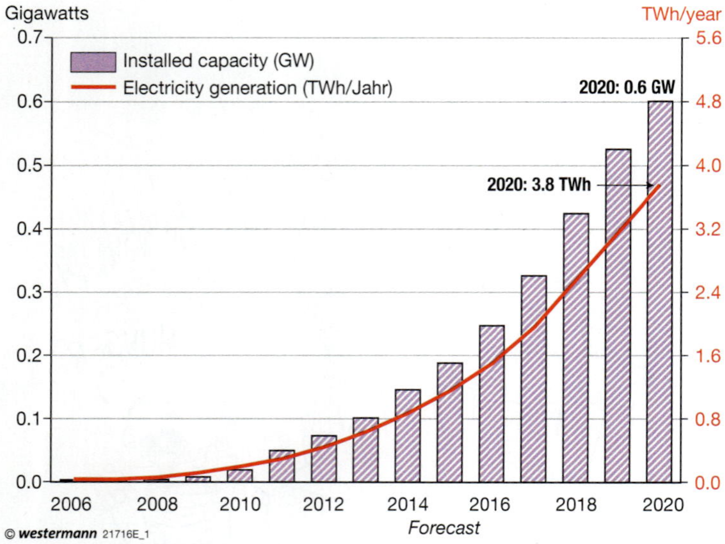

M1 Electricity from geothermal energy in Germany

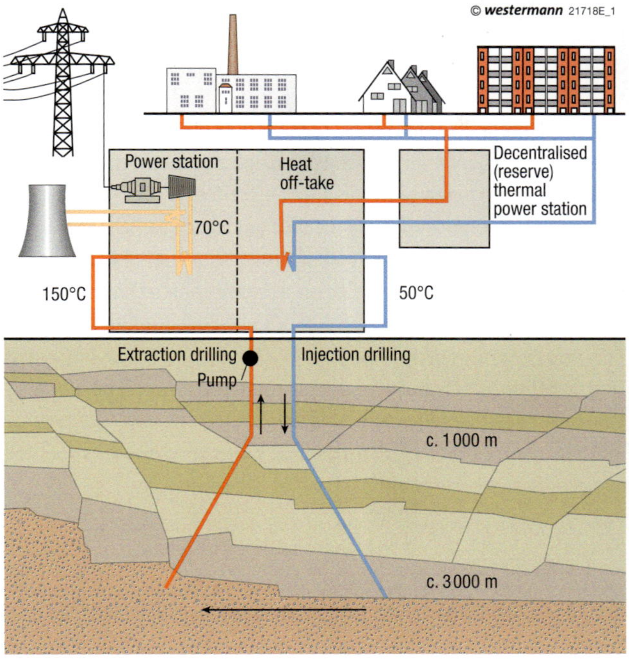

M4 Geothermal power station at Landau/Pfalz

M7 Geothermal power station at Landau/Pfalz. In 2003, thermal water with a temperature of 150°C was discovered at Landau at a depth of 3,000 m. The power station was brought into operation at the end of 2007. The question of whether the geothermal power station is really responsible for additional seismic activity is still under discussion. Capacity: electricity generation 3 MW, heat energy 6 to 8 MW.

In Wörth-Schaidt, a citizens' action group is protecting against the building of a geothermal power station. They are supported by the works council of a local company producing car components, due to fears that vibrations from the power station could disrupt production and the factory may have to close. The people in Schaidt are afraid that the power station will cause earthquakes and noise.

M5 Protests against the building of a geothermal power station

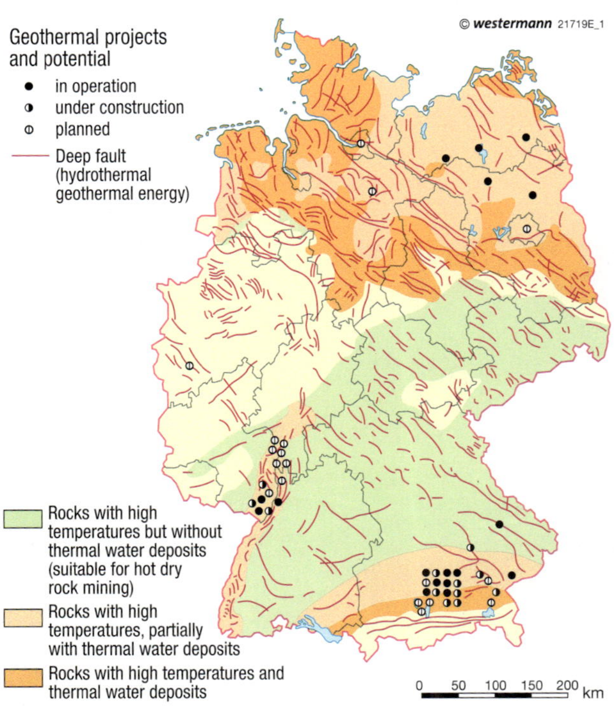

M6 Current and potential sites for geothermal energy in Germany

Two earthquakes close to the power station

In the town of Landau in Rhineland-Palatinate, there were two small earthquakes on 15 August and 14 September with magnitudes of 2.7 and 2.4 on the Richter scale. The likely cause is considered to be the geothermal power station [...]. An expert group set up by the Federal state government will clarify the matter [...].
As economics minister Hendrik Hering (SPD) reported on 3 September in reply to a question from the FDP in the state parliament, the first quake is being investigated in more detail on the basis of data from 49 measuring stations in various local and regional networks. According to preliminary results from the State Office for Geology and Mining, the earthquake's hypocentre lay some 500 metres north of the Landau geothermal power station at a depth of 2.5 to 4 km. There is also further evidence of a connection between the earthquake and the use of geothermal energy at Landau.
But this connection cannot be proved with certainty at the current state of knowledge, because the Upper Rhine Rift is a region where frequent small earthquakes have been registered. As well as deep geothermal energy, the use of shallow geothermal energy in Germany has recently also faced problems of acceptance after massive damage to buildings occurred in the town of Staufen in southern Baden-Württemberg in the course of geothermal probe drilling.
The laying of geothermal probes is widely practised in order to provide heat pumps with the highest possible temperature, independently of seasonal fluctuations. At Staufen, geothermal heat from a depth of some 140 metres was to be used. But evidently the drilling cut open a massive layer of gypsum-keuper, which swells and turns to gypsum on contact with water.
The swelling displaced the ground level in the whole of Staufen's old town. The town reported on 17 September that the number of damaged houses had further increased. 217 private houses and seven municipal buildings are currently affected.
Source: www.udo-leuschner.de, 09/2009

M8 Effects of the use of geothermal energy?

Using the Earth's heat – geothermal energy

Iceland – supplier of "green energy" for Europe and the world?

Iceland – the land of fire and ice. This often-used image makes reference to natural phenomena in Iceland. Volcanic activity and glaciers mark the landscape. 37 active volcanoes, hot springs and geysers demonstrate Iceland's geothermal potential. Temperatures of 350°C are to be found relatively close to the Earth's surface. Steam raised from drill holes is already used for 90 percent of heating supplies for all households. In geothermal power stations, so much electricity is generated that hydrogen can also be produced – and hydrogen is a possible energy source of the future.

1. Explain the geothermal potential in Iceland (M6, atlas).
2. Show the importance of geothermal energy for Iceland's energy supply and economic development (M1 – M3).
3. The use of geothermal energy in Iceland is not uncontroversial. Compare the concerns of the citizens' action group against the expanded use of geothermal energy with those of the citizens' action group in Germany (M7/ M9, p. 153 M5)
4. Iceland would like to become the first "hydrogen society". Investigate the feasibility of this idea (M3).
5. ⓞ Assess the possibilities for Iceland as energy supplier for Europe.
   A Limit yourself to electricity supply (M9/ M10).
   B Consider both electricity supply and the supply of hydrogen (M3/ M9/ M10).
6. Ⓢ In the USA, a new geothermal power station was brought into operation in Nevada in 2010. Analyse the location conditions for geothermal energy in Nevada (atlas).

Geothermal energy, Hydrogen

The main energy-intensive industries are companies producing silicon metal and aluminium. Companies in this sector have settled in Iceland because it offers cheap production with low electricity prices, even though bauxite, the raw material necessary for aluminium manufacture, must be transported to Iceland from Australia and Brazil. This logistical expense shows how attractive Iceland's cheap energy is for this industry.

Source: www.eldey.de

M1 Using energy for economic development

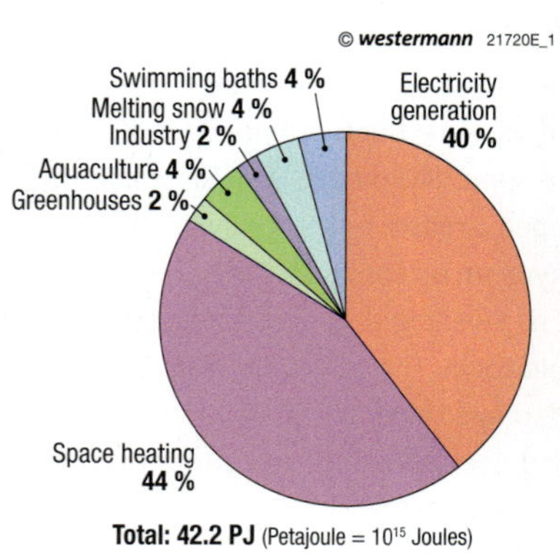

M2 Uses of geothermal energy ▷ in Iceland

Iceland aims to become the worldwide pioneer in the energy-intensive production of hydrogen, the fuel of the future. [...] The Icelanders dream of one day being able to ship their clean hydrogen all over the world.

The principle is simple but its technical implementation is complicated. With very energy-intensive electrolysis, water is split into hydrogen and oxygen with the aid of electricity.

Hydrogen is a volatile and very reactive gas which can only be stored under high pressure or in special tanks. In a fuel cell, hydrogen and oxygen produce electricity on a membrane in so-called cold combustion.

This also generates heat. The exhaust gas is steam. But an argument that is often put forward by critics against hydrogen as a energy source is its high energy consumption.

In a car, a fuel cell can drive an electric motor. In a house, a fuel cell can produce electricity and heating at the same time.

Sources: Energie im Überfluss. Island und seine natürlichen Quellen, www.fluter.de, 8.8.2006; Vulkane speien, Erde liefert Wärme, www.greenmag.de, 26.6.2010

M3 Use of geothermal energy with new hydrogen technology

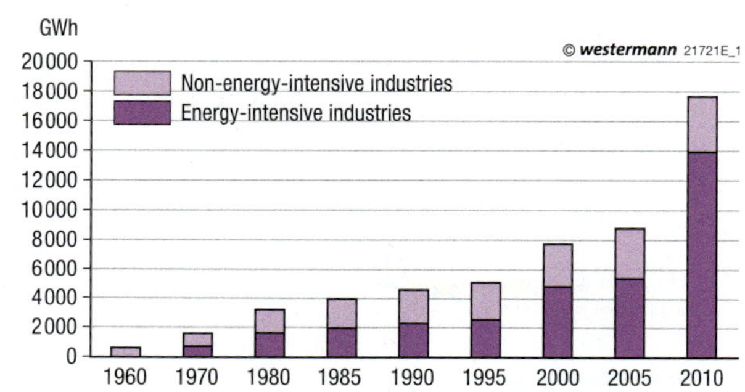

M4 Energy consumption in Iceland

M5 Geothermal power station in Iceland

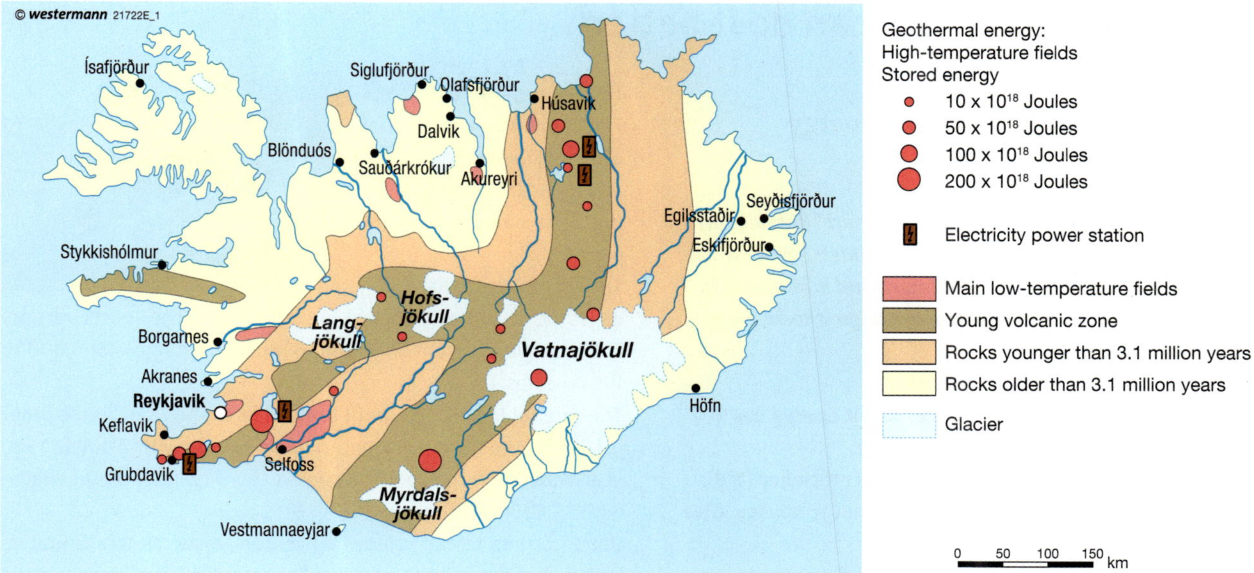

M6 Geothermal energy in Iceland

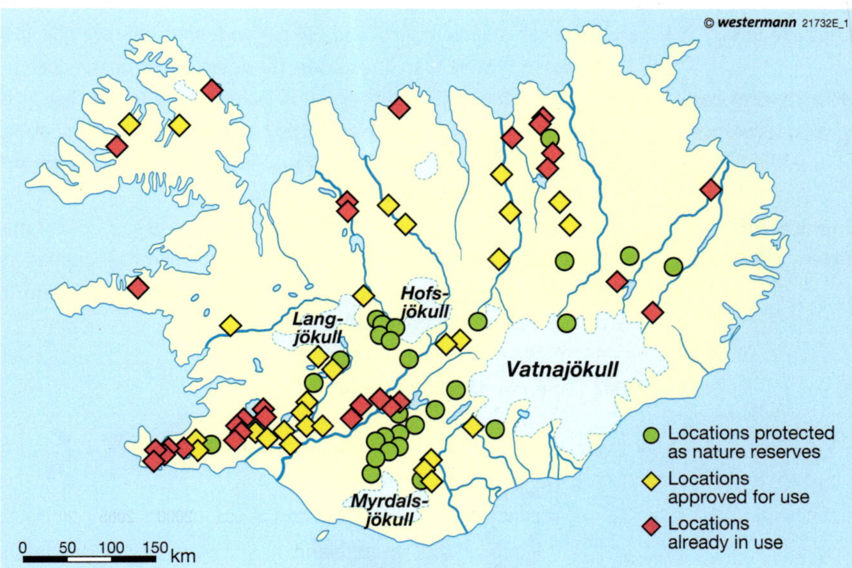

M7 Locations in the framework energy plan

- Air pollution
- Damage to health from hydrogen sulphide emissions
- Low rainfall means that the hydrogen sulphide will not be washed out of the air
- Damage to vegetation around the power station
- Drainage problems
- Less nutrition in the groundwater current due to cooling

Source: Akureyri-Wochenblatt, www.icelandreview.com, 13.09.2013

M9 Arguments of the citizens' action group against greater use of geothermal energy

Since Great Britain cannot rely on an adequate supply of imported electricity in the medium term, the longer-distance connection to Hamburg could come into consideration. [...]
The technical and economic dimensions of such a project would be [...] unique in the world. It would need to cover a distance of 1,830 km with depths of up to 1,100 m. [...] The electricity exports [could be] achieved by two high-voltage direct current cables laid in parallel. Special ships would need to be built to lay the cables.
Since the connection would pass through important fishing grounds, special precautions would also need to be taken against damage to the 12.2 m diameter cable from anchors or drag nets. And because the cable laying would only be possible in the relatively calm summer months, construction would take 6 to 10 years.

Based on: D. Bothe & E. Gläßer, Energiewirtschaft auf Island, In: Geographische Rundschau, Heft 3/2001

M8 Supply of electricity from Iceland to Germany?

M10 Possible route for laying an undersea cable from Iceland

weblink

- youtube Svartsengi-Kraftwerk

Energy efficiency – we can do it better!

Energy saving also means using energy efficiently

At the IFA consumer electronics exhibition, the word "smart" (intelligent) could be seen everywhere. There are smart TVs, smart tablets, smart washing machines and smart dishwashers. Now, the Smart Grid is intended to be the electricity grid of the future. What role can intelligent technology play in increasing energy efficiency?

1. Arrange the possibilities for more efficient energy use in a structure sketch (M1 – M3/M5/M6/M8).
2. Explain how Smart Grids increase energy efficiency (M8).
3. The future of house-building is in plus-energy houses. Give your opinion of this statement.
   Also take into account the importance of plus-energy houses in the Smart Grid (M4/M5/M7).
4. Show advantages and disadvantages of Smart Mobility, and give reasons why you would or would not use Smart Mobility (M6).
5. ⓞ Discuss the possibilities and limits for reducing energy consumption by means of modern technology, with reference to its geographical preconditions.
   A Draft a reader's letter about this.
   B Present the possibilities and limits in graphical form.
6. Ⓢ Give your opinion on increasing energy efficiency through Smart Mobility and the Smart Grid.

Smart Grid, Smart Mobility, → Energy efficiency

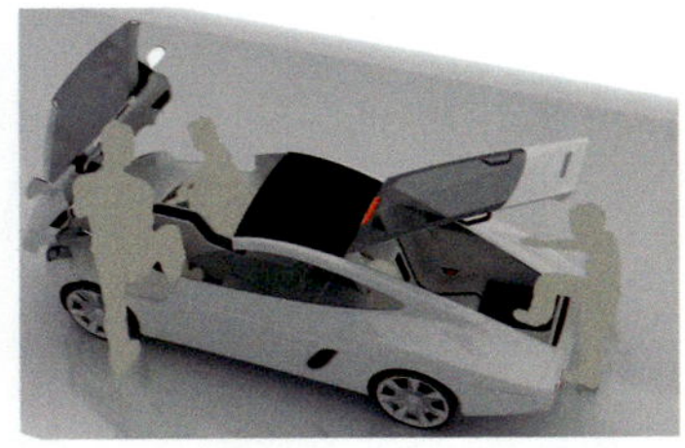

The car is called Loremo LS. The name is an abbreviation of "Low Resistance Mobile", and LS means "light and saving". The name is also the manifesto.
The car is built so that it only consumes 2.5 litres of petrol for 100 km – in the sports version. The ecological version only uses 1.5 litres. The fun [costs] 11,000 or 15,000 euros. [...]
The exterior lines of the car fit to the flow of the wind like no other [...]. Under the hood [...], a two-cylinder engine with 15 kW (20 Hp) is at work, or in the sports version a three-cylinder diesel engine with 36 kW (49 Hp).
But according to the Loremo's inventors, a switch to alternative forms of energy is quite possible. The record low fuel consumption is based on a completely new form of construction, slim and with no use of expensive high-tech materials.
[...] The car is 384 cm long and 136 cm wide, but only weighs 450 kilograms. [...] More than unusual are the doors, which do not open to the side but forwards and backwards. Even more unusual: they give access to 2 + 2 seats that are arranged back-to-back. [...] The seating arrangement is energy-saving because it provides good headroom in spite of an extremely low roof contour. [...]
The Loremo Eco takes a not-exactly-breathtaking 19 seconds for the "sprint" from 0 to 100 km/h. Its top speed of 165 km/h is capped at 140 km/h. The sports version can achieve a top speed of 220 km/h with a 9-second run up to 100 km/h.

Source: A. Ege, www.focus.de, 21.02.2006

M2 Loremo – the car of the future?

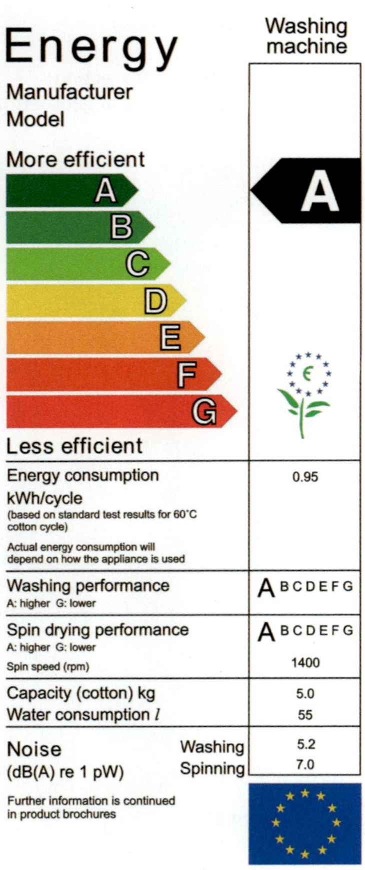

M1 Energy-saving in the household

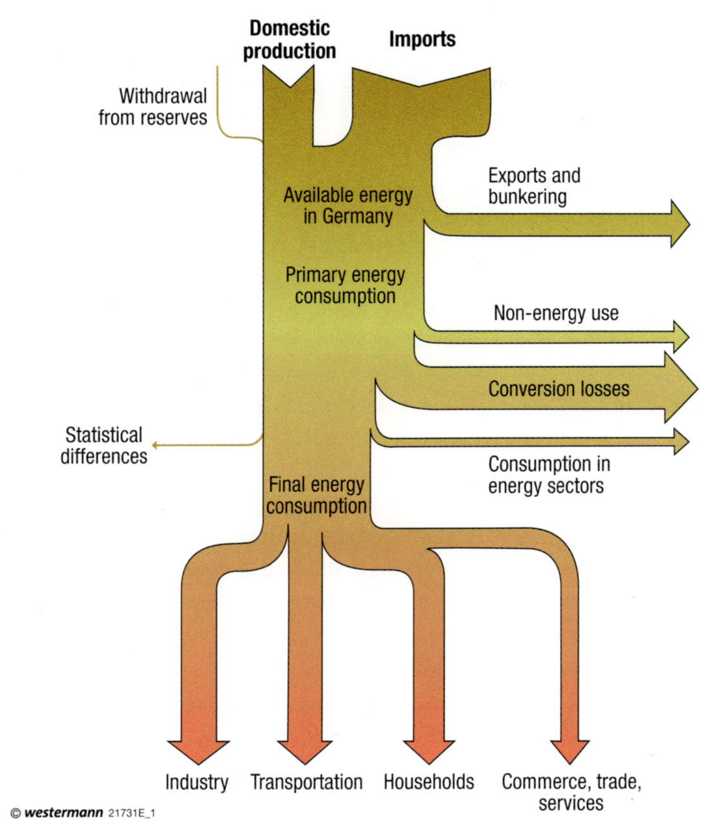

M3 Energy flow diagram for Germany

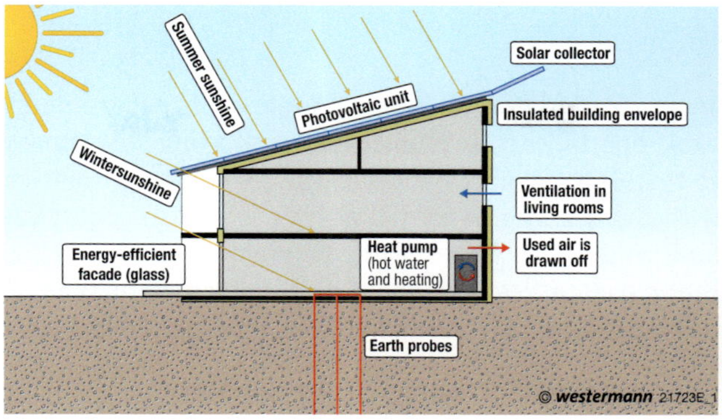

M4 Diagram of the plus-energy house near Stuttgart

M7 Plus-energy or active house near Stuttgart

Climate-conscious building has been a tradition in Germany for more than 30 years. In the early 1990s, the low-energy house was [...] the yardstick for energy-saving construction. This has been the statutory minimum standard for new buildings for over 15 years. Research and rising energy prices have led to ever-greater savings potential.
Today there are zero-energy houses, and even plus-energy or active houses.
A house is zero-energy when its energy needs can be theoretically covered by renewable energies. Over a period of a year, it produces at least the quantity of energy required to cover its needs.
The zero-energy house is often confused with the passive house. The main difference is that, in a passive house, the name is derived from passive energy use.
A passive house, as a rule, uses solar heat or the waste heat in the house to generate (passive) energy, but it still requires energy from outside.
The future of energy-efficient building will be with the plus-energy or active house. These produce more energy than they consume. With the energy that the plus-energy house generates but does not consume, an electric car could be charged, for example.

Source: www.energie-wissen.de

M5 From passive house to plus-energy house

→ Energy efficiency
Energy efficiency is a measure of the amount of energy required to achieve a given benefit. A process is efficient when a given benefit is achieved with a minimum amount of energy. The development of Smart Grids and Smart Mobility uses an approach that links energy supply with modern communication. As well as electricity, the grids also carry information. This permits supply and demand for electricity to be better coordinated.

Smart Mobility: the term stands for intelligent and sustainable mobility solutions. Smart Mobility is a solution that will make possible energy-efficient, convenient and cheap mobility. It is all about [...] optimising the use of existing mobility options by means of information and communication technologies. New kinds of mobility solution (car sharing, rent-a-bike) are being introduced. [...] But the intelligent use of these solutions (choice of the means of transport depending on the transport situation) fails on account of the varying preconditions for accessing these solutions. A Smart Mobility solution, which provides information and means of payment regardless of the kind of transport used, and which already exists to some extent abroad, is therefore not possible. If it were possible to permanently reduce the pressure to own and use cars, and if this were accepted by drivers, it could contribute to reducing traffic congestion, improving energy efficiency, increasing road safety and reducing emissions. The share of the so-called environmental network (public transport, bicycle and pedestrian traffic) would increase. In future, smartphones and developments such as tablet PCs will change information- and communication-based solutions still further. [...] Such devices will allow users to dynamically access information, also during their journey, i.e. when they need to change their planned travel method. And they can also make information available to others via smartphone GPS positioning and in connection with Facebook, Twitter and Co.

Source: S. Wolter, University of Duisburg.

M6 Smart Mobility – intelligent transport networking

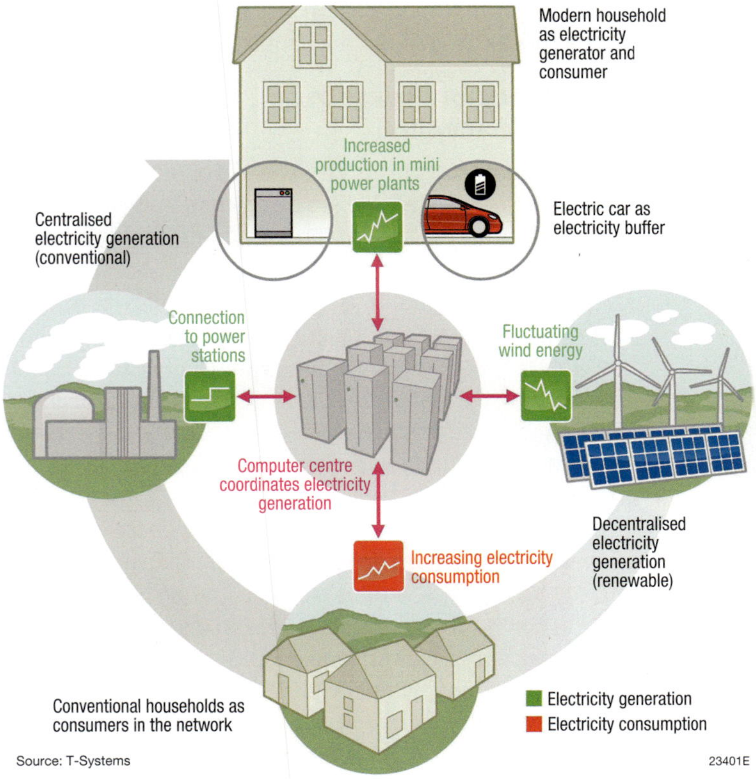

M8 How a Smart Grid works

The essentials in brief

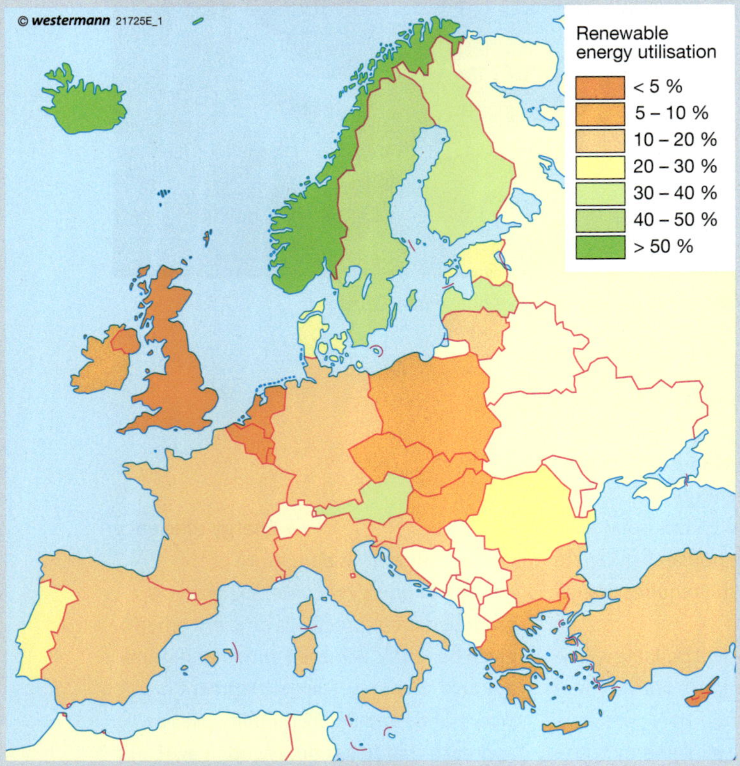

M1 Renewable energy utilisation in Europe

The conservation of resources and climate protection are among the challenges of the 21st century, which are being met in Germany by the energy transition for example. Instead of fossil energy sources, renewable energies will make a greater contribution to energy supply. But the geographical preconditions for the use of renewable energies vary, as does the geographical impact of their use.

Possibilities for the use of renewable energies

The technical prerequisites for the use of renewable energies are well advanced and are being constantly improved. In seeking opportunities for the use of renewable energies, a country must naturally take into account the available geographical preconditions. If hydroelectric power is to be used, there must be a sufficient supply of water as well as the greatest possible relief energy. Norway, for example, is a country where these preconditions exist. For the use of solar energy, the duration and intensity of solar radiation are important. Although solar collection units can be slanted to catch the angle of incidence of the sun's rays, areas near the tropics are also favoured for the use of solar energy because of their cloudless skies. For the use of wind power, not only the strength of the wind is decisive, but also winds that blow as constantly as possible. Coastal districts and areas at a higher altitude offer good conditions. Alpha Ventus is an example of an offshore wind farm which is particularly suitable on account of high and constant wind speeds.
Biomass can be used for energy generation, for example in biogas plants. Favourable locations for biogas plants are areas used for farming. But the location is not necessarily of vital importance if the problem of transporting the biomass to the biogas plant can be solved. It is a different matter with the use of geothermal energy, for which geological fault zones are a prerequisite, preferably with deposits of thermal water. Iceland is an example of a country that offers ideal conditions for the use of geothermal energy.

The transmission and storage problem

The use of renewable energies requires that the electricity they generate can be delivered to the consumer without major losses. Transmission lines have become very long, and are often not sufficiently available, as the example of Germany shows. But losses through the transmission of electricity in the power grid can be minimised by further technological developments. Another problem is the storage of energy. Pumped storage hydroelectric power stations offer storage possibilities, but solar and wind energy do not.

Usage at any price?

But the use of renewable energies also leads to problems that are debated from an economic, ecological and social perspective. The example of dams in Brazil clearly shows that they may also be associated with changes to, or even the destruction of, habitats. If populated areas are affected, the people may need to be resettled. Fauna and flora may also be affected. Changes in the landscape may result from the construction of wind farms and the installation of transmission lines, the wide-scale sowing of maize in the landscape for use as biomass, and the building of solar power plants. Ecological consequences may also arise, for example from maize cultivation or offshore wind farms. Public protests show that conflicts of interest exist not only in Germany but also in countries like Iceland and Norway.

Can renewable energies satisfy the world's hunger for energy?

Growing worldwide energy consumption makes the use of renewable energies essential, because the use of fossil energy sources is limited and the use of nuclear energy is controversial. Looking at the current importance of renewable energies for electricity supply in Europe and worldwide, considerable differences can be seen. The potential for the future is still great.

Skill Check

All the skills that you were able to acquire in this chapter are listed here.
You can judge for yourself if you have mastered each skill: ***securely, moderately or barely.***

Explanatory skills

Can I...		Unsure? Look it up on these pages
1.	describe different forms of renewable energy generation and their supply potential?	148–161
2.	describe geographical development processes that are influenced by economic and political factors (e.g. extending the use of renewable energies)?	148–161
3.	describe land use demands and conflicts as well as their possible solutions (e.g. extending the power transmission network)?	160/161
4.	systematise geographical processes and structures in terms of describing and discussing the use of renewable energies by means of a network of technical terms?	146/147

Methodological skills

Can I...		
5.	analyse different kinds of presentation and working materials (maps, images, statistical data, graphics and text) to answer questions about renewable energies?	148–161
6.	present the possibilities and limits of renewable energies orally and in writing, using the correct terminology, in a way that is problem-oriented, factually and logically structured, and relevant to the assignment, operators and materials?	148–161, 166/167
7.	substantiate written and oral statements (e.g. on the use of renewable energies) with correct and appropriate references to materials and citations?	148–161, 166/167
8.	present geographical information graphically (sketch maps, cause-and-effect networks, etc.)?	152/153

Evaluation skills

Can I...		
9.	discuss the effects of the expansion of the area of cultivation of renewable energy raw materials in connection with the need to ensure food supplies for a growing world population?	156–157
10.	evaluate different measures for reducing energy consumption in terms of their efficiency and feasibility?	162/163
11.	assess the possibilities and limits of renewable energy generation, taking into account economic interests and the requirements of climate protection?	148–161

Action skills

Can I...		
12.	present possibilities for influencing geographical processes (e.g. the building of a wind farm, increasing the efficiency of electricity supply) in my locality?	152/153, 162/163

Examination training

Use of renewable energies – the energy-autonomous village of Feldheim

1. Locate Feldheim (Treuenbrietzen) and describe the village's energy supply.
2. Explain the energy concept with reference to the natural geographical preconditions.
3. Assess the energy concept of Feldheim against the background of its sustainability and transferability to other places.

These additional materials are required to solve the assignments:

M1 Diercke Weltatlas, 2015, Map: Solar energy and geothermal energy, p. 68

M2 Diercke Weltatlas, 2015, Map: Wind energy, p. 68

M3 A selection of further atlas maps

M5 View of Feldheim

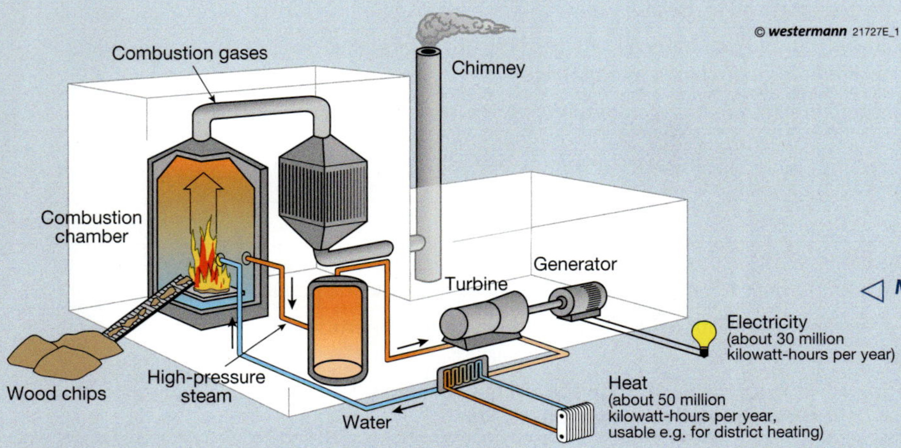

M6 How a wood chip power plant works (power-heat combination)

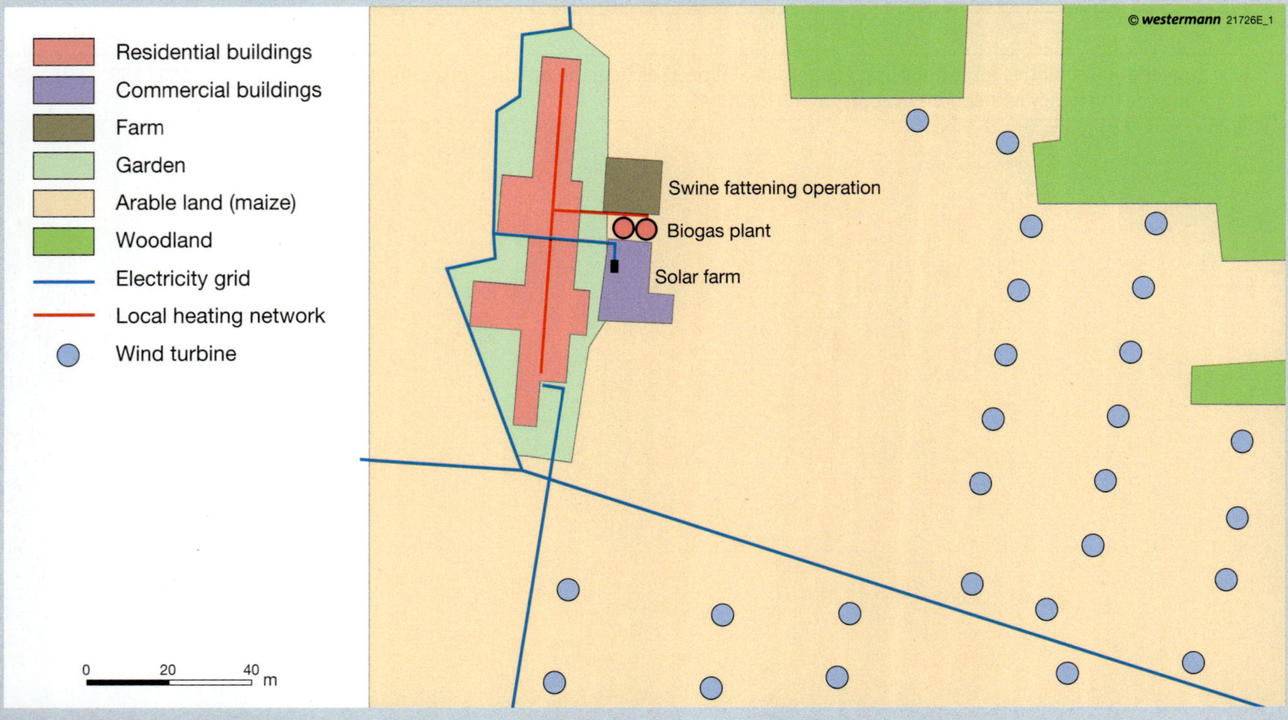

M4 Sketch map of Feldheim/Treuenbrietzen

M7 Solar plant and wind turbines at Feldheim

Feldheim in Brandenburg is reminiscent of a tiny Gallic village that aims to defeat a mighty empire. The municipality's idea that it could take its own decisions about electricity production and distribution came about in 2005. While other municipalities are buying back their electricity grids in order to take energy supply back into their own hands, Feldheim in the town of Treuenbrietzen has gone a step further. Together with a local entrepreneur, the little village has created its own electricity grid for its 150 inhabitants, connected to the German national grid. This is currently unique in Germany.
Michael Knape, the Mayor of Treuenbrietzen, recapitulates the development: "The idea of autonomy was originally limited to the newly-established solar factory, which was provided with power from the wind and sun." But the successful model caught on. "The people themselves raised the question of autonomous supply," Knape recalls. Now, power and heat are produced to a level of more than 100 percent from renewable energy plants in the village.
This participation in the economic development is central to acceptance by the population. Thus there was never any broad public rejection, even of the wind power projects. But it should not be forgotten that the profits are based on a cooperative investment, because every inhabitant of Feldheim has paid 1,500 euros for their connection to each of the heating network and the power grid. "We understood energy autonomy as a challenge, and we have mastered it," says Knape self-confidently. For electricity, the advance payment will be amortised in about seven years.
The Brandenburg-based company Energiequelle has been planning wind farm projects since 1997. Its field of business was later extended to solar energy and biogas. The firm installed a solar farm in Feldheim with 96 photovoltaic units on a former military site. The steel structures that automatically move the units to the optimum angle for the sun were manufactured locally in a newly-erected factory. This source of energy has thus created a third economic mainstay for Treuenbrietzen.
Alongside the Kohl-AG metal processing factory and the Johanniter Hospital, renewable energies are now and in future a further source of jobs and contracts for local tradesmen and building companies. Up to now, 25 new jobs have been created in a village of some 150 inhabitants.

Based on: www.kommunal-erneuerbar.de, 2010

M10 Information on the development of the energy concept

Wood chips are wood than has been cut into small pieces. Residual forest wood, low-quality wood, wood cuttings from landscape management, and old timber, are used for the production of wood chips. The burning of wood chips in power plants generates flue gases. In comparison to the burning of fossil fuels, the annual reduction in CO_2 is some 40,000 tonnes.

Source: Renewable Energy Agency

M8 Wood chips as fuel

Solarpark Selterhof (Feldheim)
Demonstration plant by the firm EQ-SYS
2.25 MWp electricity capacity
2,748 MWh annual yield

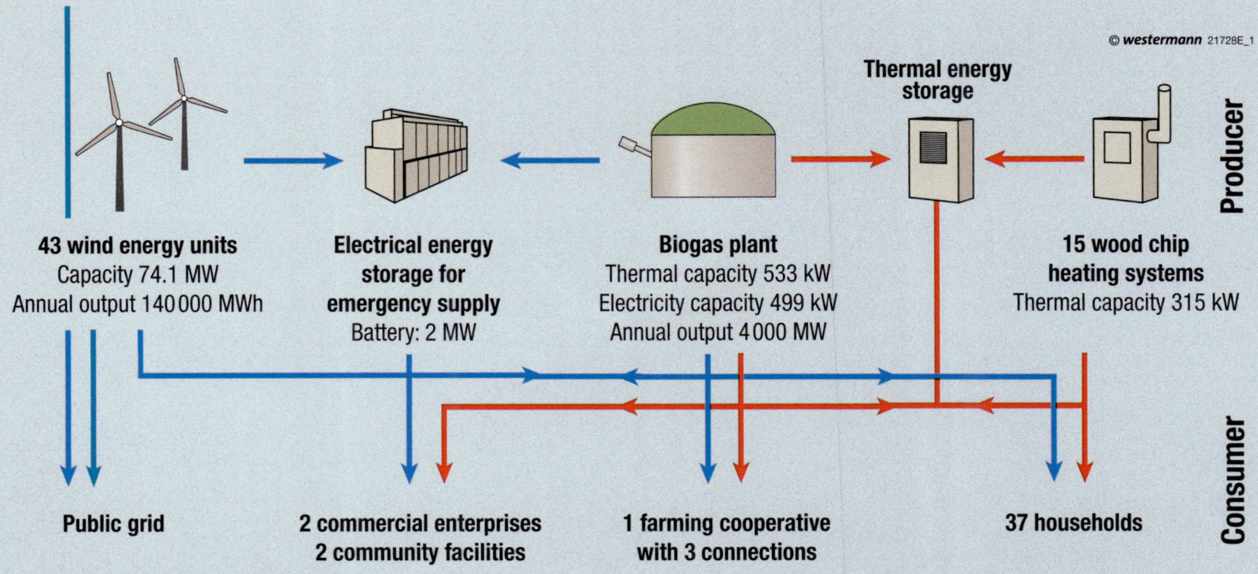

M9 The energy network at Feldheim

VII The Changing Climate

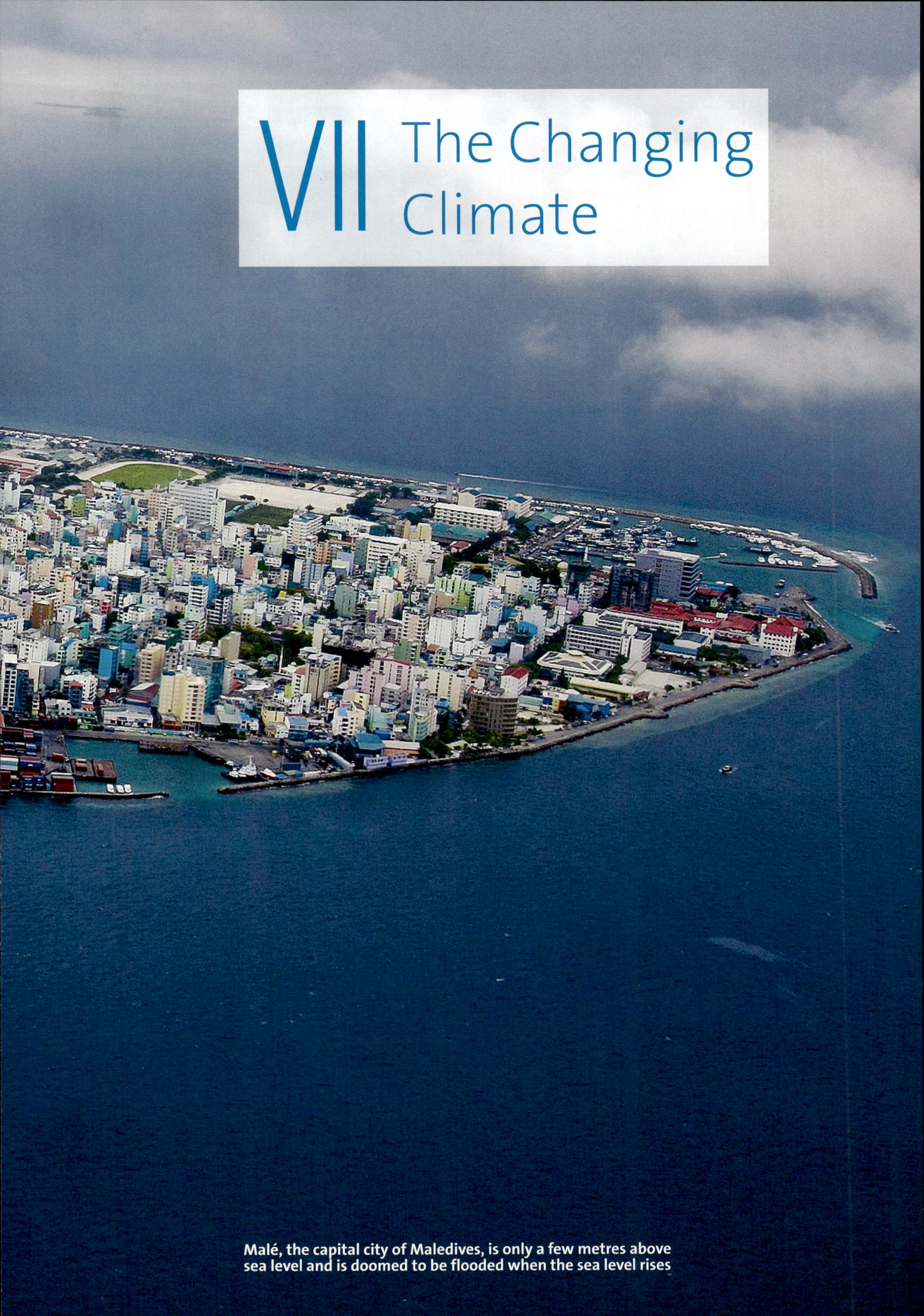

Malé, the capital city of Maledives, is only a few metres above sea level and is doomed to be flooded when the sea level rises

CHAPTER I:
Between Inhabited and Uninhabitable

M1 Retreat of the Morteratsch Glacier (Alps)

CHAPTER II:
Water –The Basis of Life

M3 Flood disaster in Australia (2012)

CHAPTER III:
Living with the Earth's Endogenous Forces

M4 Eruption of Grimsfjall volcano (Iceland). The eruption sent a column of ash nearly 20 kilometres into the air

It's getting warmer ...

Is our climate changing?

Increasingly often, the weather develops in completely unexpected ways. Many regions are threatened by periods of terrible drought, or torrential rains with floods and high tides, heat waves and forest fires, storms and storm tides. Many people see these as clear indications or harbingers of changes in the climate. But can extreme weather events be equated with climate change? Can we recognise in these extremes of weather the beginnings of a process of climate change?

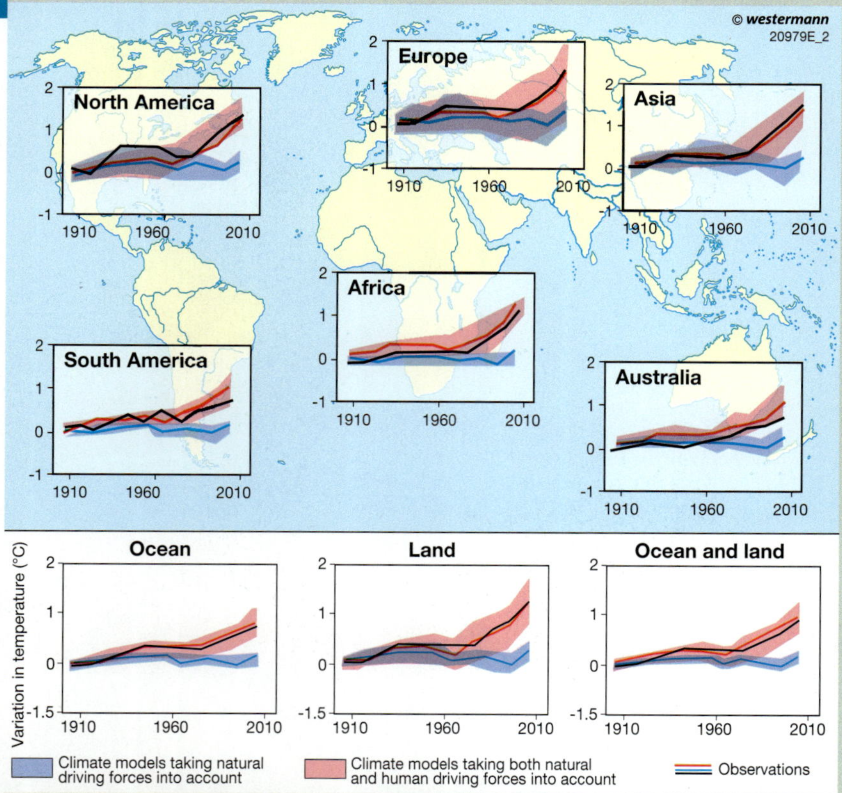

M2 Comparison of simulations of various climate models with temperature changes measured in different regions of the world

From weather to climate

What is the difference between weather and climate? When do we speak of a change in the weather, and when of a change in the climate? The two terms describe different states of the atmosphere. Weather is the state of the atmosphere at a certain time and a certain place on the Earth, and is characterised by measurable meteorological data such as air temperature, precipitation, radiation or wind direction.

By contrast, climate is defined as the average state of the atmosphere in a certain place or region or landscape zone. It is represented by overall statistical characteristics (means, extremes, frequencies and others) over a longer period of time. In general, periods of 30 years are used – so-called reference periods – but shorter time periods are also sometimes used. One speaks of a change in the climate when the difference in the value of a climatic element, such as air temperature, exceeds a given threshold value over two successive periods (covering several decades).

The complexity of climate change

Even before the publication of the fifth IPCC UNO Climate Change Report at the end of 2013, advocacy groups and government representatives were discussing how to interpret the current data and were calculating projections. The IPCC is an intergovernmental scientific council that was set up by the World Meteorological Organisation (WMO) and the United Nations Environment Programme (UNEP). It is made up of government representatives and scientists.

CHAPTER IV:
Production and Use of Fossil Fuels

M5 Open-cast mining in the Rhineland brown coal field

CHAPTER V:
New Production Technologies

M6 Extraction of oil sands in Alberta (Canada)

CHAPTER VI:
Renewable Energies

M7 Renewable energy mix in Germany

The UNO Climate Change Report brought together the results of some 10,000 scientific studies. Hundreds of scientists took part in it. They believe that climate change is primarily due to anthropogenic causes.
Against this, organisations closely associated with industry raised critical voices contradicting the interpretations put forward by the UNO's climate change panel (IPCC). They sought to convince the public that the IPCC data was too uncertain, and tried to turn opinion against the IPCC with public appearances and articles in the media.
Many people have been made aware of the subject of climate change by the media and political debate. But who understands the details of what is said and written about it? Who can judge which statements about climate change are true and which are false?
Climate change still raises many questions:
What are the consequences of the melting of glaciers and the rise in the sea level from a regional and global perspective?
Which regions will become more or less favourable habitats for mankind in future as a result of climatic warming?
Is there any connection between climate change and the increase in storm events?
What could be natural causes of climatic fluctuations?
What is the significance of mankind's role in climate change?
What actions can we take against climate change?

A multiplicity of observational data clearly shows that the Earth's climate has changed. Striking examples of climatic changes are:

→ The global mean temperature at ground level rose by 0.85°C in the period from 1880 until 2012. Each of the three past decades was warmer than all of its predecessors since 1850. In the Northern Hemisphere, the last 30-year period from 1983 to 2012 was the warmest for 1,400 years.

→ In the past 15 years, the global mean temperature has continued to increase but the speed of the rise in temperature was slower than in the preceding decades. The IPCC concludes that this finding is not evidence of a general slowing in global climate change, because such short-term changes may be primarily due to natural and internal fluctuations in the climatic system.

→ Improved measuring systems show that in the period from 1971 to 2010 the oceans stored more than 90 percent of the additional energy that was fed into the climatic system. In the uppermost 75 m of the oceans, the mean temperature rose by 0.11°C per decade from 1971 to 2010.

1. Describe what connects photos M1 to M7 with climate change.
2. Present the differences between weather and climate using concrete examples.
3. "It's getting warmer!" Describe how this development becomes apparent regionally and globally (M2).
4. Formulate hypotheses on the complex of topics around climate change.

www.westermann.de/
diercke-arbeitsmaterial

METHOD Conducting a project

The tracks of climate change

Regional and global increases in temperature have left visible tracks in many regions of the Earth. It is worthwhile to research these. The following tracks are considered to be significant by scientists and the media:

- → **Track 1:** Melting of glaciers – pages 174–175
- → **Track 2:** Thawing of permafrost soil – page 176
- → **Track 3:** Rise in the sea level – page 177
- → **Track 4:** Increase in weather extremes (e.g. very hot periods, storms) – pages 178–179

A suitable way of approaching these complex and global consequences of climate change is to follow these tracks in the form of a project.

This makes it possible to research the striking examples of the impact of current changes in the climate in a multilayered and independent way. The project has a direct practical relevance, and different subject areas can be linked together and addressed in an interdisciplinary way (e.g. with physics and biology). The project as a whole is designed in such a way that, depending on the time involved, a longer time is needed for planning and implementation. In order to approach the structured project work, the individual project steps will firstly be described in detail (M3).

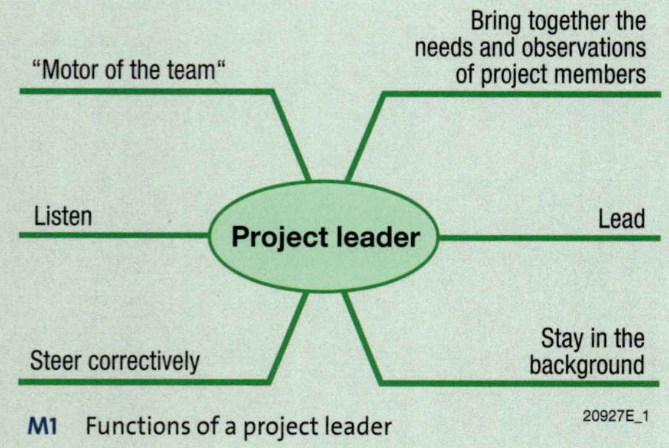

M1 Functions of a project leader

STEP 1: Introduction (working in plenum)

Structuring the topic

At the beginning of the project, the whole geography class collects ideas and sub-topics related to the individual tracks of current climate change. After an introductory brainstorming, all aspects of the tracks are collected, structured and recorded in writing in the form of a mind map.
The mind map should have main branches to which the sub-topics can be added, thus providing a good overview of the subject.
An alternative to brainstorming is working with cards. For this, each member of the class is given a maximum of three cards with the task of writing down on each card an idea or a sub-topic relating to the individual tracks of climate change in the form of keywords. It could be three sub-topics, or ideas about three different aspects of a track. Afterwards, the cards are posted on a pin board, sorted by subject and provided with titles, thus providing an overview.

Appointing a project leader

The geography class or the teacher chooses a member of the class to be the project leader. The project leader has the following tasks (M1):
- → Overall responsibility for the project
- → Coordination, control and direction of the project

The teacher is available to the project leader for expert advice during all phases of the project.

Forming groups

This project phase also includes the formation of groups. After the topic has been structured, small groups are formed. Then the class members discuss their interests within their small group and choose a track of current climate change. If more than one group chooses the same track, the topics can be allocated by drawing lots.

STEP 2: Planning (working in small groups)

In this phase of the project, a series of issues must be clarified in each of the small groups:
- → Who will undertake which task within the group? It is a good idea to distribute tasks among the group, such as e.g. timekeeper, note-taker and spokesperson.
- → What information is required in order to arrive at a comprehensive treatment of the sub-topic and the key questions?
- → What sources are available?

→ What experts/specialists can be interviewed?
→ In what time frame should the different project activities take place? It is useful to create a time schedule within the small group. An example of a time schedule for organising project work is shown in illustration M2.

STEP 3:
Implementation (working in small groups and in plenum)

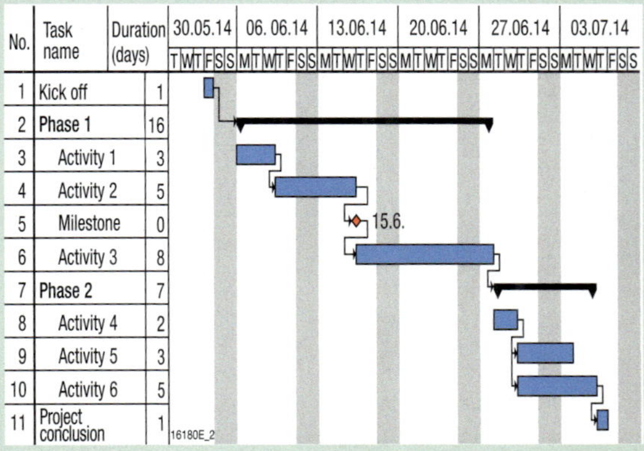

M2 Example of a schedule of project work within the small group

The small groups carry out the tasks that they discussed and that were allocated to them. Some parts of the tasks can and should also take place away from the school, for example interviewing an expert. The project leader should ensure that the small groups meet regularly.
The project leader should also keep an overview of the current status of the work. In this phase, the project leader has important functions to perform (M1). In order to document the progress of each group, the small groups should prepare short reports of their joint meetings.

STEP 4:
Presentation (working in plenum)

Before the presentation, the small groups should discuss and decide how to present the results of their project. The following forms of presentation are commonly used:
Presentations with the aid of
→ Slides
→ Wall displays, posters
→ Photographic or film documentations
→ Other media (e.g. PowerPoint, website).
In addition, the small groups should also agree on who will take on the job of preparing the presentation and who will present which parts of the topic. For the school's publicity work, the results of the project can also be published in the form of internal and external documentation (e.g. notice board, library, school magazine, yearbook, school website, newspaper).

STEP 5:
Evaluation (working in plenum)

The project work concludes with a critical reflection on the implementation of the project. This evaluation phase is very important in order to make a better estimation of the possible strengths and weaknesses of individuals for future projects in advance, and thus to optimise the workflow and results of a project.
The following questions could be asked when evaluating the project:
→ What went well during the project?
→ What problems arose, and at what stage in the project?
→ How was the role of the project leader/teacher perceived?
→ What points should be given extra attention in future projects?
→ ...

M3 Stages of a project

1. Within the project, analyse the tracks of current climate change on the following pages of this textbook with reference to the materials and current research results.
The following key tasks may be helpful for systematically researching a track:
- Examine the regional and global scope of a track of current climate change.
- Assess how far the track can be seen as an indicator of current climate change. You should also critically scrutinise the materials you are using.
- (...)
- Present your results clearly in visual form (e.g. in a digital presentation).

Information sources for researching the tracks: e.g.
- Alfred Wegener Institute
- Allianz Umweltstiftung
- Federal Ministry for the Environment, Nature Conservation and Nuclear Safety
- Germanwatch
- IPCC

PROJECT Melting of glaciers (Track 1)

Albedo, Ice sheet, Ocean currents

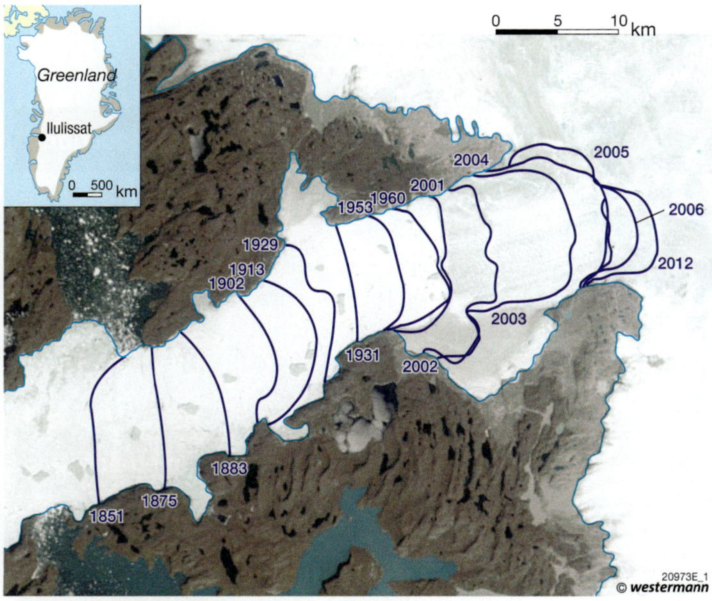

M1 The satellite photo shows how far the Jakobshavn Glacier in west Greenland (near Ilulissat) has retreated in the period from 1851 to 2012

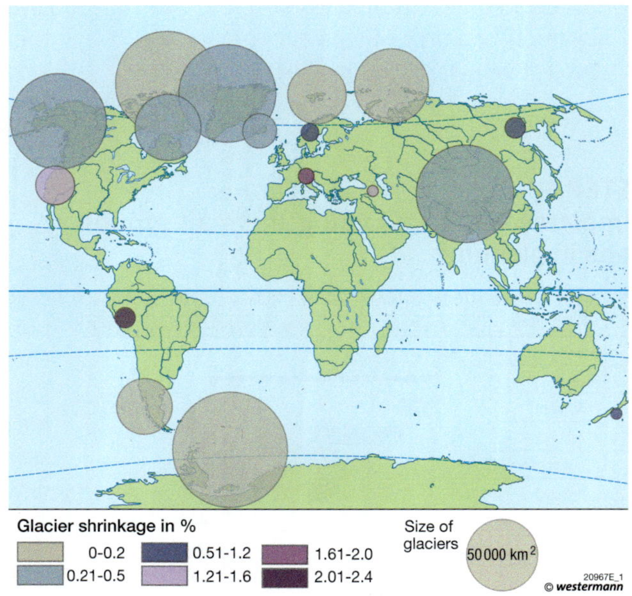

M3 Reduction in glacial mass per year worldwide.

Retreat of the mountain glaciers

One of the most clearly visible consequences of climatic warming that we have seen so far is the worldwide retreat of mountain glaciers. Glaciers represent the memory of the history of the climate; the mountain glaciers are the Earth's fever thermometer. Almost nowhere else than here can we so clearly read what is happening to our climate.

The World Glacier Monitoring Service (WGMS) concludes that the melting of the glaciers is accelerating almost everywhere in the world. In summary, the glaciers are now shrinking four times as fast as just 30 years ago – in the Andes and Alaska as much as in the Himalayas and the Alps.

In the Alps region, the glaciers have lost more than half of their mass since the beginning of the industrial revolution. While between 1975 and 2000 the glaciers of the Alps lost one percent of their volume every year, by the start of the 21st century the rate of loss had nearly tripled. With few exceptions, losses of similar dimensions can be observed in all the alpine regions of the world.

For many regions, glaciers are one of the most important sources of drinking water, because they store valuable fresh water in the winter in the form of snow and ice, and release it into the rivers in the summer in the form of meltwater. The retreat of the ice masses has a particular impact on many cities in the catchment area of the Himalayas, where the water from the glaciers feeds, among others, the seven great rivers of Asia – Ganges, Indus, Brahmaputra, Mekong, Nu Jiang, Yangtze and Huang He (Yellow River) – thus providing the water supply for hundreds of millions of people.

Source: Germanwatch, 2011

M2 Results of current research on the melting of glaciers

Arctic sea ice is disappearing as never before

With few exceptions, the global retreat of the glaciers that has occurred until now [to 2013] is continuing, and the polar ice caps are also losing mass.

From 2002 to 2011, about six times as much of Greenland's ice melted as in the previous ten years. In the period 1992 to 2001, the Antarctic ice sheet lost 30 Gigatonnes (Gt) of ice mass per year; in the period 2002 to 2011 it was 147 Gt per year – nearly five times as much.

The mean annual extent of the Arctic sea ice reduced by 3.5 to 4.1 percent per decade in the period from 1979 to 2012. For the Antarctic sea ice, a slight increase of 1.2 to 1.8 percent per decade was observed in the period 1979 to 2012.

The extent of snow cover in the Northern Hemisphere has declined since the middle of the 20th century.

Based on: IPCC, 2013

M4 IPCC Climate Change Report 2013

	Loss of mass of ice sheets		Change in sea level
	1992–2000 Gt*/year	2005–2010 Gt/year	2005–2010 mm/year
Greenland	-51 (±65)**	-263 (±30)	+0.73 (±0,08)
Antarctica	-48 (±65)	-81 (±37)	+0.23 (±0,10)
Greenland + Arctic	-100 (±92)	-344 (±48)	+0.96 (±0,13)
Glaciers + ice caps			+0.72 (±0,18)

* Gigatonnes; **The mass losses and changes in sea level are subject to considerable annual fluctuations (numbers in brackets) Source: AWI

M5 Overview of loss of mass of the ice sheets in Greenland and Antarctica

Will the North Pole become ice-free in summer?

What effect will this global warming have on the Arctic sea ice? Will the North Pole be ice-free in summer at some point?

Prof. R. Gerdes: The warming will lead to a further reduction in the extent and thickness of the ice. This trend will be accentuated by feedback effects: the further the ice cover shrinks, the more the open ocean will be made visible.

But because the ocean – in contrast to ice – is a very good absorber of solar radiation, it will warm up additionally. This warming will then again lead to an even greater reduction in the Arctic ice cover. But at the moment it is difficult to predict when we will have an ice-free North Pole in the summer months. Our climate models do not give a clear answer to this question. [...]

It is clear that temperatures will continue to rise. [...] What could happen to the polar region in the worst case?

Prof. Lochte: It could lead to faster rates of melting of the Greenland ice sheet, which would then have an impact on the sea level – with global consequences. And if the sea ice continues to shrink as it has been doing, the ecosystem there will be changed. [...] But it will also bring opportunities. It is already being discussed whether the North-East and North-West Passages can be used for shipping in future, and whether mineral resources can be extracted from the Arctic Ocean such as oil, gas and metals.

If the ice in the Arctic melts, will this have an impact on the sea level?

Prof. R. Gerdes: Not directly. After all, the sea ice floats, so it displaces exactly the amount of water that corresponds to its weight. If this ice melts, it has no immediate effect on the sea level. But there are possible indirect consequences – not on the global sea level but on its regional distribution. Where the ice melts, it can affect the ocean currents and thus the regional distribution of the water level.

It is not easy to predict how the ocean currents might change as a result of reducing ice cover. But if it really results in an impact on the wide-scale circulation of currents in the Atlantic, including the Gulf Stream, it could lead to a substantial rise in the sea level on the coasts of North America and Europe.

M6 Except from an interview with climatologists (Alfred Wegener Institute)

M8 A calving glacier face

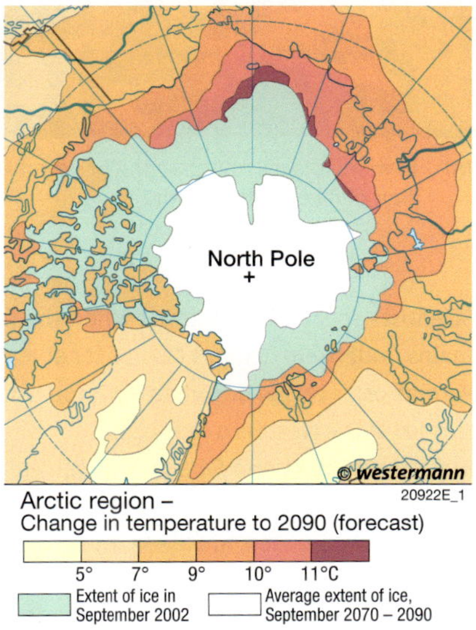

M9 North polar region – change in temperature to 2090 (projection)

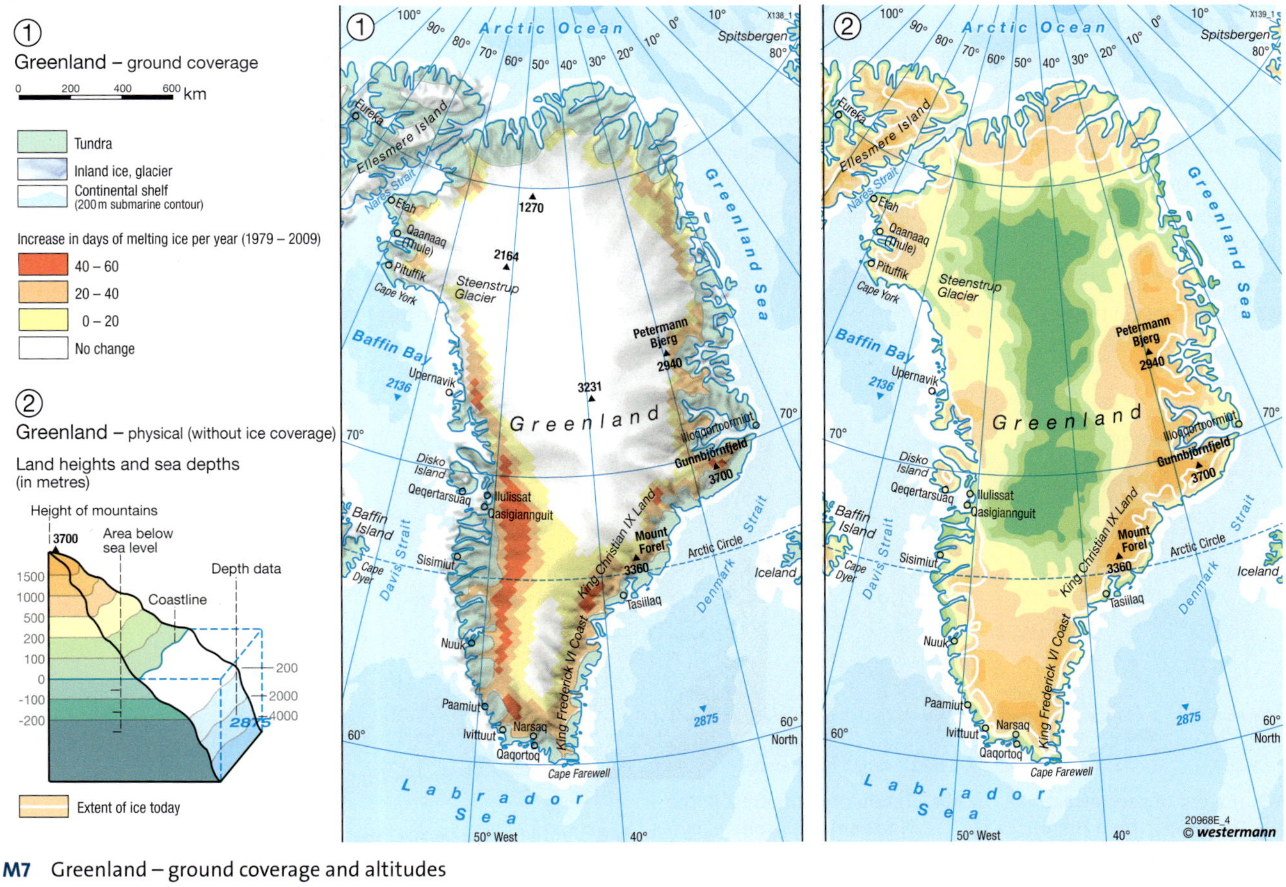

M7 Greenland – ground coverage and altitudes

PROJECT Thawing of permafrost soil (Track 2)

Permafrost, Tundra

M1 In many places, the thawing of permafrost leads to the formation of lakes in the tundra

Scientists speak of permafrost when the temperature of a soil remains under zero degrees Celsius for at least two successive years. Permafrost is very widespread – it is found on 24 percent of the land surface in the Northern Hemisphere.

In parts of Alaska and Siberia, which were not covered by glaciers during the last Ice Age, massive permafrost deposits with a high ice content have formed. In Siberia, the permafrost can reach a depth of up to 1,400 m.

In the last few decades, it is not only the atmosphere that has warmed up. The temperature of the upper layer of permafrost has also generally increased.

This is proved by recordings in drill holes and other measurements. Parts of today's permafrost are very subject to thawing, because their shape is discontinuous and irregular and they have temperatures approaching zero degrees Celsius.

It is to be expected that the extent and mass of the permafrost will become smaller in future. By the year 2100, there is a danger that it may shrink even below its minimum level of 5,000 to 9,000 years ago, when temperatures reached their hitherto maximum levels in the Holocene period.

Source: www.weltderphysik.de

M2 Permafrost in the past and future

The warming and ultimately the thawing of the permafrost could trigger processes that accelerate thawing still further as well as processes that could slow it down.

In many parts of Siberia, the thawing of the permafrost produces lakes. The resulting erosion processes eat away at the landscape. Water and heat penetrate into the soil more quickly, perhaps resulting in accelerated thawing of the permafrost. On the other hand, in some places the meltwater is drained away faster through erosion channels.

The processes of change working on the permafrost are made much more complicated by the simultaneous change in vegetation resulting from climatic warming. In various decades, the spread of bush towards the north has had contradictory effects on the permafrost. In winter, bushes foster the accumulation of thicker snow cover, the powerful insulating effect of which reduces the cooling of the soil in the winter. In summer, on the other hand, bushes produce more shade than the low tundra vegetation and thus have a cooling effect that stabilises the permafrost.

The processes are finely detailed and highly dynamic. Simulating them presents a major challenge to scientists.

It is estimated that 1,672 Gigatonnes of organic carbon are stored in the tracts of permafrost. This is more than double the quantity of carbon contained in all the vegetation on Earth. Even if only a small proportion of this were released, the atmospheric carbon content – in bound form, i.e. methane or carbon dioxide – would increase significantly. And this does not yet take into account the permafrost located on the sea floor in Arctic shelf regions (shallow areas of sea near the coasts). This was formed in the last Ice Age, when the sea level was many metres lower and the relevant areas were part of the land masses. It is not known how much carbon could be released from this submarine permafrost if the processes of warming were to continue.

If the soil material begins to thaw, it may result in a positive feedback in the climate. If deep permafrost layers melt, relatively fresh, "deep frozen" organic soil material is made available for decomposition by microorganisms. This decomposition ultimately leads to the release of the greenhouse gases carbon dioxide and methane. Their release would amplify the warming of the atmosphere.

Source: www.weltderphysik.de

M5 When the permafrost thaws

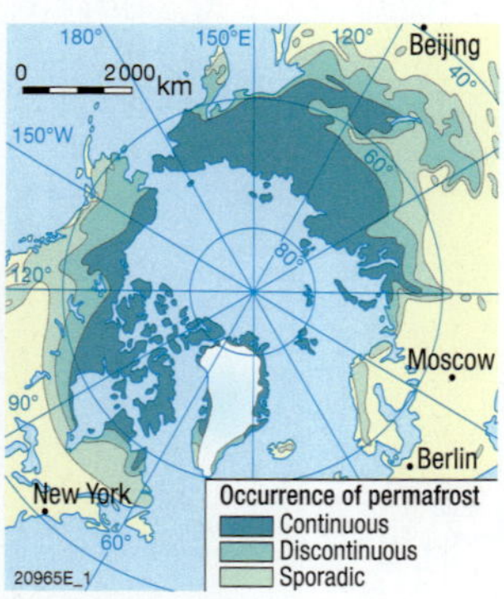

M3 Extent of permafrost in the Northern Hemisphere today

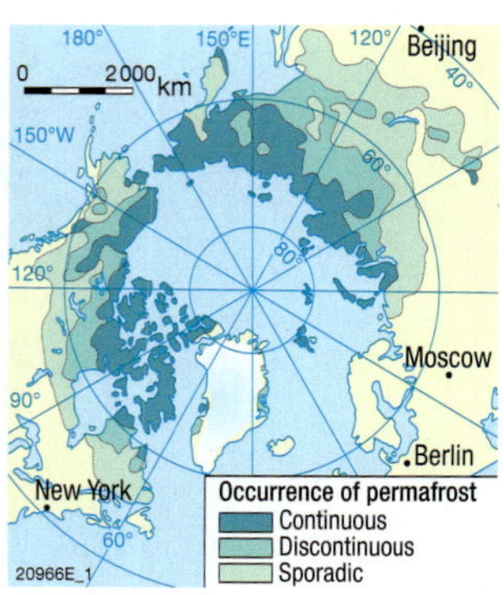

M4 Extent of permafrost in the Northern Hemisphere (by 2050)

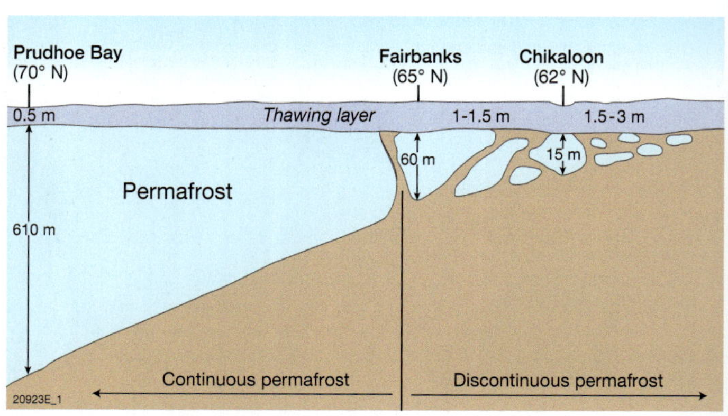

M6 Profile of the permafrost zone in Alaska between 60° and 70° n. Br.

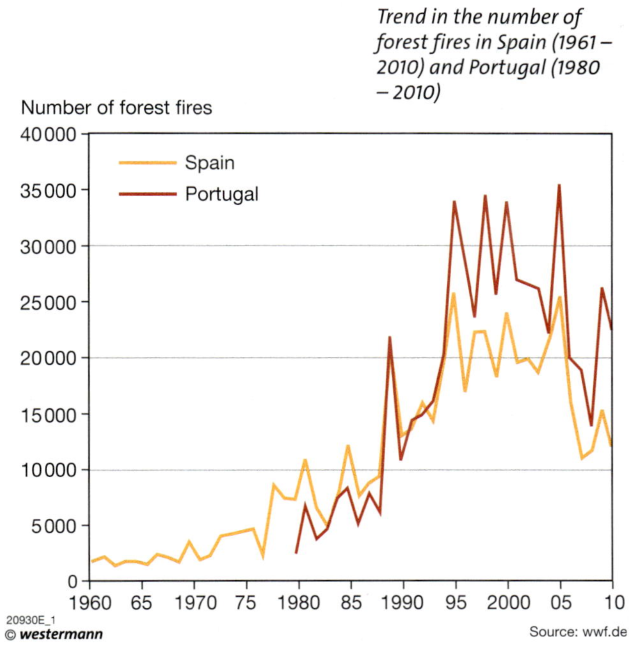

Small-scale fires have occurred in the Mediterranean region since ancient times; they are part of the natural dynamics or are used as an instrument for managing the natural resources. But in recent decades the number and extent of fires has increased worryingly in consequence of the socio-economic development in the Mediterranean region.

Since the Millennium, the Mediterranean countries have been confronted with a new phenomenon, so-called forest mega-fires. In extreme weather conditions, which have become more frequent as a result of climate change, real firestorms have occurred which rage with such intensity and spread so rapidly that they can no longer be brought under control. They only end when the weather conditions change or when there is nothing left to feed the fire. Associated with the human settlement of the landscape, these mega-fires cause substantial damages in the transition zones between settlements and forests, and often result in loss of life.

Based on: www.wwf.de, P. Hirschberger, Waldbrandstudie 2012

M4 Forest fires in the Mediterranean region

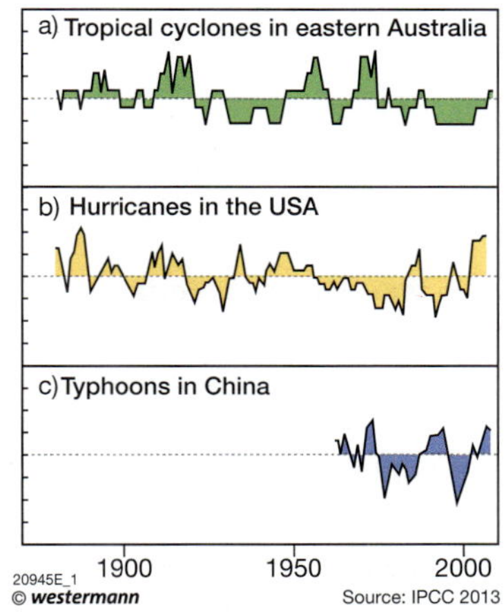

M6 Frequency of tropical cyclones affecting the land (standardised)

Another kind of extreme weather event which has increased in destructive power in recent decades is heavy tropical and winter storms such as Hurricane Katrina in the USA or Storm Kyrill in Germany. Their intensity has increased greatly, mainly on account of higher temperatures but also due to natural [...] fluctuations.

Especially in the disastrous hurricane year of 2005, the question was again raised of a possible connection between the increase in hurricanes and global warming. The intensified research work since then has focused on two main questions:

- Is it possible to discern a trend in the frequency and/or strength of tropical cyclones in the last decades and perhaps centuries?
- Is there a connection between global warming and changes in the occurrence of tropical cyclones, and how can such a connection be evaluated for the future?

The exceptionally severe hurricane season in the Caribbean and the USA in 2005 (including Hurricane Katrina) caused direct damages valued at more than 150 billion US dollars and killed more than 1,000 people in the USA alone.

In 2004, a hurricane in the Southern Hemisphere was observed for the first time, in the South Atlantic off the Brazilian coast. Even if natural fluctuations will continue to influence the intensity of heavy storm events in the coming decades, it is projected that after 2050 the effects of global warming will stand out even more clearly from those of natural fluctuations.

Based on: www.germanwatch.org

M7 Tropical cyclones and climate change

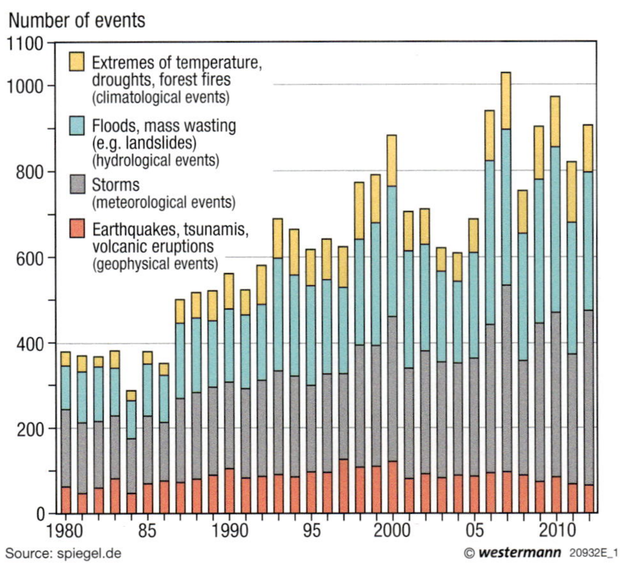

M5 Natural disasters worldwide, 1980–2012

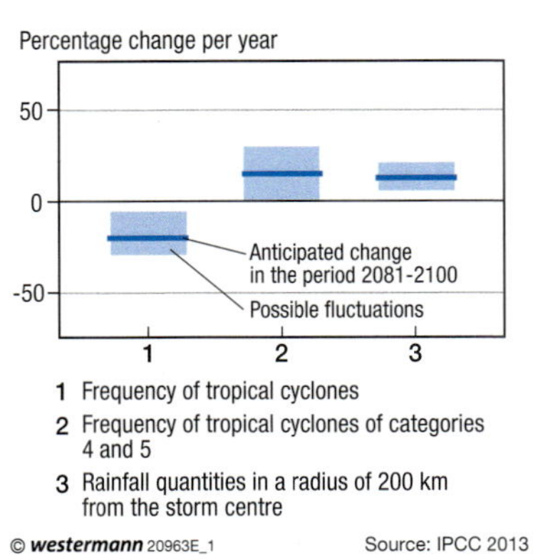

M8 ◁ Future changes in statistics of tropical cyclones. All figures show anticipated average percentage changes for the period 2081–2100 in relation to the period 2000–2019.

Causes of climate change

Looking back – natural causes of changes in climate

The climate has changed again and again in the course of the Earth's history. Phases with higher average temperatures have alternated with phases of lower temperatures. Until the very recent past, natural causes were responsible for climatic fluctuations.
But what exactly are the reasons for climatic changes in the Earth's history? This central question is very complex, and in some cases not fully explained by climatic research. This double page gives you an initial overview and points to various connections between natural events and climatic changes.

1. Characterise possible causes of changes in the climate (M2/ M3).
2. Explain the statement "Changes in the climate did not start with mankind's intervention in the climatic system" (M2–M4).
3. ⓞ **A** Investigate how far sunspots and changes in the tilt of the Earth's axis and orbital parameters could have an influence on the climate (M1/ M5, internet).
   B Explain the causes of changes in the climate in earlier epochs of the Earth's history that were brought about by volcanic eruptions, plate tectonic movements and meteorite strikes (M6–M9, internet).
4. Ⓢ Research how temperature data in the past is reconstructed, and explain why this data is uncertain.

Aerosol, Orbit, Plate tectonics, Solar constant, Sunspot, Volcano

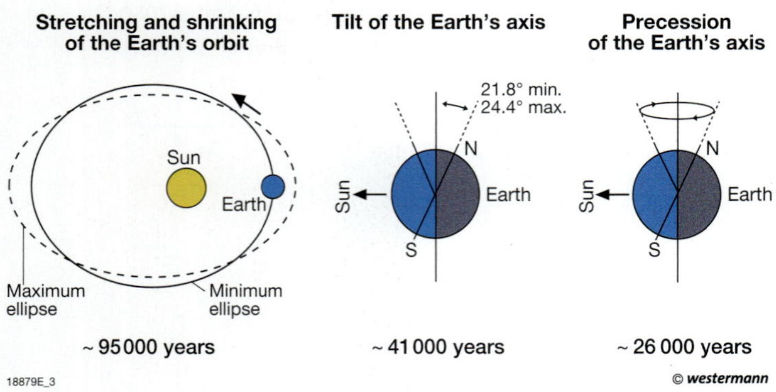

Climatic changes due to changing orbital parameters

Triggers for regularly-occurring climatic fluctuations are small changes in the Earth's radiation budget. The energy from the sun (solar constant) that reaches the Earth is not always the same. One cause is a change in the orbit of the Earth around the sun from a nearly circular to an elliptical course (eccentricity). The period in which this change occurs is some 95,000 years. In addition, the Earth's axis sways like a top in the course of approx. 26,000 years (precession). But also the change of a few degrees in the tilt of the Earth's axis (obliquity of the ecliptic) in a period of 41,000 years has an influence on the solar constant.

M1 Changes in the Earth's orbit and tilt of the Earth axis

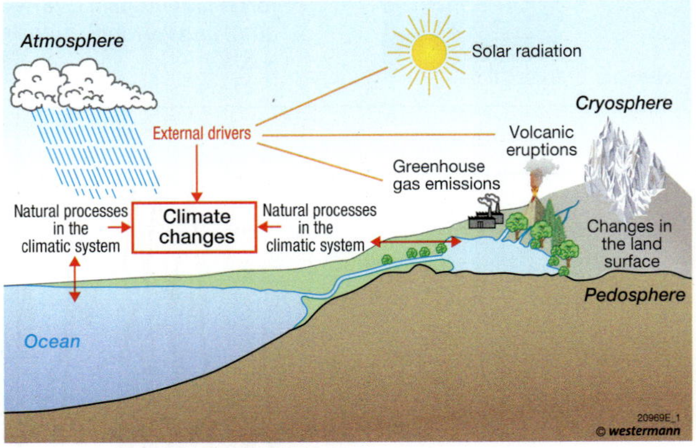

M2 Possible causes of changes in climate

External natural influences on climate

The external natural influences include fluctuations in solar activity, variations in the astronomical orbital parameters (the Earth's orbit, obliquity of the ecliptic, precession of the axis of rotation) and meteorite strikes. These strikes must be powerful enough to hurl substances as far as the stratosphere, i.e. above a height of some twelve kilometres, where they can remain for years without having any influence on the weather.

Further natural conditions

Plate tectonic processes also influence the climate, because they determine the distribution of land and sea on the Earth's surface, the characteristics of the ocean basins, the position of the continents and the formation of high mountains. Thus for example, high mountain ranges running obliquely to the main direction of current can have sufficient influence on atmospheric circulation to affect the climate.

M3 Natural causes of changes in climate

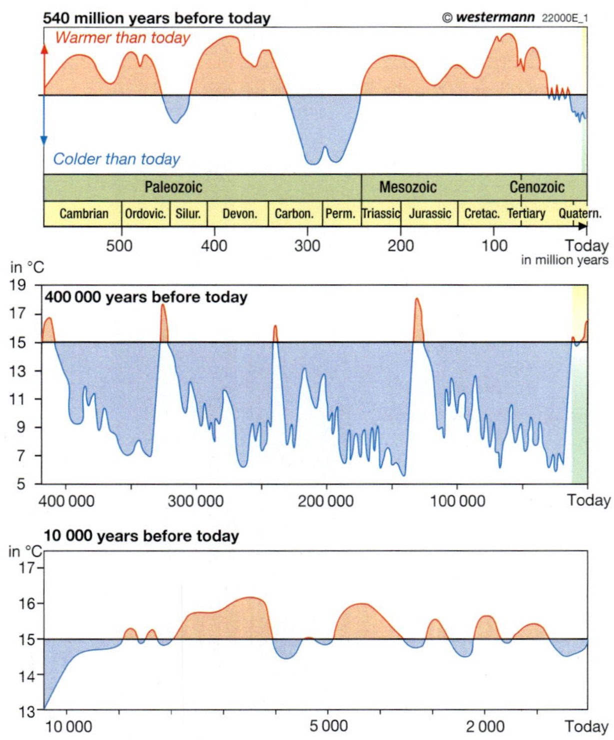

M4 Global average temperature

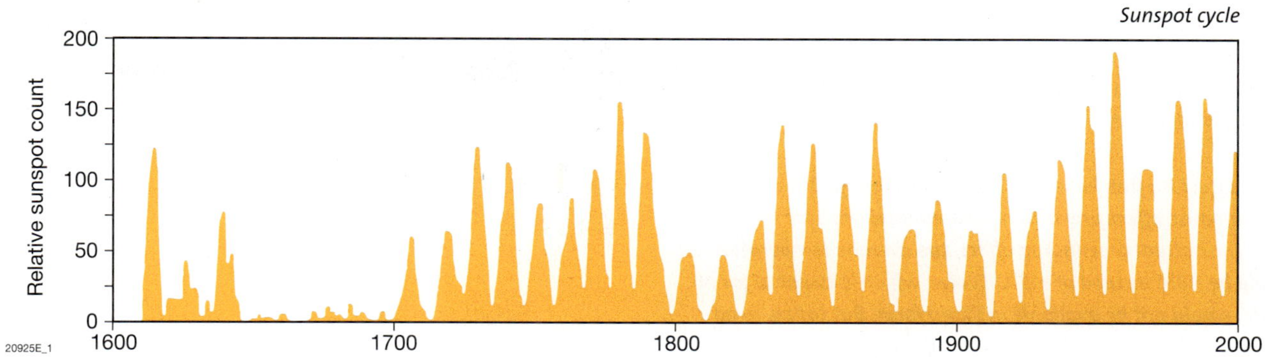

The sun itself exhibits fluctuations in the intensity of its radiation over periods of years and decades. To measure these fluctuations, the interplay of relatively cool sunspots and hot solar flares and other eruptions occurring largely synchronously is observed.

Sunspots are large formations on the sun's disc with an average diameter of some 2,000 – 50,000 km. The sunspots are visible as dark spots, from which material streams outwards. Their lifetime ranges from a few to a maximum of just under 100 days. The prevalence of sunspots fluctuates greatly in a cycle of 11.07 years.

At times of maximum activity of the cool sunspots, the intensity of radiation increases due to the simultaneous increase in flares and eruptions. The more sunspots occur, the greater is the sun's activity, ultimately leading, after a delay, to an impact on the Earth's climate.

M5 The influence of sunspots on climate change

Volcanic eruptions and their effects on the climate

How much is the 2,800 km³ of material that the Toba supervolcano on the island of Sumatra ejected some 74,000 years ago? With this material one could cover the whole of North Rhine-Westphalia with a layer 80 m thick. If the ejected material had a high sulphur content and found its way into the upper atmosphere (stratosphere), it could block out the sunlight.

Based on: W. Wiedlich, General Anzeiger Bonn, 31.05.2008

Scientists presume that the last Toba eruption is responsible for the 1,000 coldest years of the last Ice Age. That eruption led to a worldwide fall in temperatures of 5°C to 15°C. Human beings nearly died out at that time.

Based on: www.wetter-klimawandel.de

M6 The eruption of the Toba supervolcano

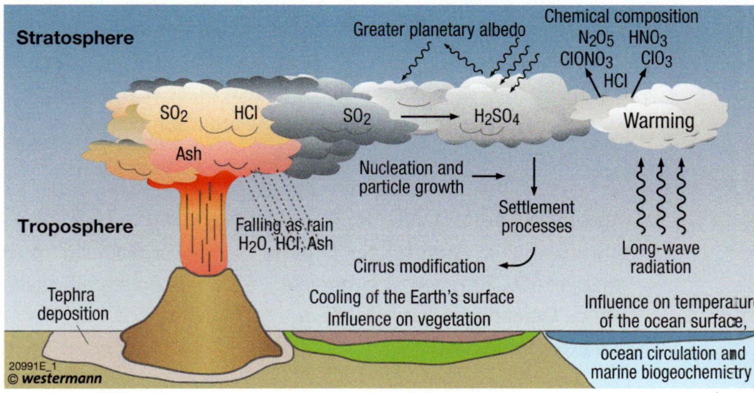

M8 The eruption of the Toba supervolcano some 74,000 years ago
(Tephra = volcanic ash; Nucleation = crystallisation, Cirrus modification = changes in the high ice clouds)

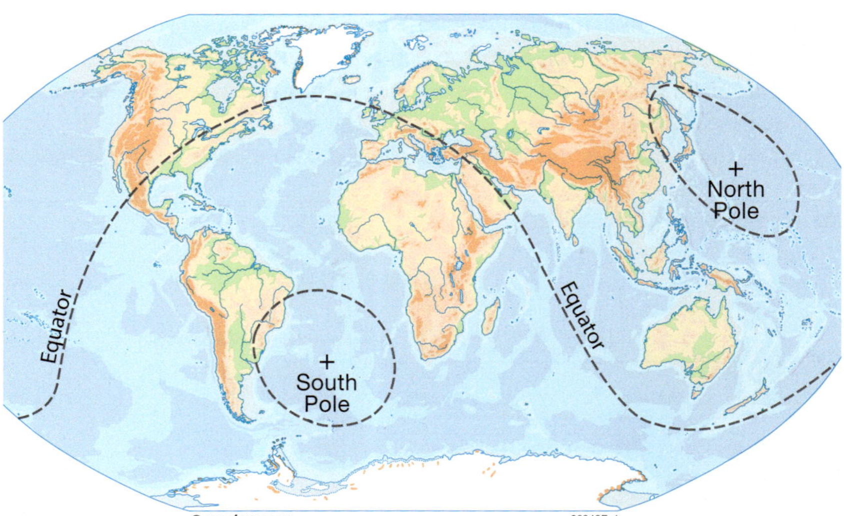

M7 World map of the Devonian (405 million years ago) and Carboniferous (350 million years ago) periods

Geological disasters and their impact on the environment

Meteorite	
Diameter	10 km
Density	4.5 g/cm³
Mass	3 billion tonnes
Speed on striking the Earth	20–60 km/sec.
Impact	
Pressure	min. 100,000 bar
Temperature	15,000°C
Impact energy (mean)	10^{30} erg = 10^{23} Joule
Crater diameter	150–300 km

For comparison: The distance from Cologne to Stuttgart is c. 300 km (in a straight line)

M9 Meteorite strike and its effects

Causes of climate change

The natural and the anthropogenically amplified greenhouse effect

The greenhouse effect is often linked with global warming. But originally the natural greenhouse effect provided a fundamental basis for life on Earth. Without the natural greenhouse effect, the average temperature on the Earth would be -18°C instead of +15°C. Water, which is essential for all processes of life, would not be available in liquid form. But today, in addition to the natural factors, mankind also contributes to the greenhouse effect and climate change (anthropogenic influence).

1. Describe the structure of the atmosphere (M1/ M4).
2. Explain the diagram of the Earth's radiation and heat balance (M2/ M5).
3. Explain the natural greenhouse effect (M1–M5).
4. Ⓞ **A** Present the influence of greenhouse gases on the Earth's climate in a causal network (M6–M11).
   B Explain the role played by greenhouse gases and aerosols in the anthropogenically amplified greenhouse effect and the influence this has on the climate (M6–M11).
5. Explain how mankind contributes to changes in the climate (M6–M11).
6. Ⓢ The melting of glaciers and the retreat of ice cover in the Arctic also has consequences for the radiation budget. Present these in a causal network.

M4 Atmosphere

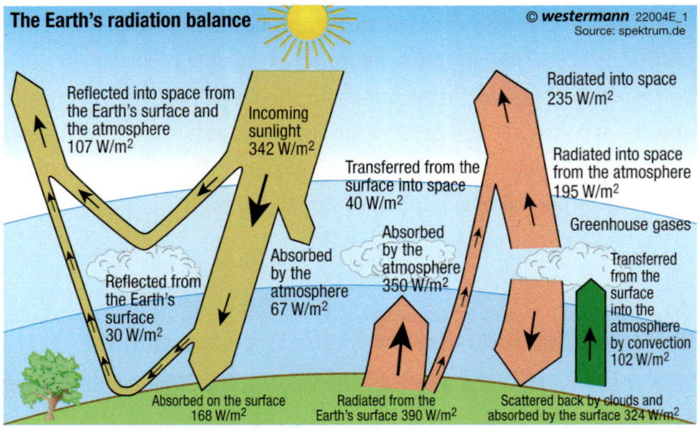

M5 Global radiation balance for the Earth-Atmosphere-Space system

Absorption, Aerosol, Albedo, Anthropogenic, Atmosphere, Climate change, CO_2, Emission, Greenhouse effect, Greenhouse gas, Radiation balance, Troposphere

The radiation from the sun that reaches the Earth is mainly short-wave radiation. Part of this is already reflected and absorbed in the atmosphere and therefore does not reach the Earth's surface. The solar radiation that reaches the Earth's surface is composed partly of direct solar radiation and a part that is dispersed from the atmosphere. When it reaches the Earth, the solar radiation is partly reflected or absorbed, i.e. converted into long-wave heat radiation. Most of the long-wave heat radiation on the warmed Earth's surface does not leave the atmosphere by a direct route.

Instead, greenhouse gases, which are also present in the atmosphere, absorb the long-wave heat radiation. In this way, heat energy is fed into the air, and part of this escapes into space. But the great majority is radiated back to the Earth. This back radiation is also called the natural greenhouse effect. The heat that is fed into the layer of air close to the ground by the heat radiation is called "sensible heat". Part of the radiation causes evaporation processes on the Earth's surface. This is not only sensible but hidden in evaporated water (latent) and retained there as latent heat.

When the steam condenses, this latent heat is released again and given out to the environment.

M2 Radiation and heat balance

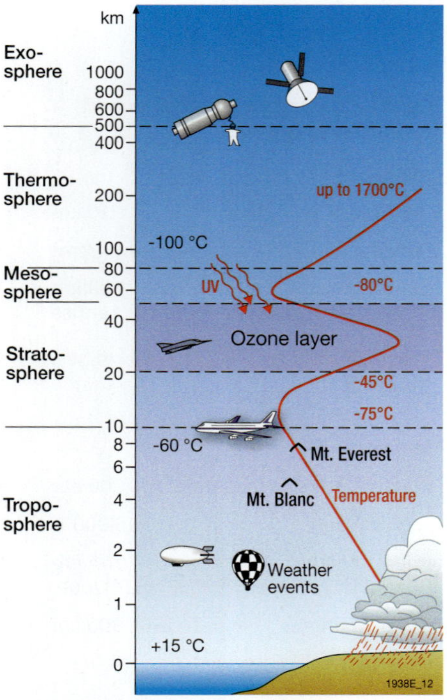

M1 The structure of the atmosphere

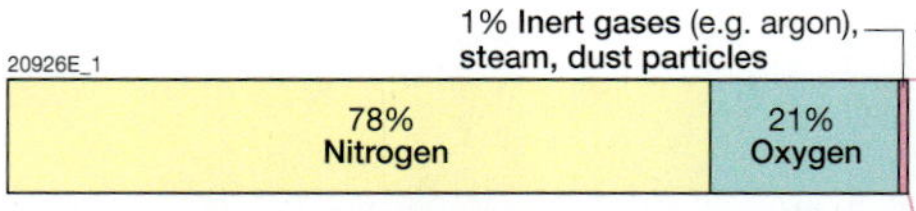

Chemical name	*Common name*	*Chemical formula*
Carbon dioxide	Carbon dioxide	CO_2
Methane	Marsh gas	CH_4
Chlorofluorocarbon (CFC)	Aerosol gas	e.g. CCL_3F
Ozone	Ozone	O_3
Nitrous oxide	Laughing gas	N_2O

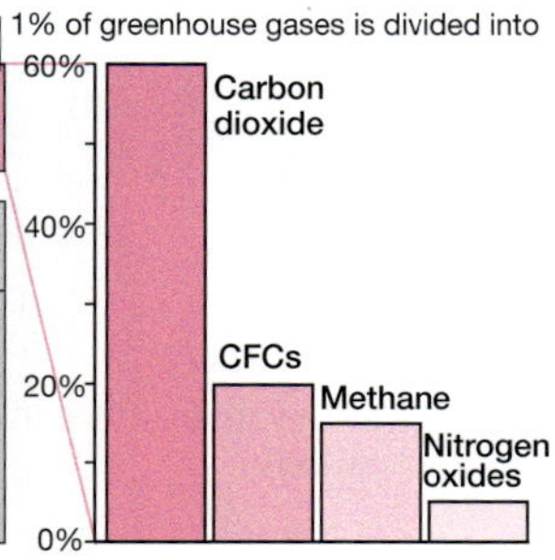

M3 Composition of air and proportion of greenhouse gases

M6 Brown coal power station in Brandenburg

Increase in emissions

Primarily the burning of fossil fuels, but also the clearance of forests, has caused the concentration of greenhouse gases in the atmosphere to increase significantly. Since 1750, the atmospheric concentrations of the greenhouse gases CO_2, CH_4 and N_2O have risen by 40 percent, 150 percent and 20 percent respectively. The concentration of carbon dioxide in the atmosphere today is higher than ever before in the past 800,000 years.

From 1750 to 2011, quantities of CO_2 totalling 545 Gigatonnes of carbon have been released by human activities. Just under half of this remained in the atmosphere and contributed to the man-made greenhouse effect. Approximately half of the rest was absorbed by the oceans, and half by the soil and vegetation.

Source: IPCC 2013

Retreat of carbon sinks

The proportion of carbon dioxide emissions that is absorbed by CO_2 sinks on land or in the sea has declined more and more in the last 50 years.

So-called carbon sinks such as rain forests or the phytoplankton in the ocean play an important role, not only in the Earth's climatic system but also in the carbon cycle, because they absorb carbon dioxide from the air and convert it into other carbon compounds. In this way they do not only remove greenhouse gases from the atmosphere, they also store the carbon for long periods of time and thus take it "out of circulation".

Forest clearance by burning to create new pasture land, and the use of the timber as fuel for heating, thus makes a double contribution to increasing the CO_2 content in the atmosphere – on the one hand through emissions, and on the other hand through the fact that the clearance of vegetation reduces the absorption of CO_2.

M7 Does mankind affect the climate?

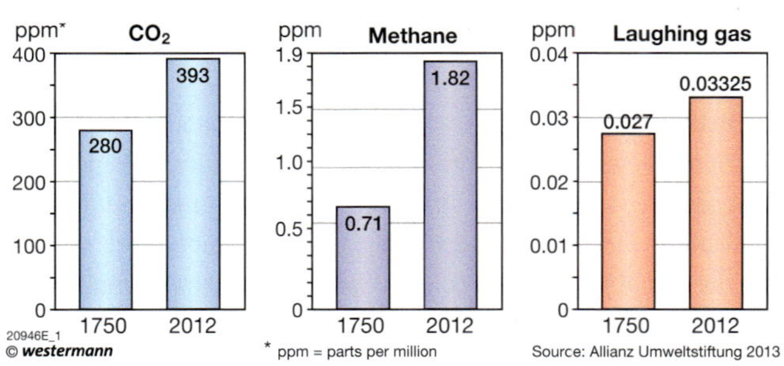

M8 Change in the composition of climate-relevant gases in the atmosphere

Greenhouse gas	CO_2	CH_4	N_2O	Ozone	CFCs
Present concentration in ppm	393	1.82	0.33	0.02	0.00052
Stay time in the atmosphere (years)	10	7-10	150	0.10	90
Present atmospheric trend (in %/year)	+0.4	+1.10	+0.25	uncertain	+5
Contribution to greenhouse effect (in %)	50	19	4	8	17
Climate impact based on one molecule of CO_2	1	32	150	2000	16,000
Anthropogenic sources	Fossil fuels, forest clearance, soil erosion	Rice cultivation, cattle rearing, waste deposits, burning of biomass, natural gas	Fossil fuels, burning of biomass, N-containing fertilisers, soil cultivation	Indirect through NO_2, CH_4, CO and carbon in the troposphere	Coolants, foaming agents, aerosols

M9 Characteristic data of greenhouse gases

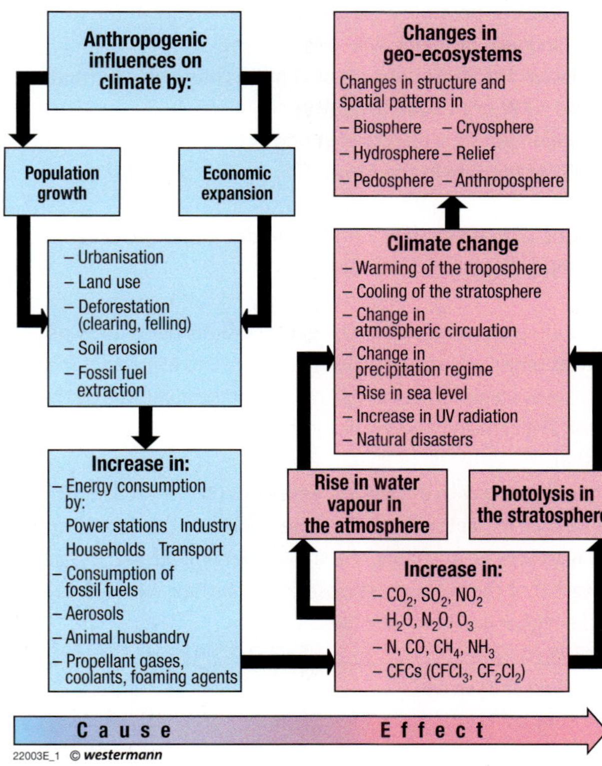

M10 Anthropogenic influences on the climate

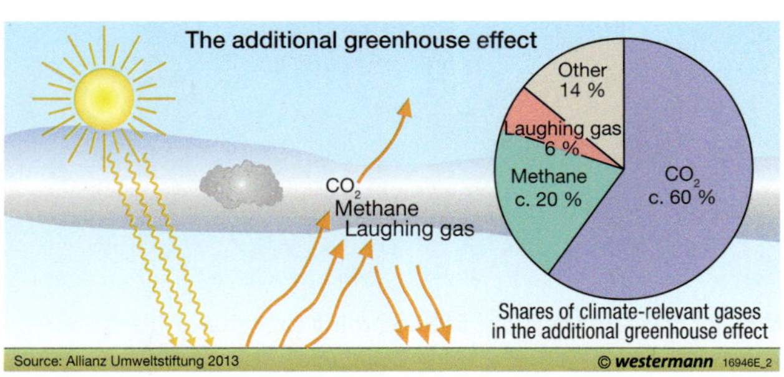

M11 The additional greenhouse effect

What will the climate be like in the future?

Hotter or colder? Scenarios for tomorrow's climate

In order to be able to estimate the future development of the climate, alongside exact climate models it is also important to determine as realistically as possible what greenhouse gas emissions are to be expected in future. Various issues play a role here:

- ***How will the world's population, living standards and the global economy develop?***
- ***What energy needs will result from this development, and how will these be met?***
- ***What technological advances are to be expected, and what impact will these have in the above-mentioned areas?***

1. Give reasons for the differences between the IPCC's various future climate scenarios (M3/ M5).
2. Looking into the future? Evaluate the chances and uncertainties of climate scenarios and projections (M3–M5).
3. ◎ **A** Discuss possible future effects of climate change on the climatic system (M1–M9, internet).
   B Present possible effects of climate change in the form of a causal network (M1–M9, internet).
4. Discuss the consequences of climate change for Germany and NRW until 2040 (M7/ M8).
5. Ⓢ Explain possible effects of climate change on summer and winter tourism in the Alps.

Aerosol, Climate change, Climate model, Emission, Greenhouse gas, IPCC, Rise in sea level, Scenario

Depending on the scenario, the glaciers could lose 15 to 55 percent (lowest emissions scenario, Scenario 1) or 35 to 85 percent (highest emissions scenario, Scenario 3) of their present volume by the end of the 21st century. It is very likely that the Arctic sea ice will retreat further.

In the scenario with the highest emissions (3), the Arctic could even be ice-free in September before the middle of the 21st century. In the Northern Hemisphere, the snow cover will retreat. It is almost certain that the permafrost close to the surface will shift in areas in the higher northern latitudes.

Based on: IPCC 2013

M1 Land and sea ice

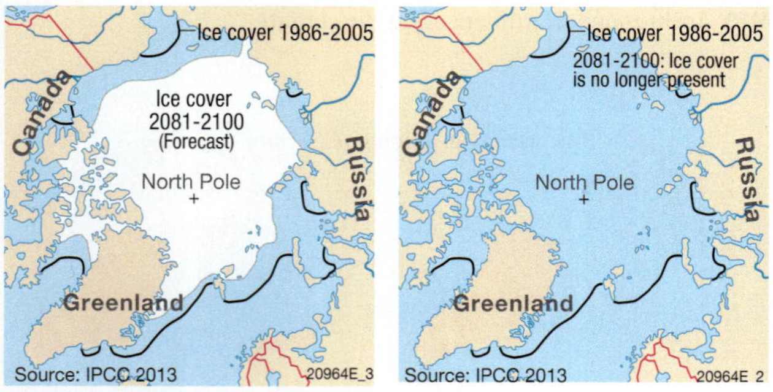

M2 Ice cover on the Arctic Sea according to Scenario 1 (left) and Scenario 3 (right)

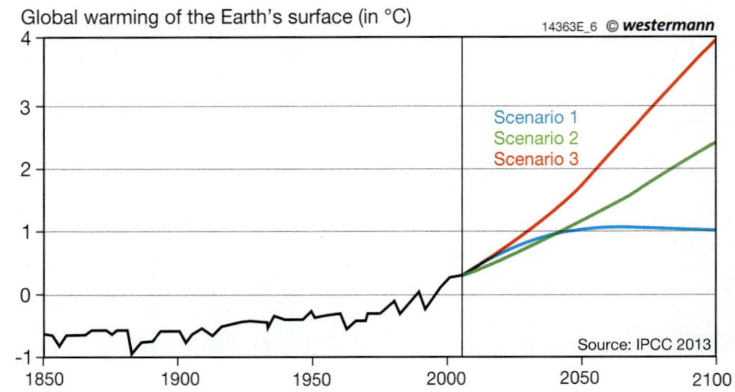

M3 Probable future development of the temperature of the Earth's surface for different emissions scenarios (see M5)

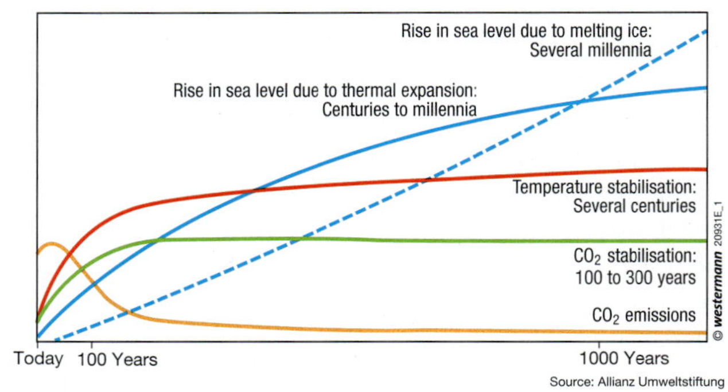

M4 Inertia of the climate

The climate of the future – climate scenarios

Scenarios developed by climatologists incorporate future trends in greenhouse gases and aerosols in the atmosphere as well as other factors influencing the climate, such as changes in land use.

In the Fifth Assessment Report of the IPCC (2013), four new scenarios [three of which are shown in M3/ M6] are used, so-called representative concentration pathways (RCP).

The RCPs cover the period from 1850 to 2100, with prolongations to the period up till 2300. In contrast to the scenarios in the two previous assessment reports, measures to reduce greenhouse gases are now taken into account.

Based on a scenario with very ambitious climate protection policies (Scenario 1) the simulations show that the mean temperature increase by the end of this century can be limited to 0.9 to 2.3°C compared to the pre-industrial era. Here, the IPCC's authors assume that global warming will probably remain below 2°C.

The simulations of three further scenarios with lower or insignificant emission reductions show temperature increases of between 1.7 and 5.4°C. All three scenarios produce increases of more than 1.5°C, and in two scenarios it is at least 2°C. In the scenario with almost unrestricted emissions (Scenario 3), temperature increases of more than 5.4°C are possible by the end of this century.

With the exception of the scenario with very ambitious climate protection policies, global warming would continue after the end of the 21st century in all scenarios. In summary, the Fifth Assessment Report indicates a range of 0.9 to 5.4°C for the mean temperature increase by the end of this century compared to pre-industrial conditions.

Based on: IPCC 2013

M5 IPCC climate scenarios from 2013

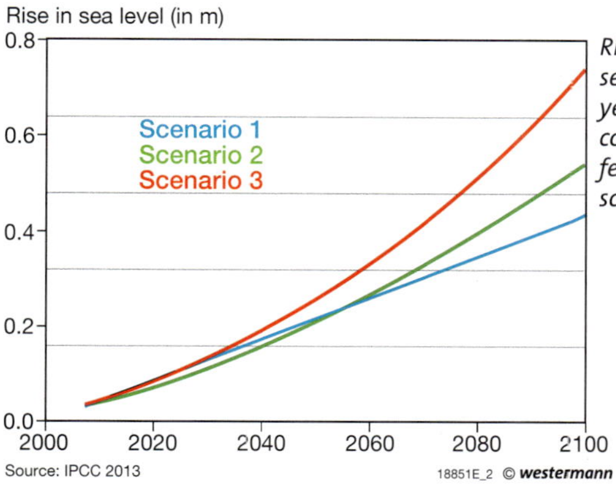

Rise in global sea level to the year 2100 according to different emission scenarios.

By the end of the 21st century, rises of a further 26 to 55 cm in the sea level are anticipated even if substantial efforts are made in the direction of climate protection (lowest emissions scenario). If emissions are not restricted, the sea level could rise by between 45 and 82 cm by the end of the century (highest emissions scenario). The IPCC does not exclude the possibility that the rise in the sea level could even be significantly greater. The latest projections for the rise in the sea level are higher than in the Fourth Assessment Report (2007), because better account has been taken of the contribution of the polar ice sheets.

Based on: IPCC 2013

M6 Rise in the sea level

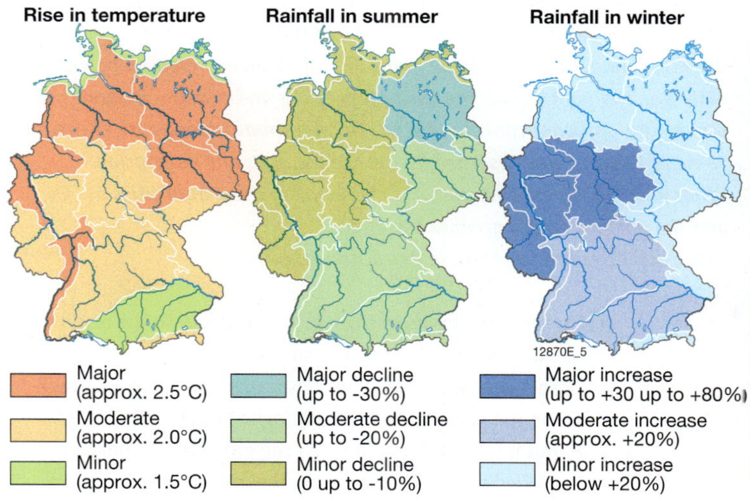

M8 Possible changes in temperature and rainfall in Germany

Feature under review	Anticipated changes	Reliability of prediction	Consequences
Temperature	1.7°C warmer than 1900, warmer winters and nights	very high	Earlier germination of plants, retreat of permafrost in the Alps
Hot periods	More frequent, more intense	very high	Health risks and stress for the biosphere, more forest fires
Alpine glaciers	Loss of 60% of area and 80% of mass	very high	Extreme fluctuations in water runoff
Rise in sea level	c. 10 cm compared to today	very high	Threat to North Sea and Baltic coasts
Precipitation	Summer drier, autumn and winter wetter, more rain than snow, higher yield from individual events	high	Greater threat of flooding
Arid and drought periods	More frequent	medium	Effects on farming, energy industry and inland shipping, higher risk of forest fires
Storms	More intense	medium	Higher risks from heavy rain, hail, squalls
Storm tides	Up to 20 cm on reaching land	high	Increased threat to North Sea coast
Extratropical (winter) storms	Trend to fewer but heavier storms, changes in wind direction	uncertain	Significant risk of damage

M7 Anticipated consequences of climate change for Germany to 2040

What exactly could happen in the future?

Ongoing unrestricted emissions of greenhouse gases would lead to further warming and changes in the volumes of glaciers, the sea level, and consequentially further components of the climatic system:

- **Extremes:** It is considered to be almost certain that more hot and fewer cold temperature extremes will occur. Very probably, heat waves will occur more frequently and last for longer. By the end of this century, heavy rainfalls will probably occur more frequently and be more intense in most areas of land in the middle latitudes and in the humid tropics.
- **Precipitation:** With increasing warming, average precipitation will further decrease in many arid regions in the middle and subtropical latitudes. In many humid regions in the middle latitudes, by contrast, increases in precipitation are to be expected in warmer conditions until the year 2100.
- **Warming of the oceans:** In all scenarios, the warming of the oceans will continue for centuries even if the emission of greenhouse gases is kept constant or reduced. The warming will penetrate from the surface of the water into the deep ocean and affect the circulation of oceanic currents such as the Gulf Stream.
- **Acidification of the oceans:** The oceans will also continue to absorb CO_2 from the atmosphere, leading to increasing acidification of the sea water.

Source: IPCC 2013

M9 Possible future consequences of climate change

What can we do against climate change?

Are there any effective recipes?

Human influence on the climate is proved in the trend of atmospheric and oceanic temperatures and in many other changes, such as for example water circulation, the melting of ice, the retreat of snow, changes in certain weather extremes, and acidification of the oceans.
Overall, from 1750 to 2013 large quantities of CO_2 were released by human activities, mainly through the use of fossil fuels and changes in land use, making a significant contribution to the man-made greenhouse effect.
Some of this CO_2 was absorbed by the oceans, soils and vegetation.
What possibilities exist for sustainably reducing these quantities of CO_2 and the anthropogenic greenhouse effect?
What activities can I undertake as a school pupil?

1. Compare the goals of the Kyoto Protocol with the results achieved so far (M1/ M2, internet).
2. Explain the system of emissions trading and give your opinion on the statement: "It is trading in hot air" (M3/ M4).
ⓞ **3. A** Put together in an overview the projects that have been carried out or are planned in the area of your school to reduce CO_2 emissions.
B Name measures that have been carried out or are planned at your school to reduce CO_2 emissions.
4. Discuss what measures you could take yourself to contribute to protecting the climate. Give reasons for your choice. For this, use the materials to determine your daily and annual CO_2 emissions (M5 – M8).
Ⓢ **5.** Prepare a short talk on the subject of "Measures to protect the climate in Germany".

CO_2, Emission, Emissions permit, Emissions trading, Kyoto Protocol, Sustainability

Goals and instruments of the Kyoto Protocol

The industrialised countries were to reduce their emissions of the six most important greenhouse gases (including CO_2 and CH_4) by at least five percent in the period 2008 to 2012 compared to the 1990 level. Differing reduction targets applied to each country depending on its economic development.
For the member countries of the EU, a common reduction target was laid down but the burden was distributed within the EU in varying ways. Thus, some countries were allowed to increase their emissions for the purpose of economic development. To a limited extent, countries were also allowed to meet their obligations through reduction measures abroad. The intention of this was to foster climate protection through international cooperation. If a country achieved greater reductions in emissions than agreed, it could also use the excess quantities for trading. Countries, as well as individual companies within a country, could sell emissions permits to one another.

M1 Reduction in greenhouse gases

1992: Rio de Janeiro (Brazil)
- Reduction in CFCs to the 1990 level by 2000.

1997: Kyoto (Japan)
- Reduction in greenhouse gases by an average of 5.2 percent from 2008 to 2012 in comparison to 1990.

2001: Bonn (Germany)
- Climate agreement by the industrialised countries; USA and China pull out and decide to go their own ways on climate matters.

2004: Milan (Italy)
- Implementation of the Kyoto Protocol.
- Determination of deadlines for reduction of CO_2.
- Resolution on emissions trading.

2008: Poznan (Poland)
- Creation of a fund to assist the poorest countries to adapt to the conditions for climate change.

2009: Copenhagen (Denmark)
- Limitation of global temperature increase to 2°C.
- No further binding resolutions.

2010: Cancún (Mexico)
- 100 billion dollar fund for developing countries from 2020.
- Kyoto Protocol members to reduce CO_2 emissions by 25 to 40 percent below 1990 levels.

2012: Doha (Qatar)
- CO_2 reduction: extension of the Kyoto Protocol.
- Minimum goal achieved: 200 states agree to the extension of the Kyoto Protocol to 2020.
  The agreement only covers 15% of global CO_2 emissions.

M2 Climate conferences and their most important results

Every country receives permits allocated by the EU, entitling it to generate a predetermined quantity of CO_2 and other greenhouse gases for a period of four years. The country initially distributes the permits free of charge to industrial companies. These allow the company to use a specified quantity of energy in the production process. If the company uses less than its permitted quantity, it can sell the unused permits nationally and internationally. The rights are purchased by companies whose greenhouse gas emissions are too high. The sale is subject to value-added tax, so that the country earns money from it. The cost of the permits is added to the price of the end product, and thus passed on to the consumer.

M3 Emissions permits

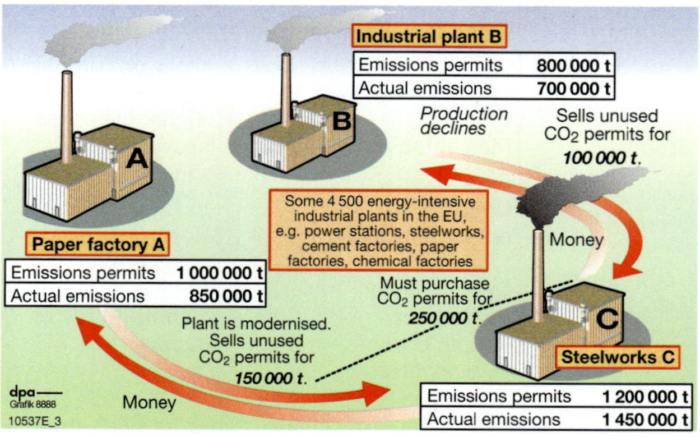

M4 EU emissions trading

Food	
Product and CO_2 emissions in g/kg of food	
Butter	23,800
Beef	13,300
Cheese	8,500
Sausage	8,000
Fried potato chips (frozen)	5,700
Tomatoes (out of season)	3,150
Eggs	1,950
Pizza (frozen)	1,250
Yoghurt	1,250
Marmalade	1,200
Milk	950
Tomatoes (in season)	350
Strawberries	300
Devices on stand-by	
Device in kg/year	
Satellite receiver	90
Music system	62
DSL router with Wi-Fi	34
DVD player	33
PC monitor	28
LCD television	7–27
Notebook	0.9

Electrical devices	
Device in g/h at maximum power	
Steam iron	15,530
Deep fryer	1,618
Kettle (2.2 kw)	1,423
Hair dryer	1,294
Toaster	712
Microwave	518
Egg cooker	259
MP3 player	8
Bath and shower	
Activity in g/min. at 40°C water temperature	
Bath (120 l)	1,560
Shower (3 min.)	520
Washing up	200
Washing hands	60
Mobility	
Means of transport in g/km per person	
Car	250
Underground	53
Bus	19
Bicycle	0

M5 CO_2 calculator

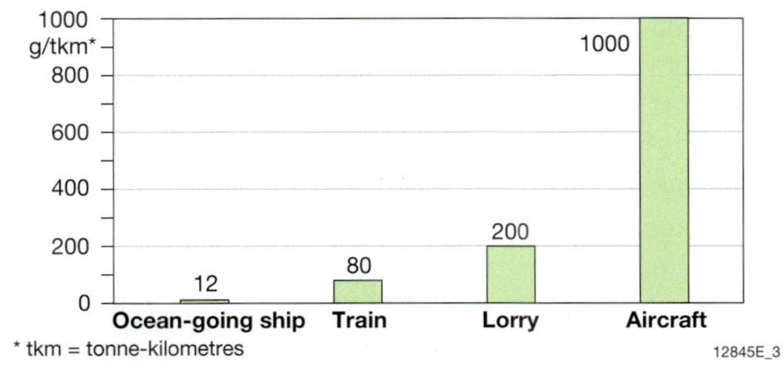

M7 Greenhouse gas emissions by mode of transport

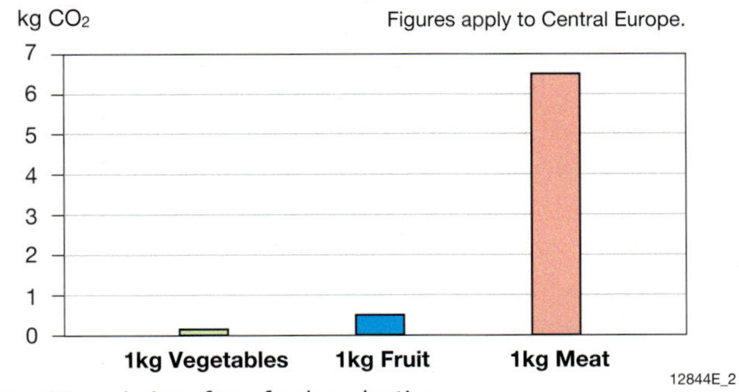

M8 CO_2 emissions from food production

What can we do as citizens?

Every individual should pay attention to the consequences of their own actions and keep themselves informed about ways of protecting the climate. Every little action that contributes to climate protection is important. Particularly effective forms of action are possible in the following three areas:

- **1. Use your political influence**
  Alliances and citizens' action groups can often have an influence on politics because they can represent either a counterweight or a partner for the government's actions. One can also exercise political influence by considering climatic and environmental aspects when deciding how to use one's vote.

- **2. Form networks**
  It is often easier to make a contribution to climate protection by working together with other activists. For example, in 1998 a private initiative was launched in the small community of Schönau in the Black Forest to invest in renewable energies. Starting with wind and solar energy, and later with co-generation and biomass, they could produce so much energy that they took over the local electricity grid and installed their own electricity plant.

- **3. Consume sustainably**
  As citizens we can enhance our quality of life and protect the climate by our consumption behaviour. These ground rules will help towards more climate-friendly and sustainable consumption:
  Inform yourself and make comparisons before deciding what to buy:
  In many fields, environment-friendly products and ways of saving energy are indicated by certification. This helps consumers to recognise particularly energy-efficient devices, for example.
  Consider the climate consequences when you buy:
  If you buy regional products, you avoid long transport distances for the products.
  Only buy what you really need:
  We often consume more than we need. Even if new technical possibilities make many products seem attractive, you should consider what functions you really need. With many devices it is the case that the more powerful it is, the more energy it uses.
  Save energy sensibly:
  It is particularly important to save in areas that are very harmful to the climate and where savings potential exists. By far the greatest proportion of greenhouse gas emissions in the field of private consumption come from car and air transport, heating and cooking.

Based on: www.germanwatch.org

M6 Measures to protect the climate

weblinks

Addresses/organisations

- **Alfred Wegener Institute**
- **Allianz Umweltstiftung**
- **Federal Ministry for the Environment, Nature Conservation and Nuclear Safety**
- **Germanwatch**
- **Greenpeace**
- **Hamburg education server**
- **IPCC**
- **WWF Switzerland**

The essentials in brief

A multiplicity of changes have occurred in the whole climatic system since the middle of the last century.

The worldwide increase in extreme climatic events and their after-effects gives rise to extensive challenges. Already today, not only the increase in temperatures in the lower atmosphere but also the warming of the oceans, the melting of glaciers, the warming of permafrost soils and the rise in sea levels have direct and noticeable effects on nature and society in both the Northern and Southern Hemispheres.

A definite result of scientific research is the recognition that the Earth's climatic system is subject to fluctuations, both over time and regionally.

More comprehensive observations show, with a high degree of certainty, that human activities are the main cause of the current climate change. Natural factors such as fluctuations in solar activity or volcanic eruptions are also measurable and also have an influence on the climate, but at the present time they only have a minor influence on long-term warming.

In particular the burning of fossil fuels, but also the clearance of forests, have resulted in a significant increase in the concentration of greenhouse gases in the atmosphere. Today, the concentration of carbon dioxide (CO_2) in the atmosphere is higher than at any previous time in the last 800,000 years.

If the present rate of emissions remained unchanged, by the middle of this century there would already be so much carbon dioxide in the atmosphere that the global mean temperature would have risen by 2°C compared to the level in the pre-industrial era.

If global warming is to be held at a given level, significant reductions in greenhouse gas emissions are necessary.

Research results show that, in a scenario with very active climate protection in the form of a significant reduction in greenhouse gas emissions, the chance exists that global warming can be kept below the level of 2°C compared to the pre-industrial era. But this will require great efforts in climate protection. And even then, a rise in the sea level is to be expected.

Since climate change is already happening, and since certain consequences such as the melting of glaciers and the rise in the sea level can no longer be prevented, we need a comprehensive adjustment strategy especially for the regions most seriously affected by the consequences of climate change (M1) and the people who live there.

With every year that we pump persistent greenhouse gases into the atmosphere unhindered, an adequately effective answer to the climate question will become more difficult and more expensive.

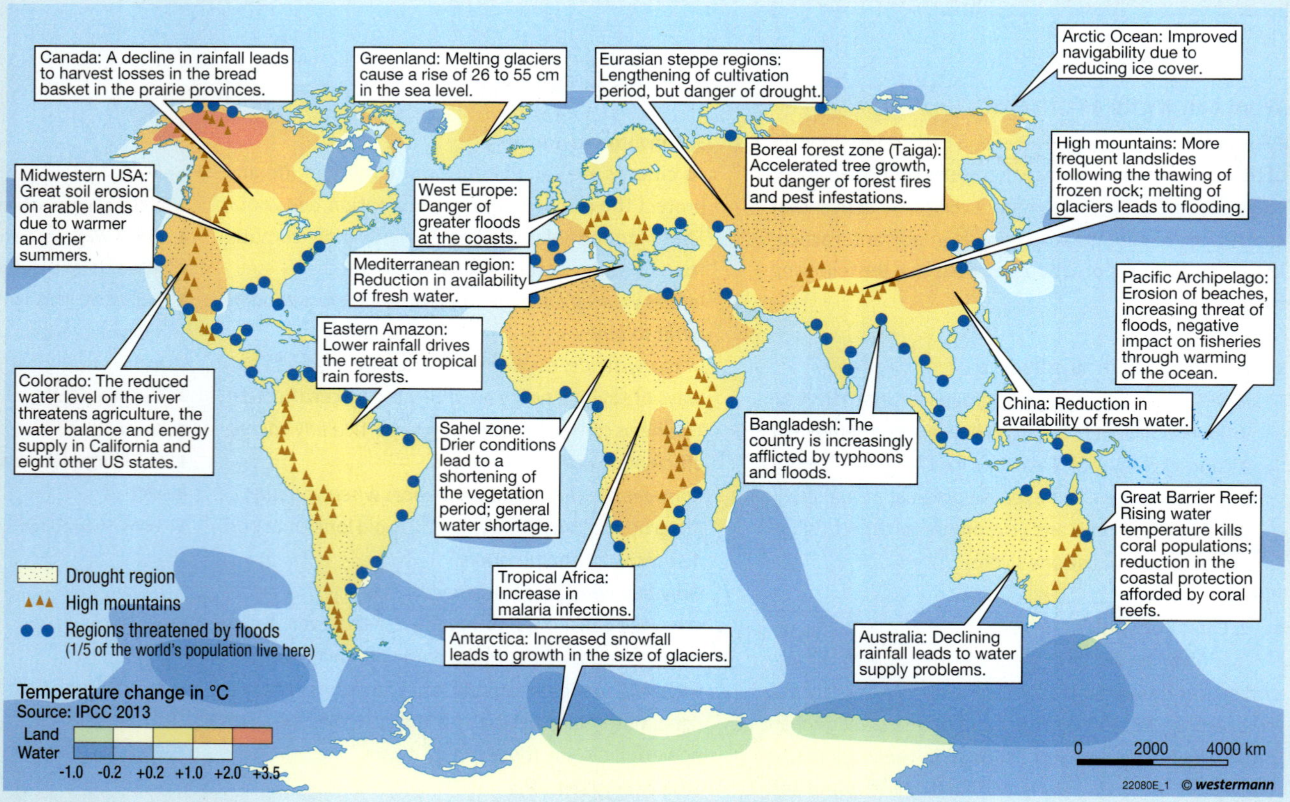

M1 Possible consequences of climate change in the course of the 21st century (IPCC, 2013)

Skill Check

All the skills that you were able to acquire in this chapter are listed here.
You can judge for yourself if you have mastered each skill: ***securely, moderately or barely.***

Explanatory skills

Can I...		Unsure? Look it up on these pages
1.	describe the trend of regional and global temperature changes?	170/171
2.	explain the connection between the warming of the Earth and the retreat of glacial ice, and discuss possible global effects arising from this?	174/175, 180–183
3.	explain the connection between the warming of the Earth and the rise in sea levels, and discuss the effects of the rise in sea levels?	177, 180–183
4.	present the possibilities for overcoming the boundaries between the inhabited and uninhabitable parts of the world arising from the consequences of climate change?	174–176
5.	explain anthropogenic influences on current changes in the climate and their possible effects, among other things on the increase in hot periods, forest fires, heavy rainfall and storm events?	178/179, 184/185
6.	explain the natural and anthropogenic causes of climate change?	180–183
7.	explain the influence of fossil fuels on climate change?	183–185

Methodological skills

Can I...		
8.	identify problem-related geographical facts and develop appropriate questions about climate change?	168–187
9.	analyse different kinds of presentation and working materials (e.g. maps, statistical data, graphics and text) and use them as the basis to answer questions about climate change?	170–187
10.	using charts, explain the structure of the atmosphere and the Earth's radiation budget?	182
11.	using suitable search strategies in the internet, research information on the tracks of climate change and analyse it in an issue-related way?	174–179
12.	present the tracks of climate change orally and in writing, using the correct terminology, in a way that is problem-oriented, structured, and relevant to the materials?	174–179
13.	substantiate statements about climate change with correct and appropriate references to materials and citations?	174–179, 182–185
14.	present geographical information graphically, for example as a causal network?	174–179

Evaluation skills

Can I...		
15.	critically scrutinise possible effects of climate change, such as e.g. the melting of glaciers and the rise in sea level?	174–179
16.	weigh up natural against anthropogenic causes of changes in climate?	180–183
17.	give my critical opinion on the validity of climate scenarios?	184/185
18.	evaluate possibilities for limiting the rise in global temperature against the background of demographic and economic developments?	186/187

Action skills

Can I...		
19.	present the results of work on climate change using appropriate terminology and with factual references?	174–179
20.	develop approaches to solutions for possible changes in climate?	186/187
21.	investigate and present opportunities for reducing CO_2 emissions in my school and in my everyday activities?	187

Index

Image Sources

Airbus S.A.S., Blagnac: 118 M6 (A. Doumenjou); Airport Development A/S, Zweigniederlassung Neuhardenberg: 151 M8 (© Sommer-Reich); akg-images, Berlin: 72 M3; alamy images, Abingdon/Oxfordshire: 113 M7 (David Wall); Alfred-Wegener-Institut für Polar- und Meeresforschung, Bremerhaven: 176 M1 (K. Piel); Alpha Ventus/ (c) 2009 Jan Oelker, all rights reserved: 152 M3; Astrofoto, Sörth: 182 M4; Aufwind-Luftbilder, Hamburg: 124 M4; Bildagentur Huber, Garmisch-Partenkirchen: 31 M6; Bildarchiv Boden-Landwirtschaft-Umwelt, Creglingen: 21 M9 (Prof. Dr. Karl Stahr); Brinkmann-Brock, Ursula, Paderborn: 14 M4, 15 M5, 31 M5 Foto o., 41 M7; Bundesanstalt für Geowissenschaft und Rohstoffe (BGR), Hannover: 139 M8; BÜNDNIS 90/DIE GRÜNEN, Berlin: 139 M6; Corbis, Berlin: 36 M2 (Everton), 137 M10 (Amy Sussman); Daldrup & Söhne AG, Ascheberg: 159 M7; Deutsche Gesellschaft für Internationale Zusammenarbeit (GIZ) GmbH: 67 M4 o. (Kampagnenmotiv IYDD 2006/SHANGHAI DGM), 67 M4 u. (Screenshot aus dem Film „An Boden verlieren...", im Auftrag des Bundesministeriums für wirtschaftliche Zusammenarbeit und Entwicklung); Deutsche Telekom AG, Bonn: 163 M8; dreamstime.com, Brentwood: 12 M3 (Ekhphoto), 16 M3 (Bidouze Stéphane); ecopix Fotoagentur, Berlin: Titel m. re.; eoVision, Salzburg: 115 M8 (U.S. Geological Survey, 2010); Europäische Kommission: 162 M1; FAO - Food and Agriculture Organization of the United Nations, Rom: 64/65 M2 m.; Focus Photo- u. Presseagentur GmbH, Hamburg: 55 M4 o. und 58 M1 (Amos Schliack); fotolia.com, New York: 13 M6 (Nouk), 22 M2 (Oleg Znamenskiy), 23 M7 (David Thyberg), 27 M4 (Lynne Richards), 50 M2 (Oleksiy Drachenko), 55 M4 u. (ferkelraggae), 68 M1 (Hendrik Schwartz), 109 M6 (S. Leyk), 118 M3 u. (Muzna Butt), 146 M1 (psdesign1), 150 M5 (Horst Schmidt), 153 M4 (Durrer), 175 M8 (jeffong); Getty Images, München: Titel o. li. (Mangiwau), Titel o. re. (Gallo Images/ Danita Delimont), Titel m. m. (Nat. Geographic/Roy Toft), Titel u. m. (The Image Bank/Keren Su), Titel u. re. (Photographer's Choice/Jens Kuhfs), 7 VII und 168/169 (Mohamed Abdulla Shafeeg), 77 M5 (Christopher Pilitz/The Image Bank), 100/101 M2 (AFP/Stringer), 118 M3 o. (Popperfoto), 119 M13 (Marwan Naamani/AFP), 127 M8 (Sean Gallup), 133 M4 (Christopher Goodwin), 156 M4 (Luis Acosta/AFP); Getty Images - Lonely Planet Images, München: 33 M8 (Porges/Arnold); Glücksman, Edvard/geolog.egu.eu: 167 M7; Graner, Tilmann, Sondershausen: 40 M3; Greenpeace Deutschland e. V., Hamburg: 139 M5; Hübscher, Heinrich, Lüneburg: 147 M5; Institut für Gebäude- und Solartechnik/TU Braunschweig: 163 M7; iStockphoto.com, Calgary: Titel o. m. (SusanneB), Titel m. li. (Vibelmages), 6 u. l. und 10/11 (Brosa), 7 V und 130/131 (jonmullen), 24 M2 (jonaldm), 26 M1 Foto (photoquest7), 26 M2 (ChrisBoswell), 29 M8 (sara_winter), 35 M5 (nonimatge), 64/65 M2 li. (Alexander Ryabintsev), 119 M10 (Alex Jeffries), 119 M9 (HeiFi), 120 M1 (DanBrandenburg), 123 M5 (SpVVK), 7 VI und 144/145 (robas), 147 M8 (Rhoberazzi), 157 M6 (Purdue9394), 160 M5 (brytta); Jakisch, Wilfried, Beeskow: 30 M3; Janzen, Jörg, Berlin: 34 M2 (aus: GR 12/2012); Jenni Energietechnik AG, Oberburg: 123 M7 (Orlando Eisenmann); Kalthoff, Frankfurt/M.: 74 M1; Klaer, W., Mainz: 43 T-Savanna und W-Savanna; Klein, Dorothee, Köln: 70 M3; Klohn, Werner Prof. Dr., Vechta: 32 M3, 33 M4; Kreuzberger, Norma, Lohmar: 69 M6, 154 M2; laif, Köln: 6 M2 und 48/49 (Michel Renaudeau/HOA-QUI), 91 M4 (Les Stone/Polaris), 7 IV und 106/107 (Vincent Prado), 146 M2 (Paul Langrock), 171 M7 (Paul-Langrock / Zenit); LOOK-foto, München: 27 M8 (Zielske); Lorenz, Wolfgang, Fotostudio, Kloster Lehnin: 166 M5; mauritius images, Mittenwald: 42 mixed forests (Mayer); Mineralogical Society of Great Britain & Ireland and The Clay Minerals Society from the ‚Images of Clay Archive' at www.minersoc.org/, Twickenham, Middlesex: 20 M1 u. M2 (reproduced with kind permission); NASA, Houston/Texas: 51 M6; NASA - Earth Observatory: 8/9 M2; Oed, Hans-Günther, Unkel: 128 M1; OKAPIA-Bildarchiv, Frankfurt/M.: 124 M2 (NAS/Biophoto Associates); Picture-Alliance, Frankfurt/M.: 6 u. III und 86/87 (dpa/Everett Kennedy Brown), 37 M8 (TASS), 50 M1 (Dave Hunt/dpa), 64/65 M2 re. (Peter Smolka/dpa), 68 M4 (dpa/M. Schrader), 69 M5 (Armin Weigel/dpa), 69 M8 (Karl-Josef Hildenbrand/dpa), 73 M5 (Okapia/Birke-Bloble), 75 M5 (dpa-Zentralbild/Wolfgang Kluge), 80 M1 u. (epa NASA), 82 M1 (dpa), 88 M1 (AP/Gwadar local government office), 90 M2 (Michael Laughlin), 92 M3 (South Florida Sun-Sentinel/landov/Michael Laughlin), 98 M2 (Abaca/E. De Malglaive), 99 M6 (dpa-Infografik), 99 M8 (dpa-Infografik), 118 M1 (DPPI Media/Frederic Le Floch), 118 M2 (dpa/Arno Burgi), 120 M4 (ZUMA Press/mr7), 127 M12 (dpa/Schachtanlage Asse), 127 M9 (dpa/Jochen Lübke), 133 M5 (dpa-Zentralbild/Michael Reichel), 133 M6 (dpa/Andreas Gebert), 135 M5 (empics/The Canadian Press/Jeff McIntosh), 141 M8 (dpa/Photo Itar-Tass), 147 M6 (dpa-Zentralbild/eruoluftbild.de/Hans Blossey), 170 M1 (Keystone/Arno Balzarini), 170 M4 (epa/Egill Adalsteinsson), 171 M5 (dpa/H. Spies), 171 M6 (AP/Jeff McIntosh), 178 M1 li. (dpa7Martin Gerten), 178 M1 re., 183 M6, 186 M4 (dpa-Infografik); Pohle, P., Gießen: 76 M3; Reuters, Berlin: 80 M1 o. (Japan Meteorological Agency/MTSAT), 95 M4 (Handout), 140 M2 (David W. Cerny), 170 M3 (Stringer Australia); Rieke, Michael, Hannover: 37 M6; Roher, K., Passau: 42 Steppe; RWE AG, Konzernpresse/www.rweimages.com, Essen: 114 M1, 117 M6; Schmidt, Roger, Brunsbüttel: 157 M10; Schnitzler, Michael, Wien: 14 M3; Schülerkarikatur von Johannes M. und Guilia G.: 54 M2; Schulze, W., Kreiensen: 153 M8; Shutterstock.com, New York: Titel u.li. (Anderl), 46 M2 (Thomas Nord), 82 o. (justasc); sportlink & cie AG, Luzern: 162 M2; Statkraft Germany GmbH, Düsseldorf: 149 M7; Strohbach, Dietrich, Berlin: 33 M7; TopicMedia Service, Putzbrunn: 13 M4 (A. N. T.); U.S. Department of Defense: 119 M11 (Adam Randolph); U.S. Geological Survey/Pacific Gas & Electric Company: 105 M6 (Cooperative Research & Development); ullstein bild, Berlin: 60 M1 (Lineair), 89 M3 (imagebroker.net/Manfred Bail); UNCCD Secretariat, Bonn: 67 M3; United Nations Environment Program (UNEP), Nairobi Kenia: 51 M4; vario images, Bonn: 124 M1 li. (U. Baumgarten); VBHG Verband bergbaugeschädigter Haus- und Grundeigentümer e.V., Herten: 111 M6; Wendorf, Monika, Hannover: 33 M5; Willeke, Matthias, Hannover: 39 M7 (aus: PG 12/2011); World Vision Deutschland, Friedrichsdorf: 63 M3; www.gazprom.com: 122 M2; www.gegen-gasbohren.de, Nordwalde: 139 M9; Zeitbild Verlag und Agentur für Kommunikation GmbH, Berlin: 9 M4.